T0292344

THEORETICAL MECHANICS

AN INTRODUCTORY TREATISE

ON THE

PRINCIPLES OF DYNAMICS

THEORETICAL MECHANICS

AN INTRODUCTORY TREATISE

ON THE

PRINCIPLES OF DYNAMICS

WITH APPLICATIONS AND NUMEROUS EXAMPLES

BY

A. E. H. LOVE, M.A., D.Sc., F.R.S.

SEDLEIAN PROFESSOR OF NATURAL PHILOSOPHY IN THE UNIVERSITY OF OXFORD.
HONORARY FELLOW OF QUEEN'S COLLEGE, OXFORD. FORMERLY FELLOW
AND LECTURER OF ST JOHN'S COLLEGE, CAMBRIDGE

THIRD EDITION

CAMBRIDGE
AT THE UNIVERSITY PRESS
1921

CAMBRIDGE
UNIVERSITY PRESS

University Printing House, Cambridge CB2 8BS, United Kingdom

Cambridge University Press is part of the University of Cambridge.

It furthers the University's mission by disseminating knowledge in the pursuit of education, learning and research at the highest international levels of excellence.

www.cambridge.org
Information on this title: www.cambridge.org/9781107593459

First edition 1897
Second edition 1906
Third edition 1921
First paperback edition 2015

A catalogue record for this publication is available from the British Library

ISBN 978-1-107-59345-9 Paperback

PREFACE

THE book, of which this is the third edition, is intended as a text-book of Dynamics, for the use of students who have some acquaintance with the methods of the Differential and Integral Calculus. Its scope includes the subjects usually described as Elementary Dynamics, Dynamics of a Particle, and Dynamics of a Rigid Body moving in two dimensions. It also includes an attempt to trace the logical development of the theory.

Within the chosen range of subject-matter there are many topics, such as the oscillation of a pendulum through wide angles, which it would be inappropriate to omit although they are difficult for a beginner. Articles dealing with such topics are marked with an asterisk. A student reading the subject for the first time, and without the guidance of a teacher, is advised to confine his attention to the unmarked Articles and the unmarked collections of Examples inserted in the text. Many of these Examples are well-known theorems, and some of them are referred to in subsequent demonstrations. Collections of miscellaneous Examples are appended to most of the Chapters. It is hoped that these will be useful to teachers and to students engaged in revising their work. A few of them, which I have not found in Examination papers, are taken from the well-known collections of Besant, Routh, and Wolstenholme.

The works which were most useful to me in regard to matters of theory, when I was writing the first edition of this book, were Kirchhoff's *Vorlesungen über mathematische Physik* (*Mechanik*), Pearson's *Grammar of Science*, and Mach's *Science of Mechanics*. The last ought to be in the hands of all students who wish to follow the history of dynamical ideas. In regard to methods for the treatment of particular questions, I am conscious of a deep obligation to the teaching of Mr R. R. Webb.

For the second edition the book was largely re-written and entirely re-arranged. In the present edition few changes have been made in the text. New Articles, which have been added, are marked with the letter A, thus "102 A." The number of miscellaneous Examples has been reduced considerably, but it is hoped that the most interesting have been retained, and that these are still sufficiently numerous.

A. E. H. LOVE.

OXFORD.
April, 1921.

CONTENTS

INTRODUCTION

CHAPTER I

DISPLACEMENT, VELOCITY, ACCELERATION

CHAPTER II

THE MOTION OF A FREE PARTICLE IN A FIELD OF FORCE

CHAPTER III

FORCES ACTING ON A PARTICLE

EQUATIONS OF MOTION IN SIMPLE CASES

THEORY OF MOMENTUM

WORK AND ENERGY

CHAPTER IV

MOTION OF A PARTICLE UNDER GIVEN FORCES

CHAPTER V

MOTION UNDER CONSTRAINTS AND RESISTANCES

CHAPTER VI

THE LAW OF REACTION

THEORY OF A SYSTEM OF PARTICLES

THE PROBLEM OF THE SOLAR SYSTEM

BODIES OF FINITE SIZE

APPENDIX TO CHAPTER VI

REDUCTION OF A SYSTEM OF LOCALIZED VECTORS

CHAPTER VII

MISCELLANEOUS METHODS AND APPLICATIONS

CHAPTER VIII

MOTION OF A RIGID BODY IN TWO DIMENSIONS

CHAPTER IX

RIGID BODIES AND CONNECTED SYSTEMS

MOTION OF A STRING OR CHAIN

CHAPTER X

THE ROTATION OF THE EARTH

CHAPTER XI

SUMMARY AND DISCUSSION OF THE PRINCIPLES OF DYNAMICS

APPENDIX

INTRODUCTION

1. Mechanics is a Natural Science; its data are facts of experience, its principles are generalizations from experience. The possibility of Natural Science depends on a principle which is itself derived from multitudes of particular experiences—the "Principle of the Uniformity of Nature." This principle may be stated as follows—Natural events take place in invariable sequences. The object of Natural Science is the description of the facts of nature in terms of the rules of invariable sequence which natural events are observed to obey. These rules of sequence, discovered by observation, suggest to our minds certain general notions in terms of which it is possible to state the rules in abstract forms. Such abstract formulas for the rules of sequence which natural events obey we call the "Laws of Nature." When any rule has been established by observation, and the corresponding Law formulated, it becomes possible to predict a certain kind of future events.

The Science of Mechanics is occupied with a particular kind of natural events, viz. with the motions of material bodies. Its object is the description of these motions in terms of the rules of invariable sequence which they obey. For this purpose it is necessary to introduce and define a number of abstract notions suggested by observations of the motions of actual bodies. It is then possible to formulate laws according to which such motions take place, and these laws are such that the future motions and positions of bodies can be deduced from them, and predictions so made are verified in experience. In the process of formulation the Science acquires the character of an abstract logical theory, in which all that is assumed is suggested by experience, all that is found is proved by reasoning. The test of the validity of a theory of this kind is its consistency with itself; the test of its value is its ability to furnish rules under which natural events actually fall.

The study of such a science ought to be partly experimental; it ought also to be partly historical. Something should be known of the kind of experiments from which were derived the abstract notions of the theory, and something also of the processes of inductive reasoning by which these notions were reached. It will be assumed here that some such preliminary study has been made*. The purpose of this book is to formulate the principles and to exemplify their application.

2. Motion of a particle. We have said that our object is the description of the motions of bodies. The necessity for a simplification arises from the fact that, in general, all parts of a body have not the same motion, and the simplification we make is to consider the motion of so small a portion of a body that the differences between the motions of its parts are unimportant. How small the portion must be in order that this may be the case we cannot say beforehand, but we avoid the difficulty thus arising by regarding it as a geometrical point. We think then in the first place of the motion of a point.

A moving point considered as defining the position from time to time of a very small part of a body will be called a "particle."

Motion may be defined as *change of position taking place in time.*

In regard to this definition it is necessary to attend to two things: the measurement of time, and the determination of position.

3. Measurement of time. Any instant of time is separated from any other instant by an *interval.* The duration of the interval may be measured by the amount of any process which is effected continuously during the interval. For the purposes of Mechanics it is generally more important that time should be conceived as measurable than that it should be measured by an assigned process.

The process actually adopted for measuring time is the average rotation of the Earth relative to the Sun, and the unit in terms of which this process is measured is called the "mean solar second." In the course of this book we shall generally assume that time is measured in this way, and we shall denote the measure of the time

* Historical accounts are given by E. Mach, *The Science of Mechanics* (Translation), Chicago, 1893, and by H. Cox, *Mechanics*, Cambridge, 1904.

which elapses between two particular instants by the letter t, then t is a real positive number (in the most general sense of the word "number") and the interval it denotes is t seconds.

4. Determination of position. The "position of a point" means its position relative to other points. Position of a point relative to a set of points is not definite until the set includes four points which do not all lie in one plane. Suppose O, A, B, C to be four such points; one of them, O, is chosen and called the *origin*, and the three planes OBC, OCA, OAB are the faces of a trihedral angle having its vertex at O. (See Fig. 1.) The position of a point P with reference to this trihedral angle is determined as follows:—we draw PN parallel to OC to meet the plane AOB in N, and we draw NM parallel to OB to meet OA

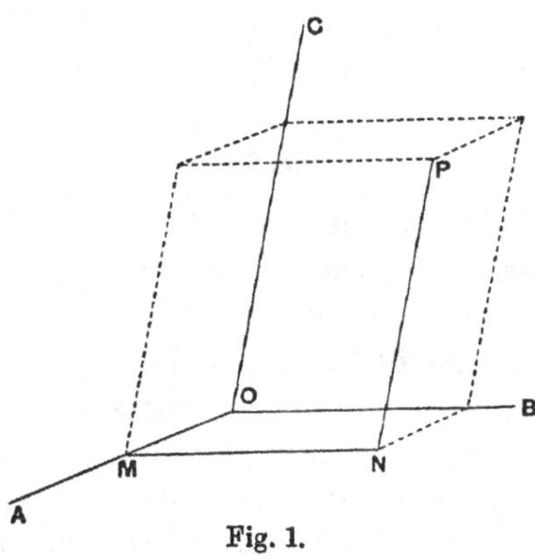

Fig. 1.

in M; then the lengths OM, MN, NP determine the position of P. Any particular length, *e.g.* one centimetre, being taken as the unit of length, each of these lengths is represented by a number (in the general sense), viz. by the number of centimetres contained in it. It is clear that OP is a diagonal of a parallelepiped and that OM, MN, NP are three edges no two of which are parallel. The position of a point is therefore determined by means of a parallelepiped whose edges are parallel to the lines of reference, and one of whose diagonals is the line joining the origin to the point.

It is generally preferable to take the set of lines of reference to be three lines at right angles to each other, then the faces of the trihedral angle are also at right angles to each other; sets of lines so chosen are called *systems of rectangular axes,* and the planes that contain two of them are *coordinate planes**.

It is clear from Fig. 2 that a set of rectangular coordinate planes divide the space about a point into eight compartments, the particular trihedral angle $OABC$ being one compartment. The lengths OM, MN, NP of Fig. 1, taken with certain signs, are

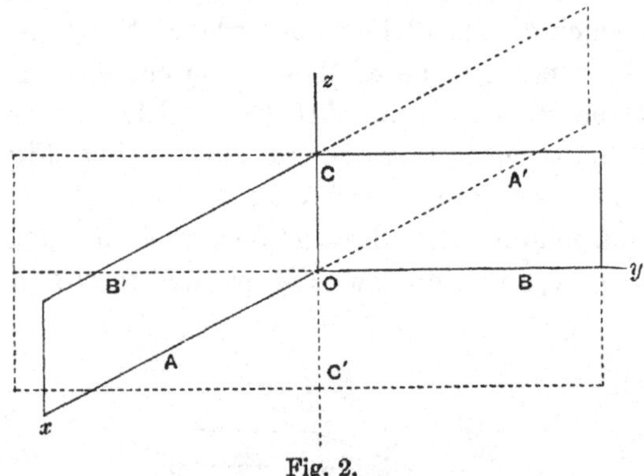

Fig. 2.

called the *coordinates* of the point P, and are denoted by the letters x, y, z. The rule of signs is that x is equal to the number of units of length in the length OM when P and A are on the same side of the plane BOC, and is equal to this number with a minus sign when P and A are on opposite sides of the plane BOC, and similarly for y and z.

Axes drawn and named as in Fig. 2 are said to be "right-handed." If the letters x and y are interchanged the axes are left-handed. In most applications of mathematics to physics right-handed axes are preferable to left-handed axes†. To fix ideas we may think of the compartment in which x, y, z are all positive as being bounded by two adjacent walls of a room and the floor of the room. If we look towards one wall with the other wall on the left-hand, and name the intersection of the walls the axis of z, the intersection of the floor with the wall on our left the axis of x, and the intersection of the floor with the wall in front of us the axis of y, the axes are right-handed. An ordinary, or right-handed, screw, turned so as to travel in the positive direction of the axis of x (or y, or z), will rotate in the sense of a line turning *from* the positive direction of the axis of y (or z, or x) *to* the positive direction of the axis of z (or x, or y). The senses of rotation belonging to the three screws are indicated in Fig. 3.

* We shall, in the course of this book, make use of rectangular coordinates only.

† In the course of this book the axes will be taken to be right-handed unless a statement to the contrary is made.

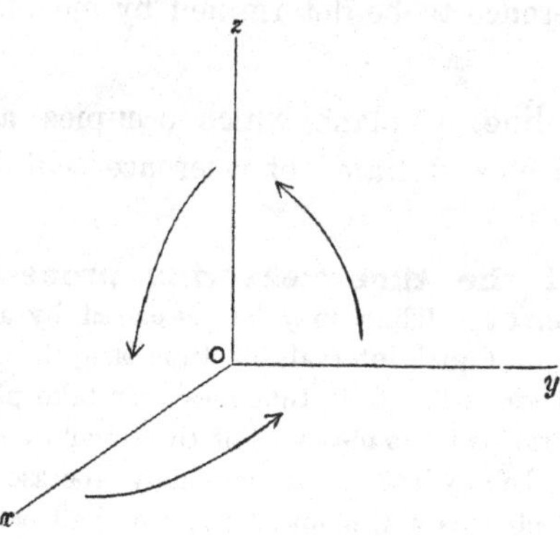

Fig. 3.

5. Frame of reference. A triad of orthogonal lines OA, OB, OC, with respect to which the position of a point P can be determined, will be called a *frame of reference.*

To determine a frame of reference we require to be able to mark a point, a line through that point, and a plane through that line. Suppose O to be the point, OA a line through the point, AOB a plane through the line. We can draw on the plane a line at right angles to OA meeting it in O, and we can erect at O a perpendicular to the plane. The three lines so determined can be a frame of reference.

In practice we cannot mark a point but only a small part of a body, for example we may take as origin a place on the Earth's surface; then at the place we can always determine a particular line, the vertical at the place, and, at right angles to it, we have a particular plane, the horizontal plane at the place; on this plane we may mark the line which points to the North, or in any other direction determined with reference to the points of the compass, we have then a frame of reference. Again we might draw from the place lines in the direction of any three visible stars, these would determine a frame of reference. Or again we might take as origin the centre of the Sun, and as lines of reference three lines going out from thence to three stars.

When we are dealing with the motions of bodies near a place on the Earth's surface, for example, the motion of a train, or a cannon-ball, or a pendulum, we shall generally take the frame of reference to be determined by lines which are fixed relatively to the Earth, and we shall generally take one of these lines to be the vertical at the place. When we are dealing with the motion of the Earth, or a Planet, or the Moon, we shall generally take the frame of reference to be determined by means of the "fixed" stars.

A point, or line, or plane which occupies a fixed position relatively to the chosen frame of reference will be described as "fixed."

6. Choice of the time-measuring process and of the frame of reference. Time may be measured by any process which goes on continually. Equal intervals of time are those in which equal amounts of the process selected as time-measurer take place, and different intervals are in the ratio of the measures of the amounts of the process that take place in them. In any interval of time many processes may be going on. Of these one is selected as a time-measure; we shall call it the *standard process.* "Uniform processes" are such that equal amounts of them are effected in equal intervals of time, that is, in intervals in which equal amounts of the standard process are effected. Processes which are not uniform are said to be "variable." It is clear that processes which are uniform when measured by one standard may be variable when measured by another standard. The choice of a standard being in our power, it is clearly desirable that it should be so made that a number of processes uncontrollable by us should be uniform or approximately uniform; it is also clearly desirable that it should have some relation to our daily life. The choice of the mean solar second as a unit of time satisfies these conditions. So long as these conditions are not violated, we are at liberty to choose a different reckoning of time for the purpose of simplifying the description of the motions of bodies.

The choice of a suitable frame of reference, like the choice of the time-measuring process, is in our power, and it is manifest that some motions which we wish to describe will be more simply describable when the choice is made in one way than when it is made in another. We shall return to this matter in Chapter XI.

CHAPTER I

DISPLACEMENT, VELOCITY, ACCELERATION

7. THE history of the Science of Mechanics shows how, through the study of the motions of falling bodies, importance came to be attached to the notions of variable velocity and acceleration, and also how, chiefly through the proposition called "the parallelogram of forces," the vectorial character of such quantities as force and acceleration came to be recognized. We shall now be occupied with precise and formal definitions of some vector quantities and with some of the immediate consequences of the definitions.

8. **Displacement.** Suppose that a point which, at any particular instant, had a position P with reference to any frame, has at some later instant a position Q relative to the same frame. The point is said to have undergone a "change of position" or a *displacement*. Let the line PQ be drawn. It is clear that the displacement is precisely determined by this line; we say that it is *represented* by this line. Let the line PQ drawn through P be produced indefinitely both ways, and let a parallel line be drawn through any other point, for instance through O. Then this line determines a particular direction; this is the direction of the displacement. Of the two senses in which this line may be described one, OR, is the sense from O towards that point (R) which is the fourth corner of a parallelogram having OP, PQ as adjacent sides; this is the sense of the displacement. The measure of the length of PQ is the number of units of length it contains; this number is the magnitude of the displacement. The subsequent position, Q, is entirely determined by (1) the previous position, P, (2) the direction of the displacement, (3) the sense of the displacement, (4) the magnitude of the displacement.

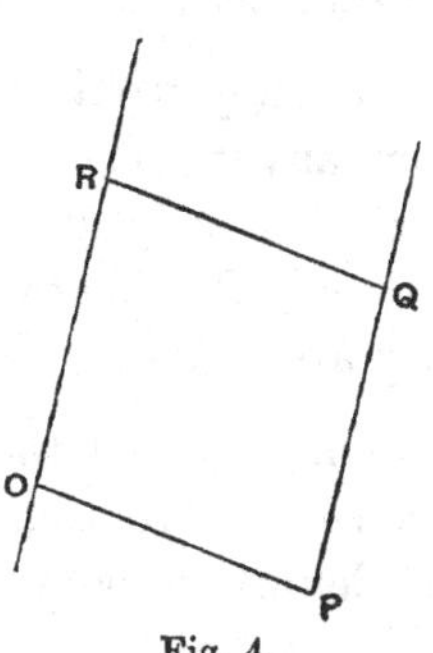

Fig. 4.

Further it is clear that exactly the same change of position is effected in moving a point from P to K by the straight line PK, and from K to Q by the straight line KQ, as in moving the point from P to Q directly by the straight line PQ. That is to say, displacements represented by lines PK, KQ are equivalent to the displacement represented by the line PQ.

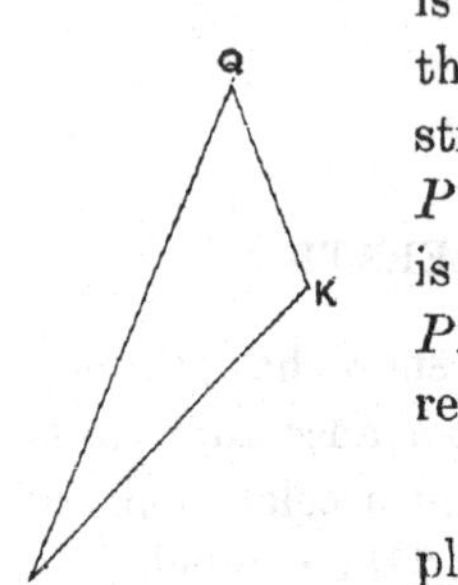

Fig. 5.

Displacement is a quantity, for one displacement can be greater than, equal to, or less than another; but two displacements in different directions, or in different senses, are clearly not equivalent to each other, even when they are equal in magnitude; and thus displacement belongs to the class of mathematical quantities known as *vectors* or *directed quantities.*

9. Definition of a vector. A vector may be defined as a directed quantity which obeys a certain rule of operation*.

By a "directed quantity" we mean an object of mathematical reasoning which requires for its determination (1) a number called the magnitude of the quantity, (2) the direction of a line called the direction of the quantity, (3) the sense in which the line is supposed drawn from one of its points, called the sense of the quantity.

Let any particular length be taken as unit of length. Then from any point a straight line can be drawn to represent the vector† in magnitude, direction, and sense. The sense of the line is indicated when two of its points are named in the order in which they are arrived at by a point describing the line.

The rule of mathematical operation to which vectors are subject is a rule for replacing one vector by other vectors to which it is (by definition) equivalent.

* The rule of operation is an essential part of the definition. For example, rotation about an axis is not a vector, although it is a directed quantity.

† The line is not the vector. The line possesses a quality, described as extension in space, which the vector may not have. From our complete idea of the line this quality must be abstracted before the vector is arrived at. On the other hand the vector is subject to a rule of operation to which a line can only be subjected by means of an arbitrary convention.

This rule may be divided into two parts and stated as follows :—

(1) Vectors represented by equal and parallel lines drawn from different points in like senses are equivalent.

(2) The vector represented by a line AC is equivalent to the vectors represented by the lines AB, BC, the points A, B, C being any points whatever.

Among vector quantities, as here defined, we note (i) displacement of a particle, (ii) couple applied to a rigid body (see Appendix to Chapter VI).

10. Examples of equivalent vectors. If AC, $A'C'$ are equal and parallel lines, their ends can be joined by two lines AA',

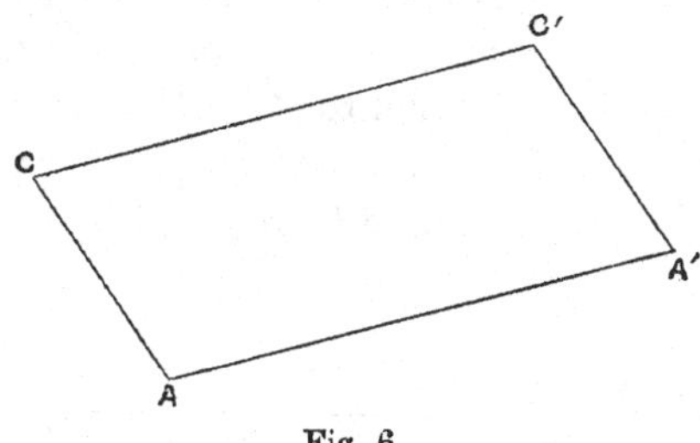

Fig. 6.

CC' which are equal and parallel; then the vectors represented by AC, $A'C'$ are equivalent; vectors represented by AC, $C'A'$ are not equivalent.

Again if A, B, C are any three points, and a parallelogram A, B, C, D is constructed having AB, BC as adjacent sides, AD

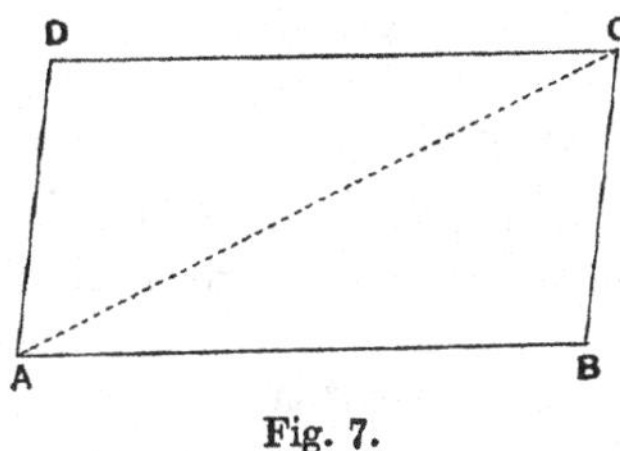

Fig. 7.

and BC are equivalent vectors. Also the vector AC is equivalent to the vectors AB, BC, or AD, DC, or AB, AD.

Further if a polygon (plane or gauche) is constructed, having AC as one side, and having any points P, Q, ... T as corners, the vector represented by AC is equivalent to the vectors represented by AP, PQ, ... TC. This is clear because by definition the vectors AP, PQ can be replaced by AQ, and so on. The statement is independent of the number of sides of the polygon, and of the order in which its corners are taken, no corner being taken more than once, provided that the points A, C are regarded as the first and last corners. [The restriction that no corner is to be taken more than once will be removed presently.]

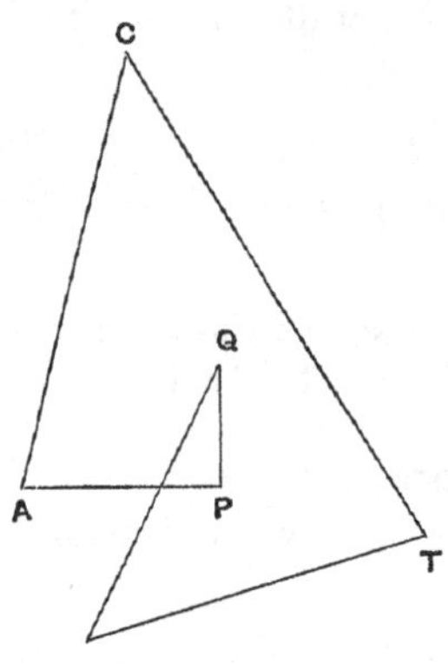

Fig. 8.

In particular, if the polygon is a gauche quadrilateral $ABDC$, a parallelepiped can be constructed having its edges parallel to AB, BD, DC, and having AC as one diagonal. Then the vector AC is equivalent to the vectors represented by the edges AB, AP, AQ which meet in A. (See Fig. 9.)

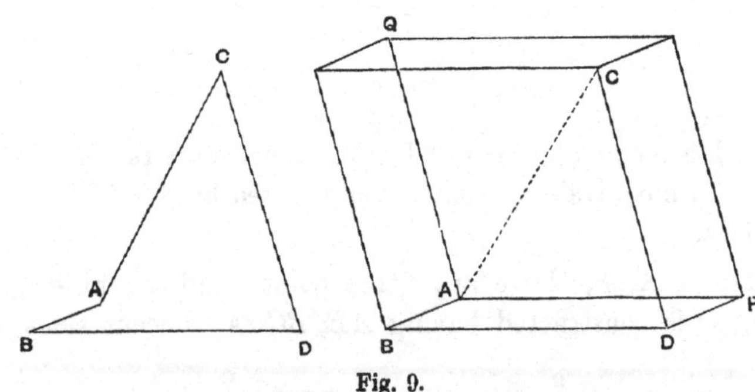

Fig. 9.

The case of this which is generally most useful is the case where the edges of the parallelepiped are the axes of reference relatively to which the positions of points are determined.

11. Components and resultant. A set of vectors equivalent to a single vector are called *components*, and the single vector to which they are equivalent is called their *resultant*.

The operation of deriving a resultant vector from given component vectors is called *composition*, we *compound* the components to obtain the resultant; the operation of deriving components in particular directions from a given vector is called *resolution*, we *resolve* the vector in the given directions to obtain the components in those directions.

It is clear from the constructions in the preceding article that we can resolve a vector in one way into components parallel to any two given lines which are in a plane to which the vector is parallel, and again we can resolve the vector in one way into components parallel to any three given lines not in the same plane.

When the directions of the component vectors are at right angles to each other the components are called *resolved parts* of the resultant vector in the corresponding directions.

Thus, if we take a system of rectangular coordinate axes, any vector parallel to a coordinate plane, *e.g.* the plane of (x, y), can be resolved into components parallel to the axes of x and y, these are the resolved parts of the vector in the directions of the axes of x and y.

Again, if we take a three-dimensional system of rectangular axes, any vector can be resolved into components parallel to the axes of x, y, and z, and these are the resolved parts of the vector in the directions of these axes.

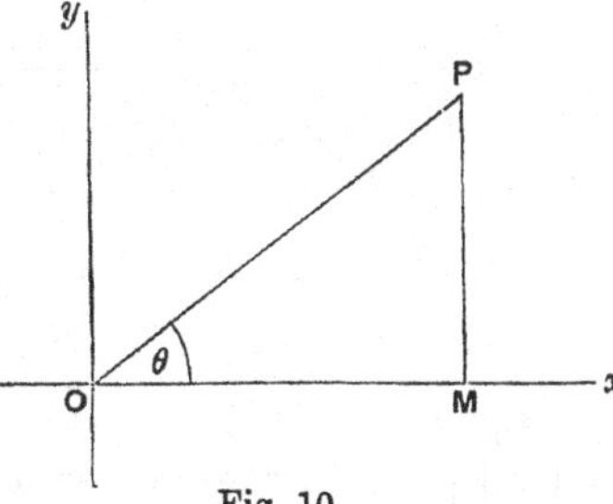

Fig. 10.

In the former case (Fig. 10) we take OP to represent the vector, and draw PM at right angles to Ox, then OM and MP represent the resolved parts of the vector parallel to the axes. If R is the magnitude of the vector represented by OP, and θ, ϕ the angles* between the lines OP and Ox, Oy, then $R\cos\theta$ and

* In Fig. 10 $\cos\phi$ is $\sin\theta$, but it is easy to draw a figure, *e.g.* Fig. 11, which makes it appear that $\cos\phi$ is $-\sin\theta$. With the usual conventions in regard to the signs of trigonometrical functions we shall always have

$$\cos\phi = \sin\theta$$

provided that θ is the angle traced out by a line OP starting from Ox and turning round O in the direction from Ox to Oy.

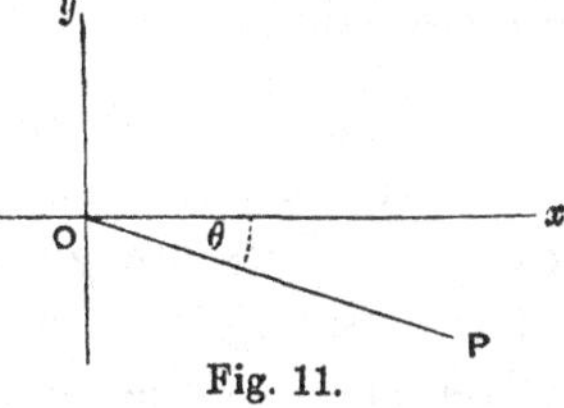

Fig. 11.

$R \cos \phi$ are the magnitudes of the resolved parts respectively, and these are the *projections* of OP on the axes.

More generally, we take OP to represent the vector, and construct a parallelepiped with O and P as opposite corners and with its faces parallel to the coordinate planes, then the resolved parts of the vector in the directions of the axes are numerically equal to the projections of OP on the axes. If R is the magnitude of the vector represented by OP, and if l, m, n are the cosines of

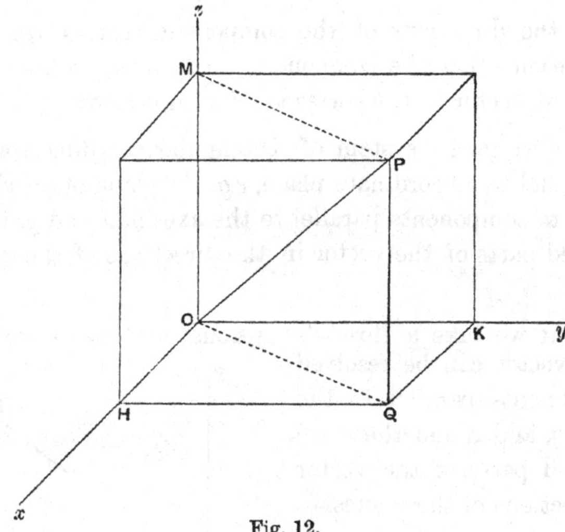

Fig. 12.

the angles which OP makes with Ox, Oy, Oz respectively, the resolved parts in these directions are Rl, Rm, Rn respectively.

This rule determines the senses as well as the magnitudes of the resolved parts; thus, when $\cos \theta$, in the first case, and l, in the second case, are negative, the resolved part parallel to the x axis is in the negative direction of that axis, *i.e.* in the direction xO produced.

It is clear from this rule that, when the magnitudes and signs of the resolved parts of a vector in the directions of three mutually rectangular lines are given, the vector is uniquely determinate, that is to say there is one and only one vector which has given resolved parts parallel to three such lines.

The construction in the former of these cases is a construction for the resolved parts of a vector parallel and perpendicular to a line. As before, let OP be a line representing the vector, and OA a line parallel and perpendicular to which the vector is to be resolved. Draw PM at right angles to OA. Then the vector is equivalent to vectors represented by OM, MP, and the magnitudes of these are respectively $R \cos \theta$ and $R \sin \theta$, where R is the magnitude of the vector to be resolved, and θ is the angle between its direction and OA.

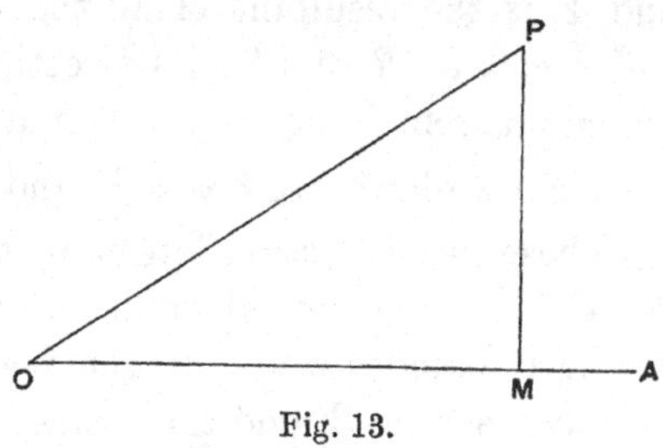

Fig. 13.

The vector represented by MP is the resolved part of the vector represented by OP at right angles to the line OA.

12. Composition of any number of vectors. I. Consider first the case where all the vectors are parallel to a plane, and take it to be the plane of (x, y). Let OP_1, OP_2, $\ldots OP_n$ be lines representing the vectors, (supposed to be n in number,) in magnitude, direction, and sense, and let $\theta_1, \theta_2, \ldots \theta_n$ be the angles which the lines OP_1, OP_2, $\ldots OP_n$ make with Ox, *i.e.* the angles traced out by a revolving line turning about O from Ox towards Oy. Let $r_1, r_2, \ldots r_n$ denote the magnitudes of the vectors.

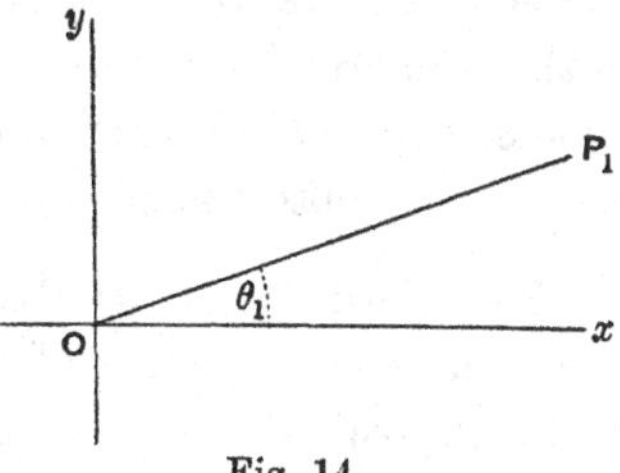

Fig. 14.

Then the vector represented by OP_1 may be replaced by vectors $r_1 \cos \theta_1$ parallel to Ox, and $r_1 \sin \theta_1$ parallel to Oy, and similarly for the others.

All the resolved parts parallel to Ox are equivalent to a single vector X parallel to Ox given by

$$X = r_1 \cos \theta_1 + r_2 \cos \theta_2 + \ldots + r_n \cos \theta_n = \Sigma (r \cos \theta).$$

All the resolved parts parallel to Oy are equivalent to a single vector Y parallel to Oy given by

$$Y = r_1 \sin \theta_1 + r_2 \sin \theta_2 + \ldots + r_n \sin \theta_n = \Sigma (r \sin \theta).$$

The vector whose resolved parts parallel to Ox and Oy are X and Y is the resultant of all the vectors. Let the magnitude of this vector be R, and let its direction and sense be those of a line going out from O and making an angle ψ with Ox.

Then we have $R\cos\psi = X$, and $R\sin\psi = Y$.

These two equations determine the magnitude R and the angle ψ. R is the numerical value of $\sqrt{(X^2+Y^2)}$, and ψ is that one among the angles whose tangents are Y/X for which the sine has the same sign as Y and the cosine has the same sign as X.

II. Consider the more general case where the vectors are not parallel to a plane. Let $r_1, r_2, \dots r_n$ be the magnitudes of the vectors, and call any one of these numbers r. Let l, m, n be the cosines of the angles which the line representing this vector in direction and sense makes with the axes Ox, Oy, Oz. Then this vector may be resolved into rl, rm, rn parallel to the lines Ox, Oy, Oz, and the whole set of vectors is equivalent to a vector whose resolved parts parallel to the axes are X, Y, Z, where $X = \Sigma rl$, $Y = \Sigma rm$, $Z = \Sigma rn$, the summations extending to all the vectors of the set. The resultant is therefore a vector whose magnitude, R, is the numerical value of $\sqrt{(X^2+Y^2+Z^2)}$, and such that the line representing it in direction and sense makes with the axes Ox, Oy, Oz angles whose cosines are X/R, Y/R, Z/R.

13. Vectors equivalent to zero. When the magnitude of the resultant of any set of vectors is zero the set of vectors is said to be equivalent to zero. Thus two equal vectors parallel to the same line, and in opposite senses, are equivalent to zero.

It is clear that the sum of the resolved parts, in any direction, of a set of vectors equivalent to zero is equal to zero.

Again vectors parallel and proportional to the sides of a closed polygon, and with senses determined by the order of the corners when a point travels round the polygon, are equivalent to zero.

This last statement enables us to do away with the restriction (Art. 10) that in the resolution of a vector into components parallel to the sides of a polygon not more than two sides of the polygon may meet in a point.

14. Components of displacement. Let x, y, z be the coordinates of a moving point at any particular instant with reference to any particular frame, x', y', z' the coordinates of the point at a subsequent instant, with reference to the same frame, then $x'-x$, $y'-y$, $z'-z$ are the components, parallel to the axes, of a vector quantity which is the displacement of the point. (Cf. Art. 8.)

15. Velocity in a straight line. Consider in the first place a point moving in a straight line, *e.g.* one of the lines of reference, and let s be the number of units of length it passes over in t units of time. Then it may happen that the two numbers s and t have a constant ratio whatever number we take for t. The point is then said to move *uniformly* in the line, and the fraction $\frac{s}{t}$ is defined to be the measure of its velocity. A point moving uniformly describes equal lengths in equal times.

Again consider the case where the point moves in a straight line, but the number of units of length passed over in any interval of time does not bear a constant ratio to the number of units of time in the interval. In this case there will be equal intervals of time in which the point describes unequal lengths; in the one of two equal intervals in which it describes the greater length we should say it was moving faster, in the other, in which it describes the shorter length, we should say it was moving more slowly. We have thus an idea of velocity of a point not moving uniformly, and we seek to make it precise.

For a point moving in a straight line we may define the *average velocity in any interval of time* to be the fraction

$$\frac{\text{number of units of length described in an interval}}{\text{number of units of time in the interval}}.$$

When the point is not moving uniformly this fraction is a variable number, which has a definite value when the measure of the interval is given and the first instant of the interval is given. Taking the first instant of the interval always the same, and taking for the measure of the interval a series of diminishing numbers, we obtain a series of fractions, which approach a limiting value as the measure of the interval is indefinitely diminished. This limiting value is defined to be the velocity of the point at the first instant of the interval. We might in the same way define the velocity of a point at the last instant of an interval.

We can now define the velocity of a point moving in a straight line at any instant. It is *the limit of the average velocity in an interval of time beginning or ending at the instant, the interval being diminished indefinitely.*

The two limits are in general the same; when they are different we call them the velocity *just after* the instant and the velocity *just before* the instant respectively.

Let t be the measure of the interval of time which has elapsed since some particular instant, chosen as the origin of time, and suppose that at the end of this interval the point has described a length s measured from some particular point in the line of its motion. We say that the point is at s at time t. In the same way suppose that it is at s' at time t'. Then in the interval $t'-t$ it describes a length $s'-s$, and its average velocity in the interval is $\frac{s'-s}{t'-t}$. The number s is a function of the number t, and the limit of the fraction just written is the number known as the differential coefficient of s with respect to t. The velocity of the moving point is accordingly measured by $\frac{ds}{dt}$.

The number $s'-s$ is the measure of the displacement of the point during the interval $t'-t$. When the velocity is uniform it is measured by the displacement in a unit of time. If the unit of time were replaced by a smaller unit the displacement in it would be replaced by a shorter length, and this length would measure the velocity in terms of the new unit of time. However short an interval is taken for the unit of time the length described in it measures the velocity in terms of it. When we wish to recall this fact, and to bring it into connexion with the definition of variable velocity we say that the latter is measured by "the rate of displacement per unit of time," but we must not attach to this phrase any other meaning than that which has just been explained, *i.e.* the phrase means nothing but the limit of the fraction

$$\frac{\text{number of units of length described in an interval}}{\text{number of units of time in the interval}}$$

when the interval is diminished indefinitely.

16. Velocity in general. When the point is not moving in a straight line it will have a component of displacement in any interval $t'-t$ parallel to each of the three axes of reference. Let these components be $x'-x$, $y'-y$, $z'-z$. Then each of the fractions $\frac{x'-x}{t'-t}$, $\frac{y'-y}{t'-t}$, $\frac{z'-z}{t'-t}$ has a limit, and these limits are, as above, the rates of displacement per unit time parallel to the axes. They are defined to be the component velocities parallel to the axes. As before x, y, z are functions of t, and the component velocities parallel to the axes are

$$\frac{dx}{dt}, \frac{dy}{dt}, \frac{dz}{dt}.$$

The velocity at an instant is the limit of the average velocity in an interval. This limit has a definite magnitude, and is associated with a definite straight line. At any instant the point is moving along the tangent to a curve, called its *path* or *trajectory*. The velocity is associated with this particular line, drawn in a definite sense. Let s be the arc of the curve measured from some particular point of the curve up to the position of the moving point at time t, and let s' be the corresponding arc for time t'. Then the length of the chord joining the two positions is the magnitude of the vector whose components parallel to the axes are $x'-x$, $y'-y$, $z'-z$. From the definition of s we have the equation

$$\text{Lt}_{t'-t=0}\left(\frac{s'-s}{t'-t}\right)^2 = \text{Lt}_{t'-t=0}\left\{\left(\frac{x'-x}{t'-t}\right)^2 + \left(\frac{y'-y}{t'-t}\right)^2 + \left(\frac{z'-z}{t'-t}\right)^2\right\}.$$

Thus the magnitude of the velocity of the moving point at time t is $\frac{ds}{dt}$, where s is the length of the arc of the path measured, in the sense of description of the path, from some particular point of it to the position of the moving point at time t. The magnitude of the velocity of a point is often called its *speed*, and, when it is independent of the time, the point is said to move with uniform speed whether its path is straight or curved.

It is manifest that the velocity of a moving particle can be represented in many respects by a vector, of which the components parallel to the axes are $\frac{dx}{dt}, \frac{dy}{dt}, \frac{dz}{dt}$; but the vector does not express the association of the velocity with a particular line—the tangent to the path of the particle.

17. Localized vectors. The vectors we have so far considered have no relation to any particular point, they are equally well represented by lines drawn from any point; and they have no relation to any particular line, they are equally well represented by segments of all lines parallel to their direction. They may be called *unlocalized vectors*. But it is often important to consider quantities which, in other respects, have the properties of vectors, but which have relations to particular points or particular lines.

A vector localized at a point is defined by its magnitude, direction, and sense, and also by a point and by a rule of equivalence, viz.:—two sets of vectors localized at the same point are equivalent if two sets of unlocalized vectors with the same magnitudes, directions, and senses are equivalent.

There is in general no rule of equivalence for vectors localized at different points.

A vector localized in a line is a vector localized at any point in a particular line, which is in the direction of the vector, with the additional rules of equivalence, (i) Two vectors localized in the same line are equivalent if they have the same magnitude and the same sense, (ii) Two vectors localized in lines which meet are equivalent to a single vector localized in a line.

All the constructions in the previous Articles apply to vectors localized at points and to vectors localized in lines, provided that all components and resultants are localized at the proper points or in the proper lines. In particular a vector localized at a point is equivalent to components (or resolved parts) of the same magnitudes, directions, and senses as if it were unlocalized, provided that these components and resolved parts are localized at the same point; also a vector localized in a line is equivalent to components (or resolved parts) of the same magnitudes, directions, and senses as if it were unlocalized, provided that these components and resolved parts are localized in lines which meet in a point on the line of the resultant.

Thus a vector localized at O may be represented (as in Fig. 12) by a line OP, and is equivalent to vectors localized at O and represented by lines OH, OK, OM; and a vector localized in the line OP, having the same magnitude and sense, is equivalent to vectors localized in any three lines parallel to Ox, Oy, Oz, meeting in a point on OP, and having the magnitudes and senses of OH, OK, OM.

The differences between the three classes of vectors may be expressed thus:—

A vector (unlocalized) is equivalent to any parallel vector of equal magnitude and like sense. Thus the line representing the vector may be drawn from any point.

A vector localized in a line is equivalent to any vector of equal magnitude and like sense localized in the same line. The line representing it may be drawn from any point in a particular line, and is a segment of that line.

A vector localized at a point is not equivalent to any other single vector. The line representing it must be drawn from the point.

A vector localized in a line is clearly determined by its components parallel to three given lines and by one point of the line, in particular the line in which it is localized is thereby determined.

As examples of vectors localized in lines we may cite (i) velocity of a moving particle, (ii) force applied to a rigid body (Chapter VI). Force applied to a particle is an example of a vector localized at a point (Chapter III).

18. Formal definition of velocity. We may now define the velocity of a moving point to be a vector, localized in a line through the position of the point, whose resolved part in any direction is the rate of displacement of the point in that direction per unit of time.

19. Measurement of velocity. The measure of any particular velocity is a number expressing the ratio of the velocity to the unit velocity.

The unit velocity is that with which a point describes one unit of length uniformly in each unit of time.

The number expressing a velocity is the ratio of a number expressing a length to a number expressing an interval of time. It therefore varies inversely as the unit of length and directly as the unit of time.

Velocity is accordingly said to be a quantity of one dimension in length, and of minus one dimension in time; or its dimension symbol is LT^{-1}, where L stands for length, and T for time.

20. Moment of localized vector. The reason for defining velocity as a localized vector is that special significance is found to attach to a certain quantity called the "moment of the velocity." We shall attend at present to the cases of vectors localized in lines that lie in a plane and vectors localized at points in a plane, and having their directions parallel to the plane*. We define the moment of such a vector about a point in the plane as follows:—

Draw a line L' in the direction of the vector, so that if the vector is localized in a line that line is L', and if the vector is localized at a point the line L' passes through the point. The moment of the

* A more general discussion will be given in Chapter III.

vector about a point O is the product, with a certain sign, of the magnitude of the vector and the perpendicular to L' from O. The rule of signs is this: Draw a line L through O at right angles to the plane containing O and L', and choose a sense of description of this line; then, if the senses of L and the vector are the same as those of translation and rotation in an ordinary right-handed screw, the sign is +, otherwise it is −.

The rule of signs may also be stated thus: Let a watch be placed in the plane of O and L', so that a line drawn from the back to the face is in the sense of L; when the sense of the vector is opposite to that of the motion of the hands the sign is +, otherwise it is −.

21. Lemma. *The moment about a point O of a vector localized at a point A is identical with the moment about O of the resolved part of the vector at right angles to OA.*

Let θ be the angle which the direction of the vector makes with the line AO, and draw ON at right angles to the line of the vector. The magnitude of the resolved part of the vector at right angles to AO is $R \sin \theta$, where R is the magnitude of the vector. The perpendicular from O on the line of the vector is the line ON, and it is equal to $OA . \sin \theta$.

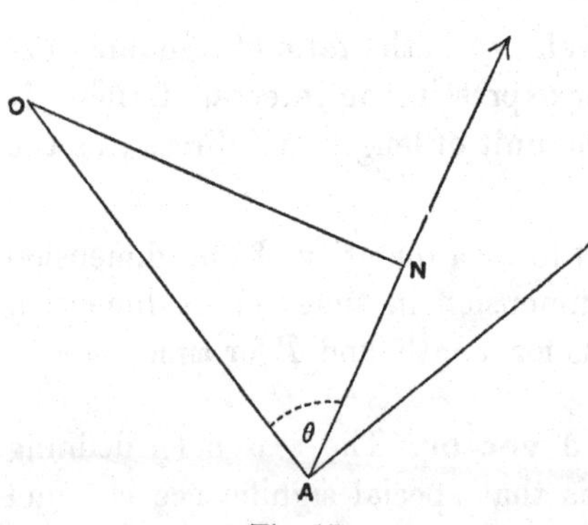

Fig. 15.

Now moment of R about $O = R . ON$

$= R . OA \sin \theta$

$= R . \sin \theta . OA$

$=$ moment about O of resolved part at right angles to OA.

22. Theorem of moments. *The sum (with proper signs) of the moments about a point O of two vectors localized at a point A is equal to the moment of their resultant about O.*

Let P_1 and P_2 be the magnitudes of the vectors, θ_1 and θ_2 the angles which the lines representing them drawn from A make with AO, R the magnitude of the resultant, ϕ the angle which the line representing it makes with AO. Then the magnitudes of the resolved parts at right angles to AO are $P_1 \sin\theta_1$, $P_2 \sin\theta_2$, and $R \sin\phi$, and we know (Article 12) that $R\sin\phi = P_1 \sin\theta_1 + P_2 \sin\theta_2$.

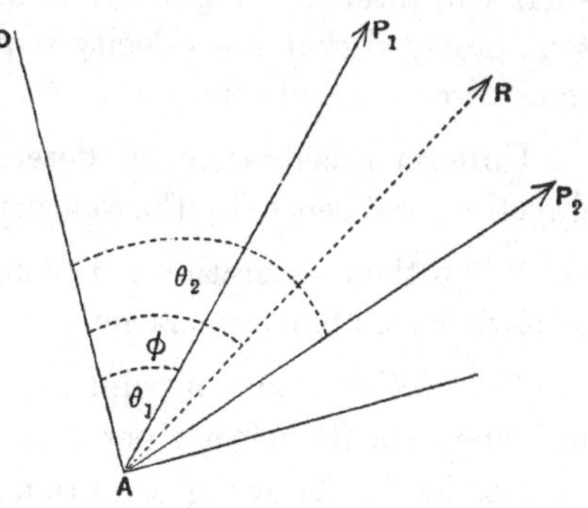

Fig. 16.

Now sum of moments of P_1 and P_2 about O

$$= OA\,(P_1 \sin\theta_1 + P_2 \sin\theta_2)$$
$$= OA \,.\, R\sin\phi$$
$$= \text{moment of } R \text{ about } O.$$

This result can be immediately extended to any number of vectors localized at a point.

It follows that, when a vector localized at a point (x_1, y_1) in the plane of (x, y), or in a line passing through this point, is specified by its components X_1 and Y_1 parallel to the axes of x and y, its moment about the origin is $x_1 Y_1 - y_1 X_1$. See Fig. 17. For example, the moment about the origin of the velocity $\left(\frac{dx}{dt}, \frac{dy}{dt}\right)$ of a particle moving in the plane of (x, y) is $\left(x\frac{dy}{dt} - y\frac{dx}{dt}\right)$, where x and y are the coordinates of its position at time t.

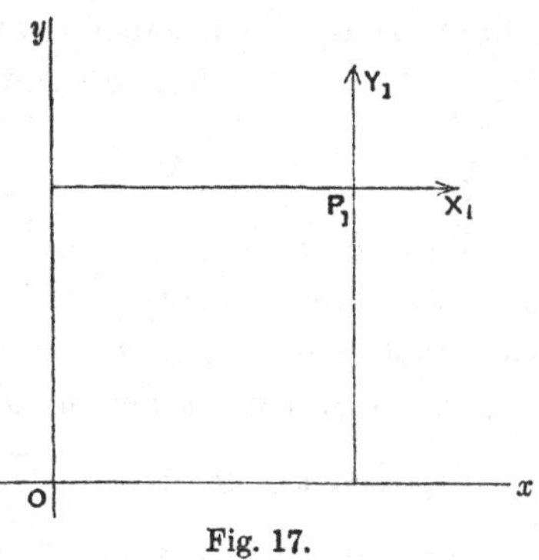

Fig. 17.

23. Acceleration. A point moving with a variable velocity, relative to any frame, is said to have an acceleration relative to that frame.

When the point is moving in such a way that its velocity

increases by equal amounts in equal intervals of time, however short the intervals may be, it is said to have a uniform acceleration, provided that the velocity acquired in every interval has the same direction and sense.

Uniform acceleration is determined, as regards magnitude, direction, and sense, by the velocity added in a unit of time.

When the acceleration is not uniform, the moving point is said to have a variable acceleration.

The acceleration of a point moving in a straight line is the rate of increase of its velocity per unit of time. This is a short way of expressing the following definition:—

Let v be the velocity of the point at time t, and v' its velocity at time t', then its acceleration is the limit of the fraction $\frac{v'-v}{t'-t}$ when the interval $t'-t$ is diminished indefinitely, or in words it is the limit of the fraction

$$\frac{\text{number of units of velocity added in an interval of time}}{\text{number of units of time in the interval}},$$

when the interval is diminished indefinitely. The number v is a function of the number t, and its differential coefficient with respect to t is the acceleration, *i.e.* the acceleration is measured by $\frac{dv}{dt}$.

When the point is not moving in a straight line it will in general have a variable velocity parallel to each of the lines of reference (coordinate axes). Let u, v, w be component velocities parallel to these axes at time t, and u', v', w' corresponding components at time t', then the fractions $\frac{u'-u}{t'-t}$, $\frac{v'-v}{t'-t}$, $\frac{w'-w}{t'-t}$ have limits when the interval $t'-t$ is diminished indefinitely, and these limits are the differential coefficients $\frac{du}{dt}$, $\frac{dv}{dt}$, $\frac{dw}{dt}$. The vector which has these components parallel to the axes is defined to be the acceleration of the point, or in other words *we define the acceleration of a moving point to be the vector, localized in a line through the point, whose resolved part in any direction is the rate of increase of the velocity in that direction per unit of time.*

24. Measurement of acceleration. The measure of any particular acceleration is the number expressing the ratio of the acceleration to the unit acceleration.

The unit acceleration is that uniform acceleration with which a moving point gains a unit of velocity in a unit of time.

The number expressing an acceleration is the ratio of a number expressing a velocity to a number expressing an interval of time. It therefore varies inversely as the unit of length and directly as the square of the unit of time.

Acceleration is accordingly said to be a quantity of one dimension in length and of minus two dimensions in time, or its dimension symbol is LT^{-2}.

Accelerations are not measured directly. The quantities which are measured directly are lengths and angles. By measuring angles we can estimate intervals of time, using a clock or watch, for example. The values of velocities are deduced from a knowledge of the distances described in different intervals of time. The values of accelerations are deduced from a knowledge of the values of velocities at different times.

25. Notation for velocities and accelerations. We have so frequently to deal with differential coefficients of quantities with regard to the time that it is convenient to use for them an abbreviated notation. We shall therefore denote the differential coefficient of any quantity q with regard to the time t by placing a dot over the q, thus $\dot{q}$ stands for $\dfrac{dq}{dt}$.

Now let x, y, z be the coordinates of a moving point at time t, then its component velocities parallel to the axes are denoted by $\dot{x}, \dot{y}, \dot{z}$.

Again let u, v, w be the component velocities of a point parallel to the axes, then its component accelerations are denoted by $\dot{u}, \dot{v}, \dot{w}$.

Since $\dot{u} = \dfrac{d\dot{x}}{dt}$, $\dot{v} = \dfrac{d\dot{y}}{dt}$, $\dot{w} = \dfrac{d\dot{z}}{dt}$ it is convenient to write for them $\ddot{x}, \ddot{y}, \ddot{z}$ respectively. Then $\ddot{x}$ stands for $\dfrac{d^2x}{dt^2}$ or $\dfrac{d}{dt}\left(\dfrac{dx}{dt}\right)$, and so on.

In the same way when we have to deal with any function of the time, say q, we may write $\ddot{q}$ for $\dfrac{d^2q}{dt^2}$, as we write $\dot{q}$ for $\dfrac{dq}{dt}$. Also, following the analogy of the case where q is x, y, or z, we may call $\dot{q}$ the velocity with which q increases, and $\ddot{q}$ the acceleration with which q increases.

26. Angular velocity and acceleration. Let a line, for example the line joining the positions at any time of two moving points, move so as always to be in the same plane with reference to any frame. To fix ideas we shall take the plane to be the coordinate plane of (x, y). Suppose the line to make an angle θ (measured in *radians*) with the axis x at time t, and an angle $\theta + \Delta\theta$ with the same axis at time $t + \Delta t$. Then $\Delta\theta$ is the measure of the angle turned through by the line in the interval measured by Δt, and the limit of the ratio of these two numbers is $\dot{\theta}$, the differential coefficient of θ with respect to t. This number, $\dot{\theta}$, is called the angular velocity of the line. In the same way $\ddot{\theta}$ is called the angular acceleration of the line.

27. Relative coordinates and relative motions. Let x_1, y_1, z_1 be the coordinates of a point A at time t referred to axes with origin at O, x_2, y_2, z_2 the coordinates of a second point B at the same time referred to the same axes, and ξ, η, ζ the coordinates of B at the same time referred to *parallel axes* through A. Then ξ, η, ζ are called the coordinates of B relative to A.

We have
$$\left.\begin{aligned} x_2 &= x_1 + \xi, \\ y_2 &= y_1 + \eta, \\ z_2 &= z_1 + \zeta. \end{aligned}\right\} \qquad \text{.........................(1)}$$

Let accented letters denote at time t' the quantities that correspond to unaccented letters at time t, thus let x_1', y_1', z_1' be the coordinates of A', the position of A at time t'. Then as before
$$\left.\begin{aligned} x_2' &= x_1' + \xi', \\ y_2' &= y_1' + \eta', \\ z_2' &= z_1' + \zeta'. \end{aligned}\right\}$$

By subtraction we deduce
$$\left.\begin{aligned} x_2' - x_2 &= (x_1' - x_1) + (\xi' - \xi), \\ y_2' - y_2 &= (y_1' - y_1) + (\eta' - \eta), \\ z_2' - z_2 &= (z_1' - z_1) + (\zeta' - \zeta). \end{aligned}\right\} \qquad \text{..............(2)}$$

The terms on the left are the components parallel to the axes of the displacement of B.

The terms in the first brackets on the right are the components parallel to the axes of the displacement of A.

The terms in the second brackets on the right are the components of the displacement of B relative to parallel axes with origin at A.

Thus we have the result:—The displacement of a point B relative to axes at O is compounded of the displacement of a point A relative to the same axes and the displacement of B relative to parallel axes through A.

By dividing both members of each of the equations (2) by $t'-t$ and passing to the limit when $t'-t$ is diminished indefinitely, or, what is the same thing, by differentiating equations (1) with respect to t, we find

$$\dot{x}_2 = \dot{x}_1 + \dot{\xi}, \quad \dot{y}_2 = \dot{y}_1 + \dot{\eta}, \quad \dot{z}_2 = \dot{z}_1 + \dot{\zeta},$$

and by differentiating again we find

$$\ddot{x}_2 = \ddot{x}_1 + \ddot{\xi}, \quad \ddot{y}_2 = \ddot{y}_1 + \ddot{\eta}, \quad \ddot{z}_2 = \ddot{z}_1 + \ddot{\zeta}.$$

These equations may be expressed in words as follows:—

The $\left\{\begin{matrix}\text{velocity}\\ \text{acceleration}\end{matrix}\right\}$ of B relative to axes at O is compounded of the $\left\{\begin{matrix}\text{velocity}\\ \text{acceleration}\end{matrix}\right\}$ of A relative to the same axes and the $\left\{\begin{matrix}\text{velocity}\\ \text{acceleration}\end{matrix}\right\}$ of B relative to parallel axes through A.

28. Geometry of relative motion. The geometrical view of relative motion is instructive, and leads easily to results of some importance. For shortness we shall speak of displacement, velocity, and acceleration of a point relative to a second point, meaning thereby displacement, velocity, and acceleration of the point relative to axes drawn through the second point parallel to the axes of reference.

Let A be the position at any time t of a point which moves relatively to a frame having its origin at O, and let A' be its position at time t'. From O draw OH equal and parallel to AA', and in the same sense; the vector represented by OH is the displacement of A.

Similarly let B be the position at time t of a second point referred to the same frame, and B' its position at time t'. From O draw OK equal and parallel to BB', and in the same sense; the vector represented by OK is the displacement of B.

Then the displacement of B relative to A is the vector that must be compounded with the displacement of A in order that the resultant may be the displacement of B.

Join HK. Then the vector OK is compounded of OH, HK.

Hence HK represents the displacement of B relative to A in magnitude, direction, and sense.

Now the vector HK is the resultant of HO, OK.

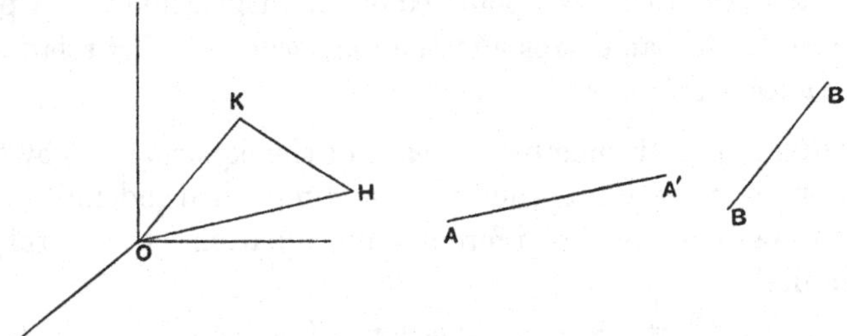

Fig. 18.

Hence to obtain the displacement of B relative to A we must compound the displacement of B with the reversed displacement of A. The resultant is the required relative displacement.

In the same way the $\left\{\begin{matrix}\text{velocity}\\ \text{acceleration}\end{matrix}\right\}$ of B relative to A is the $\left\{\begin{matrix}\text{velocity}\\ \text{acceleration}\end{matrix}\right\}$ which must be compounded with the $\left\{\begin{matrix}\text{velocity}\\ \text{acceleration}\end{matrix}\right\}$ of A in order that the resultant may be the $\left\{\begin{matrix}\text{velocity}\\ \text{acceleration}\end{matrix}\right\}$ of B.

Since the velocity of a point in any direction is the rate of increase of its displacement in that direction per unit of time, and since its acceleration in any direction is the rate of increase of its velocity in that direction per unit of time, we have the rules:—

The $\left\{\begin{matrix}\text{velocity}\\ \text{acceleration}\end{matrix}\right\}$ of B relative to A is the resultant of the $\left\{\begin{matrix}\text{velocity}\\ \text{acceleration}\end{matrix}\right\}$ of B and the $\left\{\begin{matrix}\text{velocity}\\ \text{acceleration}\end{matrix}\right\}$ of A reversed.

The compositions and resolutions described in this Article are to be effected as if the vectors involved were not localized, but the velocity and acceleration of B relative to A are to be regarded as localized in lines through B.

CHAPTER II

THE MOTION OF A FREE PARTICLE IN A FIELD OF FORCE

29. Gravity. An unsupported body near the Earth's surface generally falls towards the Earth. The differences in the behaviour of "light" bodies and "heavy" bodies are to be traced to the buoyancy and resistance of the air. When the effects due to the presence of the air are eliminated, for instance, when bodies fall in the exhausted receiver of an air pump, it is found that all kinds of bodies fall to the Earth with the same acceleration. The direction of this acceleration at any place is the "vertical at the place." The magnitude of this acceleration depends to some extent on latitude; but, in the neighbourhood of any place, it is practically constant. We call it the "acceleration due to gravity," and we denote it by the letter g. When the centimetre is the unit of length, the value of g in London is 981·2, when the foot is the unit of length the value is 32·2. The fact that bodies fall to the Earth with a constant acceleration was discovered by Galileo.

30. Field of force. A region in which a free body moves with a certain acceleration is called a "field of force." The magnitude of the acceleration is the "intensity of the field," and the direction of the acceleration is the "direction of the field." When the intensity and direction of the field are the same at all points the field is said to be "uniform."

For example, the neighbourhood of the Earth is a field of force of which the intensity near the Earth is g. We call it the "field of the Earth's gravity." If we confine our attention to a small part of the Earth's surface we may regard the field as uniform.

31. Rectilinear motion in a uniform field. Let the direction of the field be the axis of x, and let f be its intensity. A particle moving in the field parallel to the axis of x has an acceleration f. Let x_0 be the value of x at the initial position of the particle, and u its velocity (parallel to the axis of x) in this position.

Then we are given $$\ddot{x} = f,$$
with the conditions $x = x_0$ when $t = 0$, and $\dot{x} = u$ when $t = 0$

Writing v for $\dot{x}$, so that v is the velocity at time t, we are given

$$\dot{v} = f,$$

with the condition $v = u$ when $t = 0$.

Now one function of t having the constant f for its differential coefficient is the function ft, and the most general expression for a function having this differential coefficient is $ft + C$, where C is an arbitrary constant. Hence v must be of the form $ft + C$.

Putting $t = 0$, we find $u = C$, so that the constant is determined.

Hence $$v = u + ft, \text{ or } \dot{x} = u + ft.$$

Again one function of t having the function $u + ft$ for its differential coefficient is $ut + \frac{1}{2}ft^2$, hence x must be of the form $C' + ut + \frac{1}{2}ft^2$, where C' is an arbitrary constant.

Putting $t = 0$, we find $x_0 = C'$, so that the constant is determined.

Hence $$x = x_0 + ut + \tfrac{1}{2}ft^2.$$

If s is the distance described in the interval t, s is $x - x_0$, so that

$$s = ut + \tfrac{1}{2}ft^2.$$

By elimination of t between this equation and the equation $v = u + ft$, we find

$$v^2 - u^2 = 2fs.$$

In particular, the velocity acquired in moving from rest over a distance s is $\sqrt{(2fs)}$. This is described as the "velocity due to falling through s with an acceleration f."

32. Examples.

1. Prove that, when the acceleration is uniform, the average velocity in any interval of time is the velocity at the middle of the interval.

2. Obtain the formula $v^2 - u^2 = 2fs$ by multiplying both sides of the equation $\ddot{x} = f$ by $\dot{x}$ and integrating.

3. Let the distance s be divided into a great number of equal segments, and the sum of the velocities after describing those segments divided by their number, a velocity will be obtained which will have a limit when the number of segments is increased indefinitely, and this limit may be called the average velocity in the distance. Prove that, when the initial velocity is zero, this average velocity is equal to $\frac{2}{3}$ of the final velocity.

33. Parabolic motion under gravity. When a particle moving in the field of the Earth's gravity, near a place on the Earth's surface, does not move vertically, it has a component velocity in a horizontal direction. We prove that the particle describes a parabola with a vertical axis.

Let the axis of y be drawn vertically upwards, and let the plane (x, y) be the vertical plane through the initial direction of motion.

Since the acceleration parallel to the axis z is always zero, the particle does not acquire velocity parallel to this axis; and, since at time $t=0$ it has no velocity parallel to this axis, it undergoes no displacement parallel to this axis; thus the particle moves in the plane (x, y).

At time $t=0$ let the velocity of the particle be V in a direction making an angle α with the axis x (see Fig. 19).

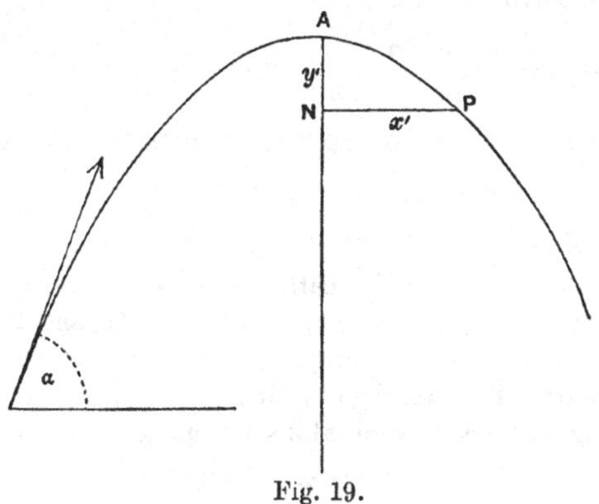

Fig. 19.

We have the equations $\ddot{x}=0,$

$$\ddot{y}=-g,$$

with the conditions that when $t=0$,

$$\dot{x}=V\cos\alpha, \quad \dot{y}=V\sin\alpha.$$

The equation $\ddot{x}=0$ shows that $\dot{x}$ is constant; and, since

$$\dot{x}=V\cos\alpha \text{ when } t=0,$$

the constant value of $\dot{x}$ is $V\cos\alpha$. Thus the horizontal component of the velocity is constant.

The equation $\ddot{y}=-g$ shows that $\dot{y}$ must be of the form

$$-gt+\text{const.};$$

and, since $\dot{y}=V\sin\alpha$ when $t=0$, $\dot{y}=V\sin\alpha-gt$.

Let the coordinates of the position of the particle at the instant when $t=0$ be x_0, y_0.

The equation $\dot{x} = V\cos\alpha$ shows that x must be of the form $Vt\cos\alpha + \text{const.}$; and, since $x = x_0$ when $t = 0$, we have

$$x = x_0 + Vt\cos\alpha.$$

The equation $\dot{y} = V\sin\alpha - gt$ shows that y must be of the form $Vt\sin\alpha - \frac{1}{2}gt^2 + \text{const.}$; and, since $y = y_0$ when $t = 0$, we have

$$y = y_0 + Vt\sin\alpha - \tfrac{1}{2}gt^2.$$

Thus the coordinates x and y are expressed in terms of t. Observing that the formula for y can be written

$$y - y_0 = \frac{V^2\sin^2\alpha}{2g} - \frac{1}{2g}(V\sin\alpha - gt)^2,$$

we can eliminate t, and obtain the equation of the path of the particle in the form

$$y - y_0 = \frac{V^2\sin^2\alpha}{2g} - \frac{1}{2g}\left(V\sin\alpha - g\,\frac{x - x_0}{V\cos\alpha}\right)^2,$$

showing that the path of the particle is a parabola with a vertical axis*.

34. Examples.

1. Prove that the vertex of the path is reached at time $V\sin\alpha/g$, that its coordinates are $x_0 + (V^2/g)\sin\alpha\cos\alpha$, $y_0 + (V^2/2g)\sin^2\alpha$, and that, if the path is referred to axes of x', y' with origin at the vertex and axis of y' drawn vertically downwards (Fig. 19), the coordinates at time t', measured from the instant of passing the vertex as initial instant, are given by the equations

$$x' = Vt'\cos\alpha, \quad y' = \tfrac{1}{2}gt'^2.$$

2. Find the length of the latus rectum of the parabolic trajectory, and determine its focus and directrix.

3. If v is the velocity at any point of the path, show that the point is at a distance $v^2/2g$ below the directrix.

4. Prove that the time until the particle is again in the horizontal plane through the point of projection is $(2V\sin\alpha)/g$. [This is called the time of flight on the horizontal plane through the point of projection.]

5. Prove that the distance from the starting point of the point where the particle strikes the horizontal through the starting point is $(V^2\sin 2\alpha)/g$. [This is called the range on the horizontal plane through the point of projection.]

6. To find the range and time of flight on an inclined plane through the point of projection. Let θ be the inclination of the plane to the horizon.

Resolve up the plane, and at right angles to it. The resolved accelerations are

$$-g\sin\theta, \quad -g\cos\theta;$$

* The result was discovered by Galileo.

the resolved initial velocities are

$$V\cos(\alpha-\theta),\quad V\sin(\alpha-\theta);$$

the resolved velocities at time t are

$$V\cos(\alpha-\theta)-gt\sin\theta,\quad V\sin(\alpha-\theta)-gt\cos\theta;$$

the distances described in time t parallel and perpendicular to the inclined plane are

$$Vt\cos(\alpha-\theta)-\tfrac{1}{2}gt^2\sin\theta,\quad Vt\sin(\alpha-\theta)-\tfrac{1}{2}gt^2\cos\theta.$$

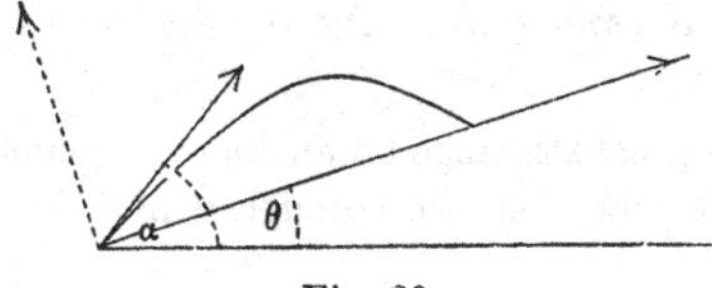

Fig. 20.

The time of flight is obtained by making the second of these equal to zero, it is

$$\frac{2V\sin(\alpha-\theta)}{g\cos\theta}.$$

The range is found by substituting this value for t in $Vt\cos(\alpha-\theta)-\frac{1}{2}gt^2\sin\theta$.

Prove that the range in question is

$$\frac{2V^2\cos^2\alpha}{g\cos\theta}(\tan\alpha-\tan\theta),$$

and that this is the same as

$$\frac{V^2}{g\cos^2\theta}[\sin(2\alpha-\theta)-\sin\theta].$$

7. Prove that, when the velocity of projection is given, the range on an inclined plane is greatest when the direction of projection bisects the angle between the plane and the vertical.

8. Show that, if a parabola is constructed having its focus at the point of projection S, its axis vertical, and its vertex at a height $V^2/2g$ above the point of projection, then the parabolic path for which the range on a line through S is greatest touches this parabola at the point where the line cuts it.

[From this it follows that all possible paths of particles moving with uniform acceleration g downwards, and starting from a point S with given velocity V, touch a paraboloid of revolution about the vertical through S having its focus at S. This paraboloid is the envelope of the trajectories of such particles.]

9. A particle is to be projected from the origin with a given velocity V so as to pass through a given point (x, y), the axes of coordinates being the same as in Art. 33. Prove that the direction of projection must make with the axis x an angle α which satisfies the equation

$$gx^2\tan^2\alpha-2V^2x\tan\alpha+(2V^2y+gx^2)=0,$$

and hence show that there are, in general, two directions in which the particle can be projected, with given velocity, from one given point, so as to pass through another given point.

[Clearly the point (x, y) must lie within the parabola $2V^2y+gx^2=V^4/g$, which is the envelope considered in Ex. 8.]

10. Prove that, in the different trajectories possible under gravity between two points A, B, the times of flight are inversely proportional to the velocities of the projectile when vertically over the middle point of AB.

11. A particle moves under gravity from the highest point of a sphere of radius c. Prove that it cannot clear the sphere unless its initial velocity exceeds $\sqrt{(\frac{1}{2}gc)}$.

12. Prove that the greatest range on an inclined plane through the point of projection is equal to the distance through which the particle would fall during the time of flight.

35. Motion in a curved path. When the motion of a body, treated as a particle, is observed, the things that can be observed are the positions of the particle at different times. The aggregate of these positions constitutes the path of the particle. For example, the path may be a circle, and equal arcs may be described in equal times. In such cases we have the mathematical problem of deducing the acceleration of the particle from the observations, that is to say the problem of determining the direction and intensity of the field of force. Conversely we may set before ourselves the problem: Given the acceleration of the particle, to determine its path and its positions at different times. The solutions of such problems are facilitated by a theorem of kinematics to which we proceed.

36. Acceleration of a point describing a plane curve. Let a particle move in the plane of (x, y).

Let v be the velocity at any point P of the path, v' the velocity at a neighbouring point Q, and $\Delta\phi$ the angle QTA between the

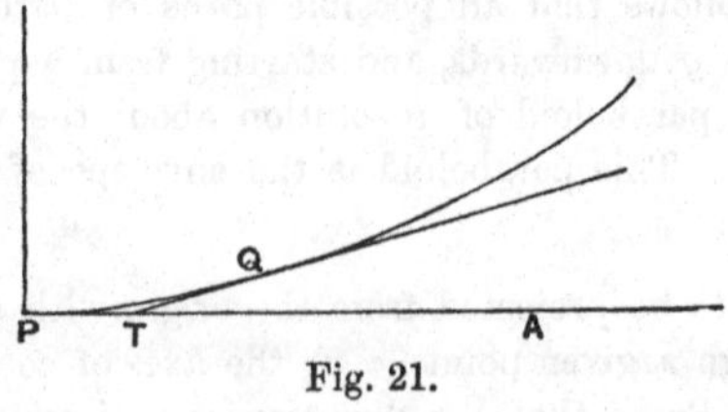

Fig. 21.

tangent at P and the tangent at Q. Also let Δt be the time taken by the particle to move from P to Q, and let Δs be the length of the arc PQ.

The velocity at Q can be resolved into components $v' \cos \Delta\phi$ in the direction of the tangent at P and $v' \sin \Delta\phi$ in the direction of the normal at P.

Hence the acceleration in the direction of the tangent at P is the limit of $\dfrac{v' \cos \Delta\phi - v}{\Delta t}$ when Δt is diminished indefinitely. Now

$$\frac{v' \cos \Delta\phi - v}{\Delta t} = \frac{v' - v}{\Delta t} - v' \frac{1 - \cos \Delta\phi}{\Delta t},$$

and
$$\frac{1 - \cos \Delta\phi}{\Delta t} = \frac{2 \sin^2 (\frac{1}{2}\Delta\phi)}{(\Delta\phi)^2} \frac{\Delta\phi}{\Delta t} \Delta\phi.$$

The limits of the three factors of this expression are $\frac{1}{2}$, $\dot{\phi}$, zero. Hence the above limit is $\dfrac{dv}{dt}$ or $\dot{v}$. Since we have

$$\frac{dv}{dt} = \frac{dv}{ds} \cdot \frac{ds}{dt} = v \frac{dv}{ds},$$

we may write $v \dfrac{dv}{ds}$ for the component acceleration parallel to the tangent, and we may also write $\ddot{s}$ for it, since v is $\dot{s}$.

Again the acceleration in the direction of the normal at P is the limit of $\dfrac{v' \sin \Delta\phi}{\Delta t}$, and this is the same as the limit of

$$v' \frac{\sin \Delta\phi}{\Delta\phi} \frac{\Delta\phi}{\Delta s} \frac{\Delta s}{\Delta t},$$

and the limits of these factors in order are v, 1, $\dfrac{1}{\rho}$, v, where ρ is the radius of curvature of the curve at P. Thus the acceleration in the direction of the normal drawn towards the centre of curvature is $\dfrac{v^2}{\rho}$ or $\dfrac{\dot{s}^2}{\rho}$.

37. Examples.

1. A particle describing a circle of radius a with velocity v has an acceleration v^2/a along the radius directed inwards.

If the radius vector drawn from the centre to the particle turns through an angle θ in time t, the acceleration of the particle has components $a\dot{\theta}^2$ along the radius (directed towards the centre) and $a\ddot{\theta}$ along the tangent in the sense of increase of θ.

2. Verify the result that, in parabolic motion of a projectile under gravity, the value of v^2/ρ at any point of the path is equal to the resolved part along the normal to the path of an acceleration equal to g.

3. Assuming this result, and that the horizontal component of the velocity is constant, deduce the result that the path is a parabola.

4. Interpret the formula v^2/ρ for the normal component acceleration so as to show that the velocity, at any point P, of a particle describing a curved path, in any field of force, is equal to that due to falling through one quarter of the chord of curvature at P, drawn in the direction of the field, with an acceleration equal to the intensity of the field at P.

38. Simple harmonic motion. A point moving in a straight line in such a way that its displacement from a fixed point at time t can be expressed in the form

$$a \cos(nt + \epsilon),$$

where a, n, ϵ are any real constants, is said to have a "simple harmonic motion."

Let the straight line be the axis of x, and the fixed point the origin. Then we have

$$x = a \cos(nt + \epsilon),$$

and therefore

$$\ddot{x} = -n^2 x.$$

We shall now show that, if the acceleration is connected with the displacement by an equation of the form

$$\ddot{x} = -\mu x,$$

where μ is a positive constant, the motion is simple harmonic motion.

Multiply both sides of the equation by $\dot{x}$. Observe that

$$\ddot{x}\dot{x} = \frac{d}{dt}(\tfrac{1}{2}\dot{x}^2), \text{ and } x\dot{x} = \frac{d}{dt}(\tfrac{1}{2}x^2),$$

so that

$$\frac{d}{dt}(\tfrac{1}{2}\dot{x}^2 + \tfrac{1}{2}\mu x^2) = 0.$$

Hence $\dot{x}^2 + \mu x^2$ is constant. Since $\dot{x}^2 + \mu x^2$ is necessarily positive, we may take the constant value of it to be μa^2, where a is real, and may be taken to be positive. Then

$$\dot{x}^2 = \mu(a^2 - x^2).$$

Since $\dot{x}^2$ is necessarily positive, this equation shows that the value of x cannot be greater than a or less than $-a$. We may therefore introduce a real variable θ, in place of x, by the equation

$$x = a \cos\theta.$$

Then the equation $\dot{x}^2 = \mu(a^2 - x^2)$ becomes

$$\dot{\theta}^2 = \mu,$$

which gives

$$\theta = \pm(t\sqrt{\mu} + \epsilon),$$

where ϵ is an arbitrary constant. We have thus obtained the complete primitive of the equation $\ddot{x} = -\mu x$ in the form

$$x = a \cos(t\sqrt{\mu} + \epsilon).$$

This equation represents a simple harmonic motion. The motion is periodic, that is to say, it repeats itself after equal intervals of time. The period is $2\pi/\sqrt{\mu}$. In the formula for x the constant a is called the "amplitude" of the motion, and the constant ϵ determines the "phase" of the motion.

On putting

$$a \cos\epsilon = A, \qquad -a \sin\epsilon = B,$$

the formula for x becomes

$$x = A \cos(t\sqrt{\mu}) + B \sin(t\sqrt{\mu}),$$

which is another form of the complete primitive of the equation $\ddot{x} = -\mu x$.

Let the moving point have at time $t=0$ a position denoted by x_0 and a velocity denoted by $\dot{x}_0$. In the formula last written put $t=0$, then $x_0 = A$. Again differentiate both sides of the formula with respect to t, and in the result put $t=0$, then $\dot{x}_0 = B\sqrt{\mu}$. Hence the formula may be written

$$x = x_0 \cos(t\sqrt{\mu}) + \frac{\dot{x}_0}{\sqrt{\mu}} \sin(t\sqrt{\mu}).$$

Simple harmonic motion may be regarded as the type of to-and-fro, or oscillatory, motion. Oscillatory motions can generally be described either as simple harmonic motions or as motions compounded of simple harmonic motions in different directions.

39. Composition of simple harmonic motions. We consider the case where the moving particle has a simple harmonic motion of period $\frac{2\pi}{\sqrt{\mu}}$ parallel to each of the axes of x and y, the acceleration in each case being directed towards the origin.

We have the equations

$$\ddot{x} = -\mu x,$$
$$\ddot{y} = -\mu y,$$

and we deduce that x and y must be given by equations of the form

$$x = A \cos(t\sqrt{\mu}) + B \sin(t\sqrt{\mu}),$$
$$y = C \cos(t\sqrt{\mu}) + D \sin(t\sqrt{\mu}),$$

where A, B, C, D are arbitrary constants depending on the initial conditions, viz. A and C are the coordinates, and $B\sqrt{\mu}$, $D\sqrt{\mu}$ the resolved velocities at the instant $t=0$.

Solving the above equations for $\cos(t\sqrt{\mu})$ and $\sin(t\sqrt{\mu})$, we have

$$(AD-BC)\cos(t\sqrt{\mu})=Dx-By,\quad (AD-BC)\sin(t\sqrt{\mu})=Ay-Cx;$$

eliminating t, we find

$$(Dx-By)^2+(Ay-Cx)^2=(AD-BC)^2,$$

so that the path of the moving point is an ellipse whose centre is the origin, and whose position with reference to the origin and axes is fixed. The whole motion is clearly periodic with period $\frac{2\pi}{\sqrt{\mu}}$.

Let us change the axes to the principal axes of the ellipse, and suppose the moving point to be at one extremity ($x=a$) of the major axis at the instant $t=0$, then at this instant $x=a$, $y=0$, and, since the point is moving at right angles to the major axis, $\dot{x}=0$. Let $\dot{y}=b\sqrt{\mu}$ at this instant. Then we must have at time t

$$x=a\cos(t\sqrt{\mu}),\quad y=b\sin(t\sqrt{\mu}).$$

Thus $2b$ is the minor axis, and $t\sqrt{\mu}$ is the eccentric angle at time t.

The point therefore moves so that its eccentric angle increases uniformly with angular velocity $\sqrt{\mu}$.

40. Examples.

1. Prove that, if a point N moves on a fixed diameter of a circle (Fig. 22) so that its acceleration is given by the equation $\ddot{x}=-\mu x$, a point P on the circle, whose projection on the fixed diameter is N, describes the circle with constant speed.

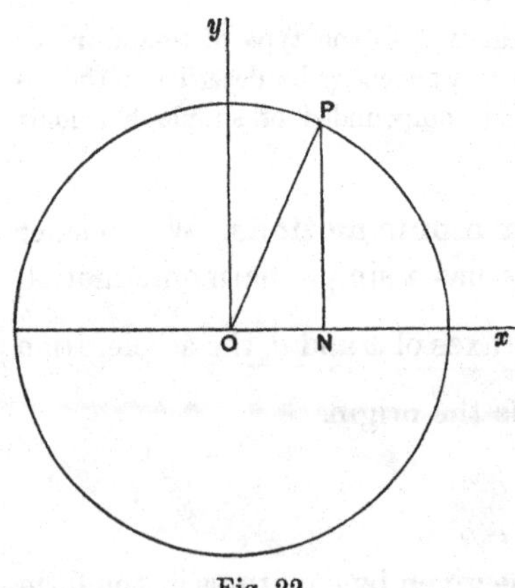

Fig. 22.

2. Prove that, when the equation is $\ddot{x}=\mu x$, where μ is positive, and the initial conditions are that $x=x_0$ and $\dot{x}=\dot{x}_0$ when $t=0$, then at any time t

$$x=x_0\cosh(t\sqrt{\mu})+(\dot{x}_0/\sqrt{\mu})\sinh(t\sqrt{\mu}).$$

3. Prove that when the acceleration of a particle moving in a plane is directed from the origin and is proportional to the distance the path is an hyperbola.

4. In the elliptic motion of Art. 39 prove that the velocity v at distance r from the centre is given by

$$v^2+\mu r^2=\text{const.},$$

and evaluate the constant.

5. In the hyperbolic motion of Ex. 2 prove that the velocity v at distance r from the centre of the hyperbola is given by

$$v^2=\mu r^2+\text{const.},$$

and evaluate the constant.

41. Kepler's laws of planetary motion. From a long series of observations of the Planets, and more especially of Mars, which were made by Tycho Brahe, Kepler* concluded that the motions of the Planets could be very precisely described by means of the two laws:—

(i) Every planet describes an ellipse having the Sun at a focus.

(ii) The radius drawn from the Sun to a Planet describes equal areas in equal times.

42. Equable description of areas. We consider the second of Kepler's laws, and suppose that a particle describes a plane curve in such a way that the radius vector drawn to it from a fixed point in the plane describes area uniformly. In Fig. 23 O represents the fixed point, B any fixed point on the curve, P the position of the particle at time t, r the radius vector OP, p the perpendicular from O on the tangent at P, v the velocity of the particle at P.

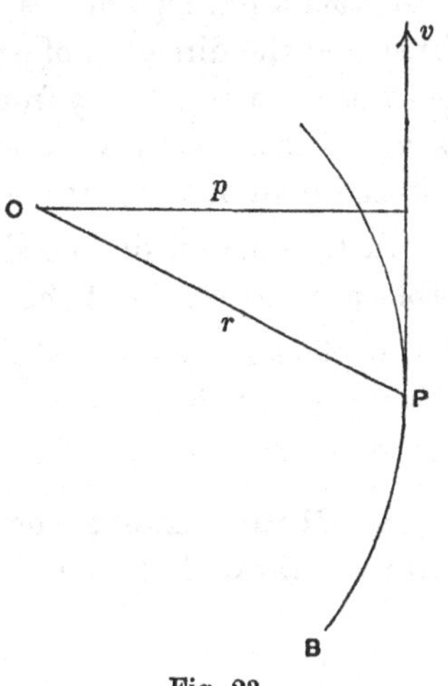

Fig. 23.

Let P' be a point on the curve near to P, Δt the time of moving from P to P', Δs the arc PP', Δc the chord PP', q the perpendicular from O to this chord. The area of the triangle POP' is $\frac{1}{2}q\,\Delta c$. Hence the rate of description of area is the limit of $\frac{1}{2}q\,\frac{\Delta c}{\Delta t}$ or $\frac{1}{2}q\,\frac{\Delta c}{\Delta s}\,\frac{\Delta s}{\Delta t}$; and this limit is $\frac{1}{2}p\dot{s}$ or $\frac{1}{2}pv$. If therefore we write

$$pv = h,$$

h is twice the rate of description of area, and the condition that the radius vector describes area uniformly is expressed by saying that h or pv is constant.

Now pv is the moment of the velocity about O. If therefore

* Joannes Kepler, *Astronomia nova...tradita Commentariis de Motibus Stellæ Martis*, 1609.

we take O to be the origin of coordinates and draw the axes of x and y in the plane of motion, we have (cf. Article 22)

$$pv = x\dot{y} - y\dot{x} = h;$$

and, since this is constant, we have $\frac{d}{dt}(x\dot{y} - y\dot{x}) = 0$, or $x\ddot{y} - y\ddot{x} = 0$, and therefore

$$\frac{\ddot{x}}{x} = \frac{\ddot{y}}{y}.$$

It follows that the direction of the acceleration is that of the radius vector, drawn from or towards the origin. We conclude that, if a particle moves in a plane path, so that the radius vector drawn to it from a fixed point describes area uniformly, it is in a field of force, and the direction of the field at any point is either directly towards or directly away from the fixed point. Such a field of force is described as "central," the fixed point being the "centre of force," and the path of the particle is a "central orbit."

In the motion discussed in Article 39 the ellipse is a central orbit, and the centre of the ellipse is the centre of force.

Kepler's second law of planetary motion may be interpreted in the statement that the Planets move in a central field of force, the centre of force being in the Sun.

43. Radial and transverse components of velocity and acceleration. Let a particle move in the plane of (x, y) and let

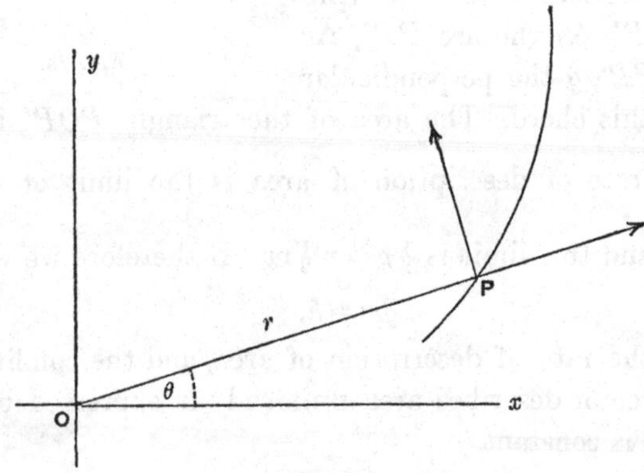

Fig. 24.

r, θ be the polar coordinates of its position at time t. It is required to express, in terms of r, θ and their differential coefficients with respect to t, the components of the velocity and acceleration in the direction of the radius vector and at right angles to it. The senses are to be those in which r and θ increase, as in Fig. 24.

Let v_1, v_2 be the required components of velocity. Then $\dot{x}$, $\dot{y}$ are the components parallel to the axes of x, y of the same velocity. We have therefore

$$v_1 \cos\theta - v_2 \sin\theta = \dot{x} = \frac{d}{dt}(r\cos\theta) = \dot{r}\cos\theta - r\dot{\theta}\sin\theta,$$

$$v_1 \sin\theta + v_2 \cos\theta = \dot{y} = \frac{d}{dt}(r\sin\theta) = \dot{r}\sin\theta + r\dot{\theta}\cos\theta.$$

Solving these equations, we find

$$v_1 = \dot{r}, \quad v_2 = r\dot{\theta}.$$

Let f_1, f_2 be the required components of acceleration. We have in like manner

$$\begin{aligned} f_1 \cos\theta - f_2 \sin\theta = \ddot{x} &= \frac{d^2}{dt^2}(r\cos\theta) \\ &= \ddot{r}\cos\theta - 2\dot{r}\dot{\theta}\sin\theta - r\ddot{\theta}\sin\theta - r\dot{\theta}^2\cos\theta, \\ f_1 \sin\theta + f_2 \cos\theta = \ddot{y} &= \frac{d^2}{dt^2}(r\sin\theta) \\ &= \ddot{r}\sin\theta + 2\dot{r}\dot{\theta}\cos\theta + r\ddot{\theta}\cos\theta - r\dot{\theta}^2\sin\theta. \end{aligned}$$

Solving these equations, we find

$$f_1 = \ddot{r} - r\dot{\theta}^2, \quad f_2 = r\ddot{\theta} + 2\dot{r}\dot{\theta} = \frac{1}{r}\frac{d}{dt}(r^2\dot{\theta}).$$

It is important to observe that the acceleration parallel to the radius vector is the resolved part along the radius vector of the acceleration relative to the frame Ox, Oy; it is not the acceleration with which the radius vector increases.

44. Examples.

1. Since the moment of the velocity about the origin is $r \,.\, r\dot{\theta}$, we verify the formulæ of Differential Calculus

$$r^2\dot{\theta} = x\dot{y} - y\dot{x} = p\dot{s}.$$

2. In a central orbit we have

$$h = r^2\dot{\theta}.$$

3. A point P describes a curve C relatively to axes through O. Prove that, relatively to parallel axes through P, O describes a curve equal in all respects to C, and that any point dividing OP in a constant ratio describes, relatively to either of these sets of axes, a curve similar to C.

45. Acceleration in central orbit. Let f be the magnitude of the central acceleration at P, and let it be directed towards P. Let r, p, ρ denote the radius vector OP drawn from the centre of force O, the perpendicular from O on the tangent at P, and the radius of curvature of the path at P. (Cf. Fig. 23 in Art. 42.)

The resolved part of the acceleration parallel to the normal at P is $f\frac{p}{r}$.

But this resolved part of the acceleration is $\frac{v^2}{\rho}$.

Hence
$$\frac{v^2}{\rho}=f\frac{p}{r}.$$

From this equation and the equation $vp=h$ we may eliminate v, and obtain the equation
$$f=\frac{h^2r}{p^3\rho}.$$

Since $\rho=r\frac{dr}{dp}$, we may also write this equation
$$f=\frac{h^2}{p^3}\frac{dp}{dr}.$$

46. Examples.

1. Show that, when the orbit is an ellipse described about the centre, the acceleration is proportional to the radius vector.

2. In the same case show that the velocity at any point is proportional to the length of the diameter conjugate to the diameter through the point.

3. Points move from a position P with a velocity V in different directions with an acceleration to a point C proportional to the distance. Prove that all the elliptic trajectories described have the same director circle.

Let the tangent at P to one of the trajectories meet the director circle in T, and let Q be the point of contact of the other tangent to this trajectory drawn from T. Prove that the trajectory in question touches at Q an ellipse having C as centre, and P as one focus, and that $2CT$ is the length of the major axis of this ellipse.

[This ellipse is the envelope of the trajectories of points starting from P with the given velocity and moving about C with the given central acceleration.]

4. Show that the central acceleration when a circle is described as a central orbit about a point on the circumference is $8h^2a^2/r^5$, a being the radius of the circle.

5. Show that the central acceleration when an equiangular spiral is described as a central orbit about its pole is proportional to r^{-3}.

6. Show that, for an ellipse described as a central orbit about any point O in its plane, the central acceleration at any point P is proportional to r/q^3, where r is the radius vector OP, and q is the perpendicular from P on the polar of O.

47. Elliptic motion about a focus. We consider now the interpretation of the first of Kepler's laws (Art. 41). Let an ellipse of semi-axes a, b be described as a central orbit about a focus S. Let S' be the second focus, e the eccentricity, $2l$ the latus rectum.

Let P be any point on the ellipse; let r and r' be the radii vectores drawn from S and S' to P; let p and p' be the perpendiculars from S and S' on the tangent at P; let C be the centre, and CD the semi-diameter conjugate to CP.

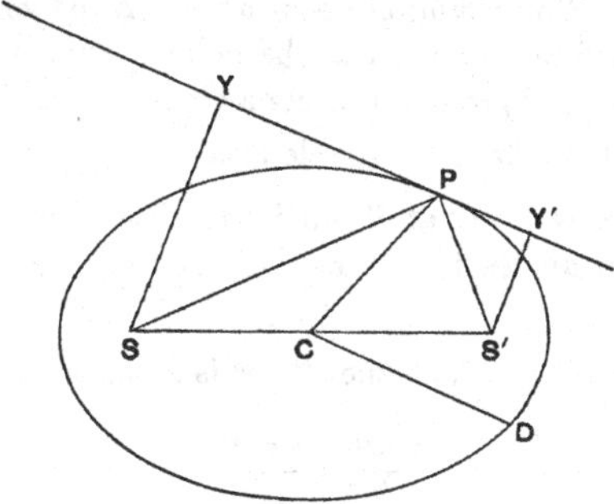

Fig. 25.

Then

$$\rho = CD^3/ab, \quad rr' = CD^2, \quad pp' = b^2, \quad r + r' = 2a, \quad b^2 = al.$$

Also, since $\angle SPY = \angle S'PY'$, we have

$$\frac{p}{r} = \frac{p'}{r'}, \text{ and therefore each of these } = \sqrt{\frac{pp'}{rr'}} = \frac{b}{CD}.$$

Now the acceleration, f, is given by

$$f = \frac{h^2 r}{p^3 \rho}$$

$$= \frac{h^2 rab}{CD^3}\left(\frac{CD}{br}\right)^3 = \frac{h^2}{r^2}\frac{a}{b^2} = \frac{h^2}{r^2 l}.$$

Thus the acceleration varies inversely as the square of the distance r, and, if we write μ/r^2 for it, we have $h^2 = \mu l$.

Accordingly Kepler's first and second laws of planetary motion may be interpreted in the statement that the field of force in which the Planets move is directed radially towards the Sun, and the intensity of the field varies inversely as the square of the distance from the Sun. The field is described as that of the Sun's *gravitation.*

48. Examples.

1. Prove that, if any conic is described as a central orbit about a focus, the acceleration is μ/r^2 towards the focus, and $\mu = h^2/l$.

Prove also that when the conic is a parabola $v^2 = 2\mu/r$, and when it is an hyperbola $v^2 = \mu(2/r + 1/a)$.

2. Prove that the velocity v at any point of the ellipse is given by the equation

$$v^2 = \mu\left(\frac{2}{r} - \frac{1}{a}\right).$$

3. Prove that in elliptic motion about a focus S the velocity at any point P is perpendicular and proportional to the radius vector from the other focus to the point W, where SP produced meets a circle centre S and radius $2a$.

[From the formula in Ex. 2, this circle is called the "circle of no velocity."]

4. Prove that the velocity at P can be resolved into two constant components, one at right angles to the radius vector SP, and the other at right angles to the major axis.

5. The periodic time in which the ellipse is described is

$$\frac{2\pi ab}{h} = \frac{2\pi a^{\frac{3}{2}}}{\sqrt{\mu}}.$$

6. To find the time of describing any arc of the ellipse.

Draw the auxiliary circle AQA'.

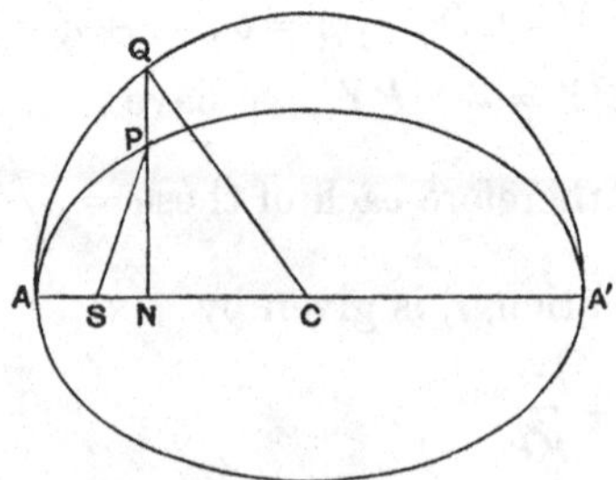

Fig. 26.

Let ϕ, $=\angle QCA$ in the figure, be the eccentric angle of P, and θ, $=\angle ASP$, the vectorial angle.

Then curvilinear area ASP = curvilinear area ANP − triangle SPN

$$= \frac{b}{a}\,(\text{curvilinear area } ANQ) - \text{triangle } SPN.$$

Now curvilinear area ANQ = sector ACQ − triangle CQN

$$=\tfrac{1}{2}(a^2\phi - a^2\sin\phi\cos\phi),$$

and triangle $SPN=\tfrac{1}{2}b\sin\phi\,(ae-a\cos\phi)$.

Hence curvilinear area $ASP=\tfrac{1}{2}ab\,(\phi-e\sin\phi)$.

Let t be the time from A to P, then, since h is twice the area described per unit of time,

$$ht=ab\,(\phi-e\sin\phi).$$

Thus
$$t=\frac{a^{\frac{3}{2}}}{\sqrt{\mu}}(\phi-e\sin\phi).$$

The quantity $\sqrt{\mu}/a^{\frac{3}{2}}$ is known as the "mean motion" and is denoted by n, so that the time in question is given by

$$nt=\phi-e\sin\phi.$$

By putting $\phi=2\pi$ we find the periodic time, as in Ex. 5.

Prove that θ is connected with ϕ by the equation

$$\cos\phi=\frac{e+\cos\theta}{1+e\cos\theta},$$ and that, if e is small,

$$\theta=nt+2e\sin nt \text{ approximately.}$$

7. (i) In an ellipse described about a focus S let ϕ_1, ϕ_2 be the eccentric angles of two points P_1, P_2, and let $\phi_1>\phi_2$. Let $\delta=\tfrac{1}{2}(\phi_1-\phi_2)$, and let σ be such that $\cos\sigma=e\cos\tfrac{1}{2}(\phi_1+\phi_2)$ and $\sin\sigma\sin\delta$ is positive. Prove that, if r_1, r_2 are the focal radii SP_1, SP_2, d is the chord P_1P_2, and t the time of describing the arc P_1P_2, then

$$r_1+r_2+d=2a\{1-\cos(\sigma+\delta)\},\quad r_1+r_2-d=2a\{1-\cos(\sigma-\delta)\},$$

and
$$nt=\chi_1-\chi_2-(\sin\chi_1-\sin\chi_2),$$

where
$$\sin\tfrac{1}{2}\chi_1=\tfrac{1}{2}\sqrt{\{(r_1+r_2+d)/a\}},\quad \sin\tfrac{1}{2}\chi_2=\tfrac{1}{2}\sqrt{\{(r_1+r_2-d)/a\}}.$$

(ii) By taking in Ex. 7 (i) both ϕ_1 and ϕ_2 to lie between $-\tfrac{1}{2}\pi$ and $\tfrac{1}{2}\pi$, prove that the point of intersection of P_1P_2 with the major axis lies on CS or CS produced according as $\sigma<\delta$ or $\sigma>\delta$; and, by passing to the limit when the ellipse becomes a parabola, prove that the time t of describing the arc P_1P_2 of a parabolic orbit about S is given by

$$6t\sqrt{\mu}=(r_1+r_2+d)^{\frac{3}{2}}\pm(r_1+r_2-d)^{\frac{3}{2}},$$

where the upper or lower sign is to be taken according as S lies within the finite area bounded by P_1P_2 and the parabola or not.

8. Two points describe ellipses of latera recta l and l' in different planes about a common focus, and the accelerations to the focus are equal when the distances are equal. Show that, when the relative velocity of the points is along the line joining them, the tangents to the ellipses at the positions of the points meet the line of intersection of the planes in the same point, and that the focal distances, r and r', make with this line angles θ and θ' such that

$$\frac{r\sin\theta}{\sqrt{l}}=\frac{r'\sin\theta'}{\sqrt{l'}}.$$

49. Inverse problem of central orbits. In regard to the problem: Given the field of force to find the orbit—we prove a general theorem as follows:—*The path of a particle moving in a central field of force is in a plane through the centre of force, and the radius vector drawn from the centre of force to the particle describes equal areas in equal times.*

At any instant, chosen as initial instant, let a plane be drawn through the tangent to the path of the particle and the centre of force. Let this be the plane (x, y), and let the centre of force be the origin. Then at the initial instant z and $\dot{z}$ vanish.

Since the acceleration is directed along the radius vector we have

$$\frac{\ddot{x}}{x} = \frac{\ddot{y}}{y} = \frac{\ddot{z}}{z},$$

or
$$y\ddot{z} - z\ddot{y} = 0, \quad z\ddot{x} - x\ddot{z} = 0, \quad x\ddot{y} - y\ddot{x} = 0.$$

Hence, by integration,

$$y\dot{z} - z\dot{y} = \text{const.}, \quad z\dot{x} - x\dot{z} = \text{const.}, \quad x\dot{y} - y\dot{x} = \text{const.}$$

The first two constants of integration vanish because z and $\dot{z}$ vanish initially. If the third also vanishes, the velocity is directed along the radius vector, and the particle moves in a straight line. We omit, for the present, the case of rectilinear motion (see Art. 54).

We may consider the equations

$$x\dot{z} - \dot{x}z = 0, \quad y\dot{z} - \dot{y}z = 0$$

as simultaneous equations to determine $\dot{z}$ and z. If $x\dot{y} - \dot{x}y$ does not vanish, these equations can only be satisfied by putting $\dot{z}$ and z equal to zero. Hence z is always zero, and the particle moves in the plane (x, y).

Since $x\dot{y} - y\dot{x}$, or the moment of the velocity, is constant, the rate of description of area by the radius vector is constant; for we saw in Art. 42 that this rate, whether constant or not, is always half the moment of the velocity about the origin.

50. Determination of central orbits in a given field. The tangential component of the acceleration of a particle describing any path can be expressed as $v\dfrac{dv}{ds}$ (Art. 36). When the acceleration is of magnitude f, and is directed towards the origin, the tangential component is $-f\dfrac{dr}{ds}$, for $\dfrac{dr}{ds}$ is the cosine of the angle between the tangent and the radius vector drawn from the origin. We have therefore the equation

$$v\frac{dv}{ds} = -f\frac{dr}{ds}. \quad \text{.........................(1)}$$

When f is a function of r, this equation can be integrated in the form

$$\tfrac{1}{2}v^2 = A - \int f dr, \quad \text{.........................(2)}$$

where A is a constant. Now, according to Art. 43, we have

$$v^2 = \dot{r}^2 + r^2\dot{\theta}^2,$$

and we have also, by Ex. 2 in Art. 44,

$$r^2\dot{\theta} = h.$$

Hence we may write

$$\dot{r} = \frac{dr}{d\theta}\dot{\theta} = \frac{h}{r^2}\frac{dr}{d\theta},$$

and equation (2) becomes

$$\frac{h^2}{r^4}\left(\frac{dr}{d\theta}\right)^2 + \frac{h^2}{r^2} = 2A - 2\int f dr.$$

If u is written for $1/r$, this equation becomes

$$\left(\frac{du}{d\theta}\right)^2 + u^2 = \frac{2A}{h^2} + \frac{2}{h^2}\int \frac{f}{u^2} du, \quad \text{..............(3)}$$

in which f is supposed to be expressed as a function of u. By this equation we can express $\dfrac{du}{d\theta}$ as a function of u, and then by integration we can find the polar equation of the path.

It is often more convenient to eliminate A from equation (3) by differentiating with respect to θ. This process gives the equation

$$\frac{d^2u}{d\theta^2} + u = \frac{f}{h^2u^2}. \quad \text{.........................(4)}$$

51. Orbits described with a central acceleration varying inversely as the square of the distance. When $f = \mu u^2$ equation (4) of Art. 50 becomes

$$\frac{d^2u}{d\theta^2} + u = \frac{\mu}{h^2} = \frac{1}{l} \text{ say,}$$

where l is a constant. To integrate this equation we put

$$u = \frac{1}{l} + w,$$

then w satisfies the equation

$$\frac{d^2w}{d\theta^2} + w = 0.$$

The complete primitive of this equation is of the form (cf. Art. 38)

$$w = A \cos(\theta - \epsilon),$$

where A and ϵ are arbitrary constants. We write e/l for A. Then the most general possible form for u is

$$u = \frac{1}{l}\{1 + e\cos(\theta - \epsilon)\}.$$

Hence all the orbits that can be described with central acceleration equal to μ/r^2 are included in the equation

$$\frac{l}{r} = 1 + e\cos(\theta - \epsilon),$$

in which e and ϵ are arbitrary constants, and l is equal to h^2/μ.

The possible orbits are conics having the origin as a focus, and the latus rectum is equal to $2l$ or $2h^2/\mu$.

According to the results of Examples 1 and 2 in Art. 48, the conic is an ellipse, parabola or hyperbola according as the velocity at a distance r is less than, equal to, or greater than $\sqrt{(2\mu/r)}$.

52. Additional Examples of the determination of central orbits in given fields.

1. If f is any function of r, any circle described about the centre is a possible orbit.

2. If $f = \mu r$ equation (3) of Art. 50 gives

$$\left(\frac{du}{d\theta}\right)^2 + u^2 = \text{const.} - \frac{\mu}{h^2u^2}.$$

Hence prove that, when μ is positive, all the possible orbits are ellipses having the centre of force as centre.

3. To find all the orbits which can be described with a central acceleration varying inversely as the cube of the distance.

If $f = \mu u^3$ equation (4) of Art. 50 gives

$$\frac{d^2u}{d\theta^2} + u = \frac{\mu}{h^2}u,$$

or

$$\frac{d^2u}{d\theta^2} + u\left(1 - \frac{\mu}{h^2}\right) = 0.$$

There are three cases according as $h^2 >, =,$ or $< \mu$.

(1) When $h^2 > \mu$, $1-\frac{\mu}{h^2}$ is positive, put it equal to n^2.

Then all the possible orbits are of the form $u = A\cos(n\theta + a)$.

(2) When $h^2 = \mu$, we have $\frac{d^2u}{d\theta^2} = 0$, so that $u = A\theta + B$, where A and B are arbitrary constants. If $A = 0$ the orbit is a circle, otherwise it is a hyperbolic spiral, as we see by choosing the constant B so as to write the above

$$u = A(\theta - a).$$

(3) When $h^2 < \mu$, $1 - \frac{\mu}{h^2}$ is negative, put it equal to $-n^2$.

Then all the possible orbits are of the form

$$u = A\cosh(n\theta + a) \quad \text{or} \quad u = ae^{n\theta} + be^{-n\theta}.$$

Putting a or b equal to zero we have an equiangular spiral.

4. Deduce the equation

$$\frac{d^2u}{d\theta^2} + u = \frac{f}{h^2u^2}$$

from the equation

$$f = \frac{h^2r}{p^3\rho}.$$

5. From the equations

$$\ddot{r} - r\dot{\theta}^2 = -f, \quad \frac{1}{r}\frac{d}{dt}(r^2\dot{\theta}) = 0,$$

which are obtained from the results of Art. 43, deduce the results

$$\dot{\theta} = hu^2, \quad \frac{d^2u}{d\theta^2} + u = \frac{f}{h^2u^2}.$$

53. Conic described about a focus. Focal chord of curvature. Let a particle be projected from a point P, with a given velocity V, in an assigned direction PT, and move in a central field of force, directed to S, and equal to μ/r^2 at a distance r from S. It describes a conic with a focus at S. Let PQ be the focal chord of curvature of this conic at P. Then, according to the result of Ex. 4 in Art. 37,

$$V^2 = 2\frac{\mu}{SP^2}\frac{PQ}{4};$$

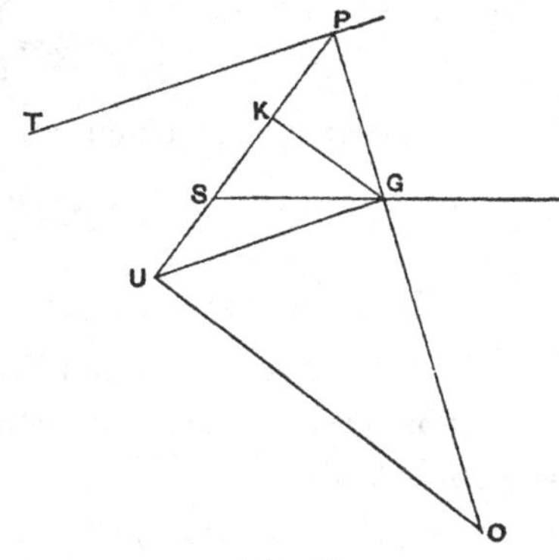

Fig. 27.

and according to the results of Exx. 1 and 2 in Art. 48, the orbit is an ellipse, parabola, or hyperbola according as V^2 is less than, equal to, or greater than $2\mu/SP$. We conclude that, when a point P, the tangent PT, a focus S, and the focal chord of curvature PQ are given, one conic, and one only, can be described, and this conic is an ellipse, parabola, or hyperbola according as $SP >$, $=$, or $< \frac{1}{4}PQ$. This may be proved directly as follows:—

Let U be the middle point of PQ. Draw PG at right angles to PT, and UG parallel to PT; draw UO and GK at right angles to SP meeting PG and SP in O and K respectively.

Then by similar triangles OPU, UPG, GPK we have

$$OP : PU = PU : PG = PG : PK.$$

Whence
$$OP = \frac{PG^3}{PK^2}.$$

Now describe a conic with focus S and axis SG to touch PT at P, G is the foot of the normal, and PK is half the latus rectum. Hence O is the centre of curvature.

Since $SG : SP =$ eccentricity, the conic is determinate and unique.

Since a semicircle on PU as diameter passes through G, we have when $SP > \frac{1}{2}PU$, $SG < SP$; when $SP < \frac{1}{2}PU$, $SG > SP$; when $SP = \frac{1}{2}PU$, $SG = SP$.

Thus the conic is an ellipse, parabola, or hyperbola according as

$$SP >, =, \text{ or } < \tfrac{1}{2}PU.$$

54. Law of inverse square. Rectilinear motion. Let a particle move in a straight line, taken as axis of x, with an acceleration directed to a fixed point of the line, taken as origin, and equal to μ/x^2 at distance x. We have the equation

$$\ddot{x} = -\frac{\mu}{x^2}.$$

Multiplying both sides of this equation by $\dot{x}$, we have an equation which may be written

$$\frac{d}{dt}(\tfrac{1}{2}\dot{x}^2) = \frac{d}{dt}\left(\frac{\mu}{x}\right),$$

and, on integrating this equation, we have

$$\tfrac{1}{2}\dot{x}^2 = \frac{\mu}{x} + C,$$

where C is an arbitrary constant.

Let the particle start from rest at the point specified by $x = 2a$, where a is positive, at the instant when $t = 0$. Then we have $C = -\mu/(2a)$, and

$$\dot{x}^2 = \mu\left(\frac{2}{x} - \frac{1}{a}\right) = \mu\frac{2a - x}{ax}.$$

This equation shows that, so long as x is positive, it is less than $2a$, and we may introduce a new variable θ, in place of x, by the equation

$$x = 2a\cos^2\theta,$$

where $\theta = 0$ when $t = 0$. The equation for $\dot{x}$ becomes

$$16a^3\cos^4\theta \,.\, \dot{\theta}^2 = \mu,$$

giving $$4a^{\frac{3}{2}}\cos^2\theta \,.\, \dot{\theta} = \pm\sqrt{\mu}.$$

Since x diminishes as t increases, $\dot{x}$ is negative and $\dot{\theta}$ is positive, so the upper sign must be taken. Then the equation gives

$$t = (a^{\frac{3}{2}}/\sqrt{\mu})(2\theta + \sin 2\theta),$$

where no constant is added because $\theta = 0$ when $t = 0$.

Here x is not expressed in terms of t but x and t are both expressed in terms of a parameter θ.

It is to be observed that, if the law of the acceleration could remain unchanged until x vanishes, or the particle reaches the origin, the velocity would become infinite. In a physically possible system, by which the acceleration could be produced, either the particle could not reach the origin or the law of acceleration would have to change. See Art. 150 and Ex. 3 in Art. 177 *infra*.

55. Examples.

1. The results obtained in Art. 54 may also be found as follows:—Let O be the point to which the acceleration is directed, A the starting point. On OA as diameter describe a circle, and let a point P describe this circle under a central force directed to O. Let N be the foot of the perpendicular from P to OA (Fig. 28). The acceleration of P is $8h^2a^2/OP^5$, where h is twice the rate at which OP describes areas about O. (See Ex. 4 in Art. 46.) The acceleration of N is the resolved part of this in the direction AO. Prove that it is $h^2/(a \,.\, ON^2)$. Observing that $ht=$ twice the curvilinear area AOP, and taking the angle AOP to be θ, deduce the results

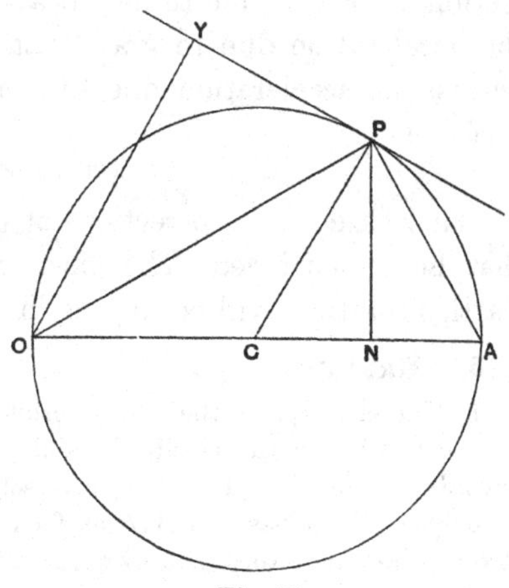

Fig. 28.

$$x = 2a\cos^2\theta, \quad t = (a^{\frac{3}{2}}/\sqrt{\mu})(2\theta + \sin 2\theta).$$

2. Prove that, if the law of acceleration remains unchanged until $x=0$, the time of moving from A to O is $\pi a^{\frac{3}{2}}/\sqrt{\mu}$.

56. Field of the Earth's gravitation. It is consonant with observations of falling bodies to state that the field of force around the Earth is central, and the acceleration of a free body in this field is directed towards the centre of the Earth. The Moon describes a nearly circular orbit about the Earth, in a period of about $27\frac{1}{3}$ days; this motion is nearly uniform, and the distance of the Moon from the Earth is about 60 times the radius of the Earth. Now the central acceleration of a particle describing a circular orbit of radius R uniformly in time T is $\frac{4\pi^2 R}{T^2}$; and, if the radius is 60 times the Earth's radius (3980 miles), and the period is $27\frac{1}{3}$ days, this acceleration, when expressed in foot-second units, is equal to $\frac{32 \cdot 1}{3600}$ approximately. Thus the Moon moves around the Earth in nearly the same way as if it were under gravity diminished in the ratio $1:(60)^2$.

From this result we conclude that the field of force around the Earth extends to the Moon, and that the intensity of this field, like that of the field around the Sun, varies inversely as the square of the distance.

For bodies in the neighbourhood of the Earth there is a correction of gravity due to height above the Earth's surface. If g is the acceleration due to gravity at the surface, and a the Earth's radius, the acceleration due to gravity at a height h above the surface is

$$ga^2/(a+h)^2.$$

There are other corrections of gravity at least as important as that here mentioned. The most important, depending upon the Earth's rotation, will occupy us in Chapter X.

57. Examples.

1. The envelope of the elliptic orbits described by particles, which start from a point P with velocity V, and move with an acceleration directed towards a point S and varying inversely as the square of the distance, is an ellipse, which has S and P as foci, and touches any of the trajectories at the point where the line drawn from P to the second focus of the trajectory meets it.

2. Show that a gun at the sea level can command $1/n^2$ of the Earth's surface if the greatest height to which it can send a shot is $1/n$ of the Earth's radius, variations of gravity due to altitude being taken into account.

3. Prove that the time in which a particle falls to the Earth's surface from a height h is $\left(\frac{2h}{g}\right)^{\frac{1}{2}}\left(1+\frac{5h}{6a}\right)$ approximately, a being the Earth's radius and $(h/a)^2$ being neglected.

MISCELLANEOUS EXAMPLES

1. Prove that the time in which it is possible to cross a road of breadth c, in a straight line, with the least uniform velocity, between a stream of omnibuses of breadth b, following at intervals a, moving with velocity V, is

$$\frac{c}{V}\left(\frac{a}{b}+\frac{b}{a}\right).$$

2. A straight line AB turns with uniform angular velocity about a point A, retaining a constant length, and a second straight line BC, also of constant length, moves so that C is always in a certain straight line through A. Prove that the velocity of C is proportional to the intercept which BC makes on the line through A at right angles to AC.

3. A point C describes a circle of radius r with angular velocity ω' about the centre O, and a point P moves so that CP is always equal to a and turns with angular velocity ω in the plane of the circle described by C. Prove that the angular velocity of OP is

$$\tfrac{1}{2}\{\omega(R^2+a^2-r^2)+\omega'(R^2-a^2+r^2)\}/R^2,$$

where R is the length of OP.

4. Two particles start simultaneously from the same point and move along two straight lines, one with uniform velocity, the other with uniform acceleration. Prove that the line joining the particles at any time touches a fixed parabola.

5. A body is projected vertically upwards with velocity v; after a time t a second body is projected vertically with velocity v' $(<v)$. If they meet as soon as possible after the instant when the first was projected

$$t=\{v-v'+\sqrt{(v^2-v'^2)}\}/g.$$

---

6. Two particles describe the same parabola under gravity. Prove that the intersection of the tangents at their positions at any instant describes a coaxial parabola as if under gravity. Prove also that, if τ is the interval between the instants when they pass through the vertex, the distance between the vertices of the two parabolas is $\frac{1}{8}g\tau^2$.

7. Prove that the angular velocity of a projectile about the focus of its path varies inversely as its distance from the focus.

8. A particle is projected from a platform with velocity V and elevation β. On the platform is a telescope fixed at elevation α. The platform moves horizontally in the plane of the particle's motion, so as to keep the particle always in the centre of the field of view of the telescope. Show that the original velocity of the telescope must be $V\sin(\alpha-\beta)\operatorname{cosec}\alpha$, and its acceleration $g\cot\alpha$.

9. A cricketer in the long field has to judge a catch which he can secure with equal ease at any height from the ground between k_1 and k_2; show that he must estimate his position within a length

$$R\{\sqrt{(1-k_2/h)}-\sqrt{(1-k_1/h)}\},$$

where $2R$ is the range on the horizontal and h the greatest height the ball attains.

10. A heavy particle is projected from a point A so as to pass through another point B; show that the least velocity with which this is possible is $\sqrt{(2gl)}\cos\frac{1}{2}a$, and that the highest point of the path is at a height $l\cos^4\frac{1}{2}a$ above A, where $AB=l$ and makes an angle a with the vertical.

11. A man travelling round a circle of radius a with speed v throws a ball from his hand at a height h above the ground, with a relative velocity V, so that it alights at the centre of the circle. Show that the least possible value of V is given by $V^2=v^2+g\{\sqrt{(a^2+h^2)}-h\}$.

12. If A and B are two given points, and C any given point on the line joining them, prove that, in the different trajectories possible under gravity between A and B, the time of flight varies as $\sqrt{(CD)}$, where D is the point in which the trajectory meets the vertical through C.

13. A gun is placed on a fort situated on a hill side of inclination a to the horizon. Show that the area commanded by it is $4\pi h(h+d\cos a)\sec^3 a$, where $\sqrt{(2gh)}$ is the muzzle-velocity of the shot, and d the perpendicular distance of the gun from the hill side.

14. It is required to throw a ball from a given point with a given velocity V so as to strike a vertical wall above a horizontal line on the wall. When the ball is projected in the vertical plane at right angles to the wall, the elevation must lie between θ_1 and θ_2. Prove that the points on the wall towards which the ball may be directly projected lie within a circle of radius

$$V^2\sin(\theta_1-\theta_2)/\{g\sin(\theta_1+\theta_2)\}.$$

15. Water issues from a fountain jet in such a manner that the velocity of emission in a direction making an angle θ with the vertical is $\sqrt{(ga\operatorname{cosec}\theta)}$, the jet being at the height h above the centre of a circular basin. Prove that, if all the water is to fall into the basin, its radius must not be less than

$$[2a\{a+\sqrt{(a^2+h^2)}\}]^{\frac{1}{2}}.$$

16. Prove that, if the sole effect of a wind on the motion of a projectile is to produce an acceleration f in a horizontal direction, the locus of points in a horizontal plane which can just be reached with a given velocity v of projection is an ellipse of eccentricity $f/\sqrt{(f^2+g^2)}$ and area $\pi v^4\sqrt{(f^2+g^2)}/g^3$.

17. A man standing on the edge of a cliff throws a stone with given velocity u, at a given inclination to the horizon, in a plane perpendicular to the edge of the cliff; after an interval τ he throws another stone from the same spot with given velocity v at an angle $\frac{1}{2}\pi-\theta$ with the line of discharge of the first stone and in the same plane. Find τ so that the stones may strike each other, and show that the maximum value of τ for different values of θ is $2v^2/wg$, and occurs when $\sin\theta=v/u$, w being the vertical component of v.

18. Two particles describe the same ellipse in the same time as a central orbit about the centre. Prove that the point of intersection of their directions of motion describes a concentric ellipse as a central orbit about the centre.

19. Particles are projected from points on a sphere of radius a with velocity $\sqrt{(gb)}$ and move with an acceleration to the centre equal to gr/a at distance r. Prove that the part of the surface on which they fall is the smaller of the two segments into which the sphere is divided by a small circle of radius b.

20. A particle P describes a rectangular hyperbola with an acceleration μCP from the centre C; a point Y is taken in CP so that $CP \, . \, CY = a^2$; prove that the rate at which P and Y separate is

$$\sqrt{\mu} CP \left(1 - \frac{a^2}{CP^2}\right)^{\frac{1}{2}} \left(1 + \frac{a^2}{CP^2}\right)^{\frac{3}{2}},$$

where $2a$ is the transverse axis.

---

21. If the acceleration of a particle is directed to a point S and varies inversely as the square of the distance, prove that there are two directions, if any, in which it can be projected from a point P with given velocity so as to pass through a point Q, and that the velocity of arrival at Q is the same for both. Prove also that the angle between one of the directions of projection and PQ is the same as the angle between the other and PS.

22. Prove that two parabolic orbits can be described about the same focus so as to pass through two given points, and that the focus lies within the finite area bounded by the line joining the given points and one parabola, and outside the finite area bounded by the same line and the other parabola.

23. Prove that the greatest radial velocity of a particle describing an ellipse about a focus is

$$2\pi ae\,(1-e^2)^{-\frac{1}{2}}/T,$$

where $2a$ is the major axis, e the eccentricity, and T the periodic time.

24. A particle describes an ellipse as a central orbit about a focus, and a second particle describes the same ellipse in the same time with uniform angular velocity about the same focus. The particles start together from the farther apse. Prove that the angle which the line joining the particles subtends at the focus is greatest when the angle described by the first particle is $\cos^{-1}\{1-(1-e^2)^{\frac{3}{4}}\}/e$, e being the eccentricity.

25. Prove that the central orbit described with acceleration $\mu/(\text{distance})^2$, by a particle projected with velocity V from a point where the distance is R, is a rectangular hyperbola if the angle of projection is

$$\operatorname{cosec}^{-1}\{V\sqrt{(V^2R^2-2\mu R)}/\mu\}.$$

26. Prove that the focal radius and vectorial angle of a particle describing an ellipse of small eccentricity e at time t after passing the nearer apse are given approximately by the equations

$$r = a(1 - e\cos nt + \tfrac{1}{2}e^2 - \tfrac{1}{2}e^2\cos 2nt),$$
$$\theta = nt + 2e\sin nt + \tfrac{5}{4}e^2\sin 2nt,$$

where $2a$ is the major axis and $2\pi/n$ is the periodic time.

Prove also that if e^2 is neglected the angular velocity about the other focus is constant.

27. Two particles describe the same ellipse in the same periodic time, starting together from one end of the major axis. One of them has an acceleration directed to a focus, and the other an acceleration directed to the centre. Prove that, if ϕ_1 and ϕ_2 are their eccentric angles at any instant, then $\phi_1 - \phi_2 = e\sin\phi_1$.

28. If the perihelion distance of a comet is $\dfrac{1}{n}$ th of the radius of the Earth's orbit, supposed circular, show that the comet will remain within the Earth's orbit for

$$(\surd 2/3\pi)(1+2/n)\surd(1-1/n) \text{ years,}$$

the comet's orbit being parabolic.

29. If the parabolic orbits of two comets intersect the orbit of the Earth, supposed circular, in the same two points, and if t_1, t_2 are the times in which the comets move from one of these points to the other, prove that

$$(t_1+t_2)^{\frac{2}{3}} + (t_1-t_2)^{\frac{2}{3}} = (4T/3\pi)^{\frac{2}{3}}, \text{ where } T \text{ is a year.}$$

30. Three focal radii SP, SQ, SR of an elliptic orbit about a focus S are determined, and the angles between them. Show that the ellipticity may be found from the equation $b\Delta = a\Delta'$, where Δ is the area of the triangle PQR, and Δ' is the area of a triangle whose sides are

$$2\surd(SQ\,.\,SR)\sin\tfrac{1}{2}QSR,$$

and two similar expressions.

---

31. A particle is projected from A with velocity $\surd(\frac{1}{2}\mu)/OA^2$ and moves with an acceleration $\mu/(\text{distance})^5$ directed to O, the direction of projection making an angle a with OA. Prove that the particle will arrive at O after a time

$$\frac{OA^3}{\surd(2\mu)}\,\frac{a - \sin a\cos a}{\sin^3 a}.$$

32. Prove that the acceleration with which a particle P can describe a circle as a central orbit about a point S is inversely proportional to $SP^2\,.\,PP'^3$, where PP' is the chord through S.

If points are taken on the orbit such that the squares of their distances from S are in arithmetic progression, the corresponding velocities are in harmonic progression.

33. Prove that any conic can be described by a particle with an acceleration always at right angles to the transverse axis and varying inversely as the cube of the distance from it.

34. A particle moves with an acceleration μy^{-3} towards the axis x, starting from the point $(0, k)$ with velocities U, V parallel to the axes of x, y. Prove that it will not strike the axis x unless $\mu > V^2k^2$, and that, in this case, it strikes it at a distance $Uk^2/(\sqrt{\mu} - Vk)$ from the origin, U, V, k being positive.

35. Prove that the acceleration towards the centre of the fixed circle with which a particle can describe an epicycloid is proportional to r/p^4, where r is the radius vector and p the perpendicular from the centre to the tangent.

36. Prove that the curve $r=a\,(1+\frac{1}{2}\sqrt{6}\cos\theta)$ is a central orbit about the origin for acceleration proportional to $r^{-4}+\frac{1}{3}ar^{-5}$.

---

37. A series of particles are describing the same curve as a central orbit about a point O with an acceleration whose tangential component is $h^2/p^2\phi'(p)$. Prove that, if the line density at any time is constant and $=\rho_0$, the line density ρ at any subsequent time t is given by

$$\phi(p)+ht=\phi(p\rho_0/\rho),$$

$\frac{1}{2}h$ being the rate of description of areas about O, and p the perpendicular from O on the tangent.

38. If inverse curves with respect to O can be described as central orbits about O with accelerations f, f', prove that

$$\frac{r^3f}{h^2}+\frac{r'^3f'}{h'^2}=\frac{2}{\sin^2\phi},$$

where h and h' are constants, r and r' are corresponding radii vectores, and ϕ is the angle r or r' makes with the tangent.

39. If f is the acceleration and $\frac{1}{2}h$ the areal velocity in a central orbit about a point O, prove that the angular acceleration a about O satisfies the equation

$$\frac{da^2}{du}-6\,\frac{a^2}{u}=8h^2u^4\,(f-h^2u^3),$$

where u is the reciprocal of the distance from O.

---

40. Prove that a body ejected from the Earth with velocity exceeding seven miles per second will not in general return to the Earth, and may leave the solar system.

41. Prove that the least velocity with which a body could be projected from the North Pole so as to meet the Earth's surface at the Equator is nearly $4\frac{1}{2}$ miles per second, and that the angle of elevation is $22\frac{1}{2}°$.

42. A stream of particles originally moving in a straight line K with velocity V is under the influence of a gravitating sphere of radius R, whose centre moves with velocity v in a straight line intersecting the line K and making with it an angle a. Prove that, if the distance of the sphere from the line is originally very great, a length

$$(2R/v) \operatorname{cosec} a \sqrt{(V^2 - 2Vv\cos a + v^2 + 2gR)}$$

of the line of particles will fall upon the sphere, g being the force per unit mass at the surface of the sphere.

CHAPTER III

FORCES ACTING ON A PARTICLE

58. The force of gravity. Consider a heavy body supported near the Earth's surface. The body may, for example, rest upon a horizontal plane, which is then the plane surface of some other body, or it may be supported by a rope or a spiral spring. In either case we should say that there was a *force* acting upon it and counteracting the force of the Earth's field. When the body is supported by a spring, the spring is stretched; if the body is supported even by a steel bar, the bar is stretched a little*, and the stretching of the bar can be observed by means of suitable instruments. If the body is supported by a man carrying it, his muscles are thrown into a state of strain, analogous to the stretching of the steel bar, and the man has a sensation of muscular effort. We should say that he exerted "force."

The operation of weighing a body in a common balance determines a certain quantity: the number of pounds or grammes which the body weighs. The number so determined is independent of the latitude and longitude of the place where the operation is performed; and it is independent also, so far as observation can tell, of the altitude of the place above, or its depth below, the mean surface of the Earth.

The stretching of a spring supporting a body can be measured; and, when the weight of the body, as determined by the common balance, is not too great, the stretching of the spring, at any definite place on the Earth's surface (*e.g.* in London), is proportional to the weight so determined. We may therefore use this stretching to determine the weight of the body, and then the body is said to be "weighed by a spring balance." The weight of the body, determined by the spring balance, is different in different

*** A steel bar, of sectional area one square inch, hanging vertically, and supporting a load of 1 ton, is extended by the fraction 0·00007 of its length, approximately.**

latitudes and at different altitudes. It is found to be proportional to the local value of g (the acceleration of a free falling body).

The primitive notion of "force" is based upon the muscular sensations of a man supporting a heavy body. The measure of force which is suggested by the above considerations is the stretching of an ideal spring supporting a heavy body*. This stretching is always proportional (i) to the weight, as determined by a common balance, (ii) to the local value of g. We are thus led to measure the force of the Earth's gravity as proportional to each of these factors.

The operation of weighing a body in a common balance teaches us how to assign to any body of sufficiently small bulk a definite constant quantity: the number of pounds or grammes which the body weighs. This quantity, or any suitable constant multiple of it, will be called the *mass* of the body. For a body which cannot be weighed in a common balance, *e.g.* a battleship, the mass may be determined by adding the masses of the several parts, each being determined by weighing in a common balance or by some equivalent method. The definition of "mass" does not cover such cases as the mass of the Earth, or Sun, or Moon. A more general definition will be given in Chapter VI. We denote the mass of a body by the letter m.

The *force of the Earth's gravity* acting upon a body† is measured by the product of the number of units of mass in the mass of the body and the number of units of acceleration in the local value of g. We denote this force by W, and write

$$W = mg.$$

59. Measure of force. Force may be defined as a certain measure of the action which one body exerts upon another. In the particular case of a body supported upon a horizontal plane, the force counteracting the force of the Earth's gravity is traced to an action of the body having the horizontal plane for part of its surface; this force is called the *pressure* of the plane upon the supported body. In the case of a body supported by a rope or

* The spring is "ideal" in as much as the extension is supposed to be proportional to the weight, however great the weight may be. An actual spring would be damaged by a sufficiently heavy weight, and it would not measure that weight correctly.

† This force is sometimes called the "weight" of the body.

spring, the force counteracting the force of the Earth's gravity is traced to an action of the rope or spring; this force is called the *tension* of the rope or spring. The force of the Earth's gravity acting upon a body is, in like manner, traced to a supposed action of the Earth upon the body.

In this last case we know that the effect of the action, if not counteracted, is to produce in the body a certain acceleration; and the measure of the force is the product, as explained above, of the mass of the body and the acceleration which it produces.

In like manner, we may say that the effect of any force on a body, when not counteracted by other forces, is to produce in the body an acceleration, and the measure of the force is the product of the measures of the mass and the acceleration. If a force P acts upon a body of mass m, it produces in it an acceleration f, and we have the formula

$$P = mf.$$

60. Units of mass and force. In the "C.G.S. system" of units, the *gramme* is the unit of mass. It is the one-thousandth part of the mass of a certain lump of platinum known as the "Kilogramme des Archives," made by Borda, and kept in Paris. The unit of force is called the "dyne." It is the force which, acting upon a body of mass one gramme, produces in it an acceleration of one centimetre per second per second.

In the "foot-pound-second system," the *pound* is the unit of mass. It is the mass of a certain lump of platinum kept in the Royal Exchequer in London. The unit of force is called the "poundal." It is the force which, acting upon a body of mass one pound, produces in it an acceleration of one foot per second per second.

In the "British engineers' system" the unit of force is the force of the Earth's gravity acting in London upon a body which weighs a pound, when weighed in a common balance. It is called a "force of one pound." The unit of mass is the mass of a body which weighs 32·2 pounds in a common balance. The mass of a body which weighs one pound in a common balance is $\frac{1}{32 \cdot 2}$ units of mass. In this system, as in the others, the unit force, acting upon the unit mass, produces in it an acceleration of one unit of length (one foot) per second per second.

In any system of units, force is a quantity of one dimension in mass, one dimension in length, and -2 dimensions in time. The dimension symbol is MLT^{-2}.

61. Vectorial character of force. In the cases which we have examined so far, either there has been a single force acting upon a body, which for definiteness we thought of as a "particle," or else the forces acting upon the body have exactly counteracted each other. In the former case, the body moves with a certain acceleration. In the latter case, it remains at rest. In the case of a heavy body supported by the tension of a cord, we may regard the Earth's gravity as producing in it the acceleration g downwards, and the tension of the cord as producing in it the acceleration g upwards. If we do this we are able to maintain in both cases the measure of force as the product of the mass and the acceleration that is produced by the force.

Consider a body supported upon a plane horizontal surface. Let the surface be gradually tilted so that the plane becomes an inclined plane. It is found that the body will begin to slide* down the plane when the plane is tilted at an angle which exceeds a certain limiting angle. If the surfaces in contact are highly polished the angle at which sliding begins is small. We might imagine the surfaces to be so smooth that sliding would take place at any inclination however small. The acceleration with which the body slides down the plane is the resultant of the acceleration g in the direction of the downward vertical and some other acceleration, f. Let α be the inclination of the plane; then the acceleration g can be resolved into two components, viz.: $g \sin \alpha$ in the direction of a line of slope drawn down the plane, and $g \cos \alpha$ at right angles to the plane. See Fig. 29. If the acceleration f is directed at right angles to the plane its amount must be $g \cos \alpha$, in the sense opposite to one of the two components of g, as shown in Fig. 29, since the body moves on the plane, and so has no acceleration at right angles to the plane. In this case, the acceleration with which the body slides

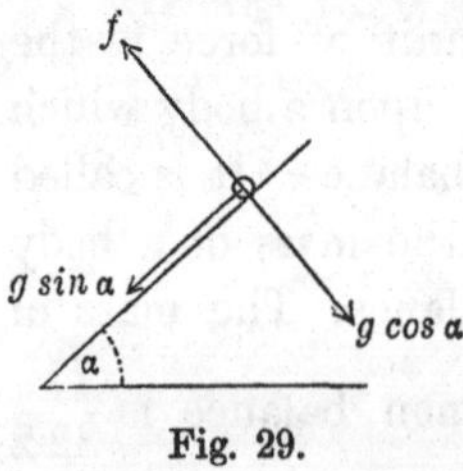

Fig. 29.

* The body should have a flat base. A solid sphere, or any body with a curved surface, placed on an inclined plane, will generally roll. We avoid for the present the complication of rolling.

down the plane is $g \sin \alpha$*, and the pressure of the plane on the body is of amount $mg \cos \alpha$, the mass of the body being m. This state of things cannot be exactly realised in practice, but it can be approximately realised when the surfaces are very smooth.

In any actual case the acceleration with which the body slides down the plane is less than $g \sin \alpha$, and the motion is said to be resisted by "friction." For the present we shall suppose that the surfaces are so smooth that the effect of friction is negligible. We have learnt that the effect of the Earth's gravity on the body is the same as that of two forces: one $mg \sin \alpha$ producing acceleration down the line of slope, and the other $mg \cos \alpha$ producing acceleration at right angles to the plane.

This result leads us to the conclusion that *force*, as a mathematical quantity, is to be regarded as a vector quantity, equivalent to "component forces" in the same way as any other vector quantity is equivalent to components.

In particular, we see that force acting on a particle ought to be regarded as what we have called a "vector localized at a point" (Art. 17), the point at which the vector is localized being the position of the particle. The line, drawn through the point, by which the vector is determined, is the "line of action" of the force. The line of action of the force and the sense of the force are the direction and sense of the acceleration which the force produces.

According to this statement any forces acting on a particle are equivalent to a single force, to be determined from the separate forces by the rules for the composition of vectors. This single force is called the "resultant" of the forces acting on the particle.

62. Examples†.

1. Find the time of descent of a particle down an inclined tube when friction is neglected and the particle starts from rest at a given point of the tube.

2. Prove that the time of descent down all chords of a vertical circle, starting at the highest point of the circle, or terminated at its lowest point, is the same.

3. Prove that the line of quickest descent from a point A to a curve, which is in a vertical plane containing A, is the line from A to the point of

* This result was used by Galileo for the determination of g.

† The results in Examples 2 and 3 were noted by Galileo.

contact with the curve of a circle described to have A as its highest point and to touch the curve. Prove also that the line of quickest descent from a curve to a point A is the line to A from the point of contact with the curve of a circle described to have A as its lowest point and to touch the curve.

4. Prove that each of the lines of quickest descent in Ex. 3 bisects the angle between the vertical and the normal to the curve at the point where it meets the curve. Hence show that the line of quickest descent from one given curve to another in the same vertical plane bisects the angle between the normal at either end and the vertical.

5. Prove that a particle projected in any manner on an inclined plane, and moving on the plane without friction, describes a parabola.

63. Definitions of momentum and kinetic reaction.

The *momentum* of a particle of mass m, moving with a velocity v, is a vector, localized in the line of the velocity, of which the sense is the same as that of the velocity, and the magnitude is the product mv.

The *kinetic reaction* of a particle of mass m, moving with an acceleration f, is a vector, localized in the line of the acceleration, of which the sense is the same as that of the acceleration, and the magnitude is the product mf.

The kinetic reaction of a particle is the same quantity as the *rate of change of momentum* of the particle per unit of time.

64. Equations of motion. The discussion of the nature of force in Articles 58—61 leads to the following statement:—

The kinetic reaction of a particle has the same magnitude, direction and sense as the resultant force acting on the particle.

This statement is to be regarded as a general principle which is suggested by the facts stated in the previous discussion and other facts of like nature. In other words it is an induction from experience. From the nature of the case it is not capable of mathematical proof. The truth of it is only to be tested by the comparison of results deduced from it with results of experiment.

The statement is expressed analytically by certain equations, which are called the "equations of motion" of the particle. They are obtained by equating the resolved part of the kinetic reaction in any direction to the sum of the resolved parts of the forces in that direction.

Let X, Y, Z be the components parallel to the axes of x, y, z of the resultant force acting on the particle, or, what comes to the same thing, the sums of the resolved parts of the forces in the directions of these axes. Let m be the mass of the particle, and x, y, z the coordinates of its position at time t. The equations of motion are

$$m\ddot{x} = X, \quad m\ddot{y} = Y, \quad m\ddot{z} = Z.$$

We have had several examples already of equations which are really equations of motion.

For example, the equations

$$\ddot{x} = 0, \quad \ddot{y} = -g$$

in Art. 33 are really equations of motion.

As a further example, consider the motion of a particle in *a central field of force*. If f is the intensity of the field, and the centre of force is the origin, and if the force is an *attraction*, it is of amount mf and is directed towards the origin; and the equations of motion are

$$m\ddot{x} = -mf\frac{x}{r}, \quad m\ddot{y} = -mf\frac{y}{r}, \quad m\ddot{z} = -mf\frac{z}{r},$$

where r denotes distance from the origin. Just as in Art. 49, these equations show that the motion takes place in a fixed plane. By means of the result of Art. 43 the equations of motion, expressed in terms of polar coordinates in the plane, can be written

$$m(\ddot{r} - r\dot{\theta}^2) = -mf, \quad m\frac{1}{r}\frac{d}{dt}(r^2\dot{\theta}) = 0.$$

Equations of Motion in simple cases

65. Motion on a smooth guiding curve under gravity. The motion of a small ring on a very smooth wire, or of a small spherical shot in a very smooth tube, can be discussed by treating the ring or shot as a particle constrained to describe a given curve, and supposing that the particle is subject not only to the force of the field, but also to a force—the pressure of the curve—directed along the normal at any point of the curve. We take the case where the curve is a plane curve in a vertical plane, and the field is that of the Earth's gravity at a place. We draw the axis of y vertically upwards, and denote by s the arc of the curve measured from some fixed point of it up to the position of the particle at the instant t, and by v the velocity of the particle in the direction of increase of s. We denote the pressure of the curve by R, and suppose that its sense is towards the centre of curvature. If the pressure really acts outwards, the value found for R will be negative.

In the left-hand figure (Fig. 30) are shown the components of the kinetic reaction along the tangent and normal. In the right-hand figure are shown the forces acting on the particle. The

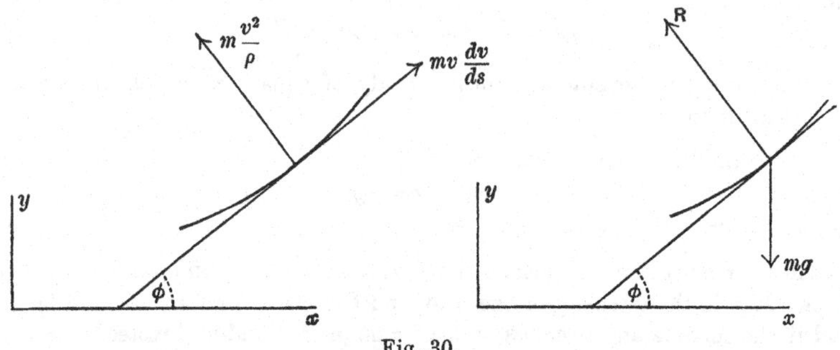

Fig. 30.

equations of motion, obtained by resolving the forces along the tangent and normal, are

$$mv\frac{dv}{ds} = -mg\sin\phi, \quad m\frac{v^2}{\rho} = R - mg\cos\phi.$$

Now $\sin\phi = \frac{dy}{ds}$, and the first of these equations becomes

$$mv\frac{dv}{ds} = -mg\frac{dy}{ds}.$$

This equation can be integrated in the form

$$\tfrac{1}{2}mv^2 = -mgy + C,$$

where C is a constant. Let v_0 be the velocity at some point (x_0, y_0) of the curve. Then $C = \frac{1}{2}mv_0^2 + mgy_0$, and the equation can be written

$$\tfrac{1}{2}mv^2 - \tfrac{1}{2}mv_0^2 = mg(y_0 - y).$$

This equation can be partially interpreted in the statement that the velocity of a particle moving under gravity without friction is always the same when it comes back to the same level*.

If the particle starts with an assigned velocity from a given point of the curve, this equation determines the velocity in any position; the equation $mv^2/\rho = R - mg\cos\phi$ determines the pressure at any point of the curve.

* This result was found by Galileo.

66. Examples.

1. When the curve is a circle, the angle ϕ of Fig. 30 is the angle which the radius of the circle drawn from the centre to the particle makes with the vertical drawn downwards. Prove that, if the particle starts from rest in a position in which $\phi = \alpha$, the velocity v in any position is given by the equation

$$v^2 = 2ga (\cos\phi - \cos\alpha),$$

where a is the radius of the circle.

Find the pressure in any position.

2. Find the greatest angle through which a person can oscillate in a swing, the ropes of which can support a tension equal to twice the person's weight.

3. When a particle moves on a smooth cycloid under gravity, the vertex of the cycloid being at the lowest point, the equation of motion, by resolution along the tangent in direction QP, may be written

$$\ddot{s} = -g\sin\theta,$$

s being the arc measured from the vertex to P, and θ the angle which the normal OP makes with the vertical. Now, by a known property of the cycloid, $s = 4a\sin\theta$, where a is the radius of the generating circle, and thus the above equation becomes

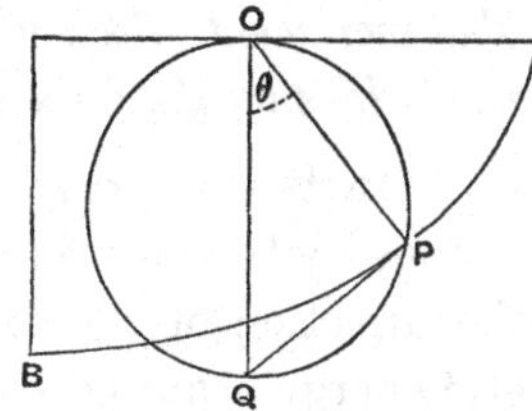

Fig. 31.

$$\ddot{s} = -\frac{g}{4a}s,$$

showing that the motion in s is simple harmonic with period $2\pi\surd(4a/g)$. Thus the time taken to fall to the vertex from any point on the curve is independent of the starting-point, and in fact is $\pi\surd(a/g)$.

[This property is known as the "Isochronism of the cycloid."]

4. Show that the time a train, if unresisted, takes to pass through a tunnel under a river in the form of an arc of an inverted cycloid of length $2s$ and height h cut off by a horizontal line is

$$\frac{s}{\surd(2gh)}\cos^{-1}\left(\frac{v^2 - 2gh}{v^2 + 2gh}\right),$$

where v is the velocity with which the train enters and leaves the tunnel.

67. Kinetic energy and work. The quantity obtained by multiplying the number of units of mass in the mass of a particle by half the square of the number of units of velocity in the velocity of the particle is called the "kinetic energy" of the particle.

The "work done" by a constant force acting on a particle is a quantity which is defined in terms of the force and the displacement of the particle. We resolve the displacement into components parallel and perpendicular to the line of action of the force. The component parallel to this line (taken in the sense of the force) has a certain magnitude, which is a number of units of length, and a certain sign. We multiply this number, with this sign, by the number of units of force in the measure of the force. The product so obtained is the work done.

In the case of a particle moving under gravity, the work done by the force of gravity is the product of the force, mg, and the distance through which the particle descends, $y_0 - y$. The equation

$$\tfrac{1}{2}mv^2 - \tfrac{1}{2}mv_0^2 = mg(y_0 - y)$$

can be expressed in words in the statement:—

The increment of kinetic energy in any displacement is equal to the work done by the force of gravity in that displacement.

68. Units of energy and work. The unit of work is the work done by the unit force in a displacement of one unit of length in the direction of the force. The unit of kinetic energy is the kinetic energy acquired by a free body on which one unit of work is done.

In the C.G.S. system of units the unit of work is called the *erg*. It is the work done by a force of one dyne acting over one centimetre.

In the foot-pound-second system the unit of work is the "foot-poundal." It is the work done by a force of one poundal acting over one foot.

In the British engineers' system the unit of work is the "foot-pound." It is the work done by a "force of one pound" acting over one foot. It is equal to the work done in the latitude of London in raising through one foot a body which weighs one pound in a common balance.

In any system of units, work and kinetic energy are quantities of 1 dimension in mass, 2 dimensions in length, and -2 dimensions in time. The dimension symbol is ML^2T^{-2}.

69. Power. An agent which does one unit of work per unit of time is said to be working up to a unit of *power*. If 550 foot-pounds of work are done per second the power is one *horse-power*. Power is a quantity having the dimensions ML^2T^{-3}. For a more extended discussion see Chapter VI.

70. Friction. Consider a body sliding down an inclined plane. Let α be the inclination of the plane. The acceleration of the body down the lines of slope is less than $g \sin \alpha$. Let f be the acceleration up the lines of slope which must be compounded with the acceleration $g \sin \alpha$ down the lines of slope in order that the resultant may be the actual acceleration of the body. The forces acting on the body are the force of gravity mg vertically downwards, the pressure $mg \cos \alpha$ at right angles to the plane, and a third force which is of magnitude mf and acts up the lines of slope. This force is called the "friction."

The body will not slide down the plane unless the inclination α exceeds a certain angle i. When $\alpha = i$, the friction just prevents motion. In this case $g \sin \alpha = f$, or $f = g \sin i$, and the friction $= mg \sin i$. In the same case the pressure $= mg \cos i$. Hence the ratio of the friction to the pressure when motion is just about to take place is $\tan i$. We write μ for $\tan i$, so that when the body is about to slide the friction is equal to the product of μ and the pressure.

It is found that, when motion takes place, the ratio of the friction to the pressure remains constant. This ratio (equal to $\tan i$ or μ) is called the "coefficient of friction." The angle i is called the "angle of friction."

The angle of friction, and the coefficient of friction, depend upon the materials of the bodies in contact and the degree of polish of the surfaces.

Whether the body moves up or down the plane, the friction acts in the sense opposite to that of the velocity, and is equal to the product of μ and the pressure.

71. Motion on a rough plane. We shall take the plane to be inclined at an angle α to the horizontal, and treat the body sliding on it as a particle moving down a line of slope. Draw the axis of x down a line of slope. The equations of motion are

$$m\ddot{x} = mg \sin \alpha - F, \quad 0 = mg \cos \alpha - R,$$

where F is the friction and R the pressure. Also we have

$$F = \mu R.$$

Hence the particle moves down the line of slope with acceleration

$$g(\sin \alpha - \mu \cos \alpha).$$

When the particle moves up a line of slope, the friction acts down the line, and the acceleration is equal to

$$g(\sin\alpha + \mu\cos\alpha)$$

down the line.

When the body slides on a horizontal plane the pressure is equal to mg vertically upwards, and the friction is equal to μmg in the sense opposite to that of the velocity.

This last result is generally taken to be applicable to the motion of a train on level rails. The resistance to the motion is taken to be proportional to the mass. The force by which the train is set in motion and kept in motion against the resistance is called the "pull of the engine." We shall consider this force further in Chapter VIII.

When there is friction the increment of kinetic energy in any displacement is less than the work done by gravity in that displacement.

72. Examples.

1. A particle is projected with a given velocity up a line of slope of a rough inclined plane. Find the height above the point of projection of the point at which it comes to rest. Supposing the inclination of the plane to be greater than the angle of friction, find the velocity with which the particle returns to the point of projection.

2. A carriage is slipped from an express train, going at full speed, at a distance l from a station, and comes to rest at the station. Prove that the rest of the train will then be at a distance $Ml/(M-m)$ beyond the station, M and m being the masses of the whole train and of the carriage slipped, and the pull of the engine being constant.

3. Prove that the extra work required to take a train from rest at one station to stop at the next at a distance l in an interval t is

$$\frac{2l}{gt^2}\frac{1}{k}\Big/\left\{\left(\frac{1}{m}+\frac{1}{n}\right)\left(\frac{1}{m}+\frac{1}{n}+\frac{1}{k}\right)\right\}$$

times the work required to run through both without stopping, where the incline of the road is 1 in m, and the resistance of the road and the brake power per unit mass are equal to the components of gravity down uniform inclines of 1 in n and 1 in k respectively.

73. Atwood's machine*. Another simple example of equations of motion is afforded by the problem of two bodies attached

* G. Atwood, *A treatise on the rectilinear motion and rotation of bodies*, Cambridge, 1784.

to a string or chain which passes over a vertical pulley. This arrangement constitutes in principle the instrument called "Atwood's machine." We shall assume that the tension of the chain is the same throughout. This amounts to assuming that there is no friction between the pulley and the chain, and that the mass of the chain is negligible in comparison with the masses of the bodies (see Chapter VI).

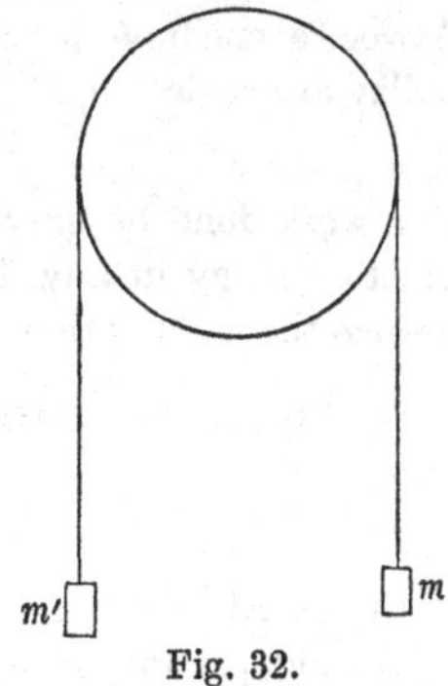

Fig. 32.

Let m, m' be the masses of the bodies, x the distance through which m has descended at time t. Then x is also the distance through which m' has ascended at time t. If m has ascended and m' descended, x is negative. Let T be the tension of the chain. The forces acting on m are mg vertically downwards, and T vertically upwards. The kinetic reaction of m is $m\ddot{x}$ vertically downwards. The equation of motion of m is therefore

$$m\ddot{x} = mg - T.$$

The forces acting on m' are $m'g$ vertically downwards, and T vertically upwards. The kinetic reaction of m' is $m'\ddot{x}$ vertically upwards. The equation of motion of m' is therefore

$$m'\ddot{x} = T - m'g.$$

By adding the left-hand, and also the right-hand, members of these equations, we find

$$(m + m')\,\ddot{x} = (m - m')\,g.$$

It follows that the heavier body descends, and the lighter ascends, with an acceleration

$$\frac{m \sim m'}{m + m'}\,g.$$

The value of g is sometimes determined by means of Atwood's machine. Various corrections have to be applied to the result. Generally the pulley turns with the motion of the chain, and the most important correction is on account of the mass of the pulley. (See Chapter VIII.)

74. Examples.

1. The kinetic energy of the two bodies in the case of the simple Atwood's machine, in which the friction and the masses of the chain and pulley are neglected, is

$$\tfrac{1}{2}m\dot{x}^2+\tfrac{1}{2}m'\dot{x}^2.$$

The work done by gravity is $mgx-m'gx$. Assuming that the increment of kinetic energy in any displacement is equal to the work done by gravity, deduce the acceleration of either body.

2. Prove that the tension of the chain is

$$\frac{2mm'}{m+m'}g.$$

3. In Atwood's machine the smaller mass m' is rigid, the mass m consists of a rigid portion of mass m' and a small additional piece resting lightly upon it. As m descends it passes through a ring by which the additional piece is lifted off. Prove that, if m starts from a height h above the ring, and if after passing through the ring it falls a distance k in the time t, then

$$g=\frac{m+m'}{2(m-m')}\frac{k^2}{ht^2},$$

the friction and the masses of the pulley and chain being neglected.

75. Simple circular pendulum executing small oscillations. A particle constrained to describe a circle in a vertical plane, without friction, is called a "simple circular pendulum." An ordinary pendulum consists of a massive body, called the "bob," suspended by a bar which can turn about a horizontal axis. When the bob is small and massive, and the bar thin, the motion of the bob, treated as a particle, approximates to that of a simple circular pendulum.

We denote the radius of the circle by l. When the radius of the circle which passes through the particle makes an angle θ with the vertical as in Fig. 33, the acceleration along the tangent to the circle is $l\ddot{\theta}$ (Ex. 1 of Art. 37). We may write down one equation of motion in the same way as in Art. 65 in the form

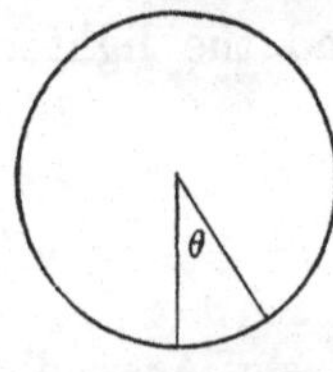

Fig. 33.

$$ml\ddot{\theta}=-mg\sin\theta.$$

If θ is very small throughout the motion, $\sin\theta$ may be replaced by θ, and we have the approximate equation

$$l\ddot{\theta}=-g\theta.$$

This equation shows that the motion in θ is simple harmonic motion of period $2\pi\sqrt{(l/g)}$. (Cf. Art. 38.)

The pendulum swings from side to side of the vertical. If it starts from rest, in a position slightly different from the position of equilibrium, it falls to this position in the time $\frac{1}{2}\pi\sqrt{(l/g)}$, passes through it, and proceeds to move away from it on the other side until its displacement is numerically equal to that at starting, and comes to rest after an interval $\frac{1}{2}\pi\sqrt{(l/g)}$ from the equilibrium position. The motion is then reversed. The time from rest to rest is $\pi\sqrt{(l/g)}$. This is known as the time of a "beat," the period $2\pi\sqrt{(l/g)}$ is the time of a "complete oscillation."

A pendulum which beats seconds is known as a "seconds' pendulum"; the time of a complete oscillation of such a pendulum is two seconds. The length of the seconds' pendulum at a place is given by the equation

$$\pi\sqrt{(l/g)} = 1.$$

Pendulum experiments afford the most exact method of determining the value of g.

76. Examples.

1. Prove that, if in London $g=981{\cdot}17$, the units being the centimetre and the second, then the length of the seconds' pendulum there is $99{\cdot}413$ centimetres.

2. A balloon ascends with constant acceleration and reaches a height of 900 ft. in one minute. Show that a pendulum clock carried with it will gain at the rate of 27·8 seconds per hour, approximately.

3. If l_1 is the length of a slightly defective seconds' pendulum which gains n seconds in an hour, and l_2 the length of another such pendulum which loses n seconds in an hour, n being small, prove that the square root of the true length of the seconds' pendulum is the harmonic mean between $\sqrt{l_1}$ and $\sqrt{l_2}$.

4. The bob of a pendulum which is hung close to the face of a cliff is attracted to the cliff with a horizontal force of intensity f. Show that the time of a beat is

$$\pi l^{\frac{1}{2}}/(g^2+f^2)^{\frac{1}{4}},$$

where l is the length of the pendulum.

5. A bead slides on a smooth circular wire of radius a, whose plane is inclined at an angle α to the vertical. Find the period of its small oscillations about the lowest point.

77. One-sided constraint. A particle may be constrained to describe a circle by means of a thread of constant length attached to the centre of the circle; or it may be inside a smooth circular cylinder. More generally a particle may be constrained to describe a curve in a vertical plane by being inside a cylinder, of which the normal section is the curve and the generators are horizontal, and not too far from the lowest generator. Or it may be outside such a cylinder, and not too far from the highest generator. In either case the constraint is "one-sided," and the particle may leave the curve. This will happen if the pressure vanishes. The particle then describes a parabola under gravity until it strikes the curve again.

Now the pressure is given, according to Art. 65, by the equation

$$R = m\frac{v^2}{\rho} + mg\cos\phi,$$

where ϕ is the angle which the tangent, drawn in a definite sense, makes with the horizontal. To make R vanish, we must have

$$\cos\phi = -\frac{v^2}{g\rho},$$

where v^2 is known. This equation determines the point at which the particle leaves the curve.

78. Examples.

1. The bob of a simple circular pendulum is projected horizontally from its equilibrium position with a velocity V. Find limits between which V must lie in order that the suspending fibre may become slack, and determine the position of the bob at the instant when the fibre becomes slack.

2. A cylinder whose section is a parabola is placed with its generators horizontal, the axis of a normal section vertical, and the vertex upwards, and a particle is projected along it in a vertical plane. Prove that if it leaves the parabola anywhere it does so at the point of projection.

3. A particle is projected from the lowest point of a vertical section of a smooth hollow circular cylinder, whose axis is horizontal, so as to move round inside the cylinder. Prove that, if the velocity is that due to falling from the highest point, the particle leaves the circle when the radius through it makes with the vertical an angle $\cos^{-1}\frac{2}{3}$.

Find the least velocity of projection in order that the particle may describe the complete circle.

4. A particle is constrained to describe a circle by means of an inextensible thread, and leaves the circle when the thread makes an angle β with the vertical drawn upwards. Prove that when it strikes the circle again the thread makes an angle 3β with the same vertical.

79. Conical pendulum. A particle can be constrained to describe a horizontal circle uniformly by the tension of a string or thread, attached to a fixed point on the vertical straight line which passes through the centre of the circle. In any position of the particle the string lies along a generator of a right circular cone having its vertex at the fixed point.

Let 2α be the vertical angle of the cone and l the length of the string. The radius of the circle is $l \sin \alpha$. Let v be the velocity of the particle and T the tension of the string. The kinetic reaction of the particle is $\frac{mv^2}{l \sin \alpha}$ directed along the radius of the circle towards its centre. The forces acting on the particle are the force of gravity, mg vertically downwards, and the tension T of the string, directed along the generator of the cone towards the fixed point. We form equations of motion by resolving vertically, horizontally along the radius of the circle, and horizontally along the tangent of the circle. Neither the kinetic reaction nor the forces have any components in the third of these directions; and we therefore have the two equations

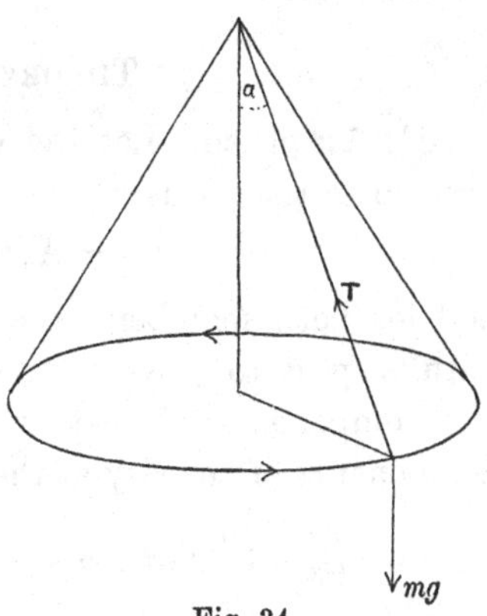

Fig. 34.

$$\frac{mv^2}{l \sin \alpha} = T \sin \alpha, \quad 0 = mg - T \cos \alpha.$$

By eliminating T we find the equation

$$v^2 = gl \frac{\sin^2 \alpha}{\cos \alpha}.$$

This equation determines the velocity with which the circle can be described when l and α are given, or the angle α when v and l are given.

80. Examples.

1. A train rounds a curve, of which the radius of curvature is ρ, with velocity v. Prove that to prevent the train from leaving the metals the

outer rail ought to be raised a height equal to $bv^2/\rho g$ above the inner, b being the distance between the rails.

[The train may be treated as a conical pendulum, in which the pressure of the rails, directed at right angles to the plane of the rails, takes the place of the tension of the string.]

2. The point of suspension of a simple pendulum of length l is carried round in a horizontal circle of radius c with uniform angular velocity ω. Prove that, when the motion is steady, the inclination α of the suspending thread to the vertical is given by the equation

$$\omega^2 (c + l \sin \alpha) = g \tan \alpha.$$

Prove also that, if $(g/\omega^2)^{\frac{2}{3}} < l^{\frac{2}{3}} - c^{\frac{2}{3}}$, the inclination can be inwards towards the axis of the circle.

THEORY OF MOMENTUM

81. Impulse. Let the equations of motion of a particle be written in the forms

$$m\ddot{x} = X, \quad m\ddot{y} = Y, \quad m\ddot{z} = Z,$$

and let both members of each of these equations be integrated with respect to t over an interval from t_0 to t_1. Let $\dot{x}_1$, $\dot{y}_1$, $\dot{z}_1$ be the components of velocity at the instant t_1, and $\dot{x}_0$, $\dot{y}_0$, $\dot{z}_0$ the components of velocity at the instant t_0. The result is

$$m\dot{x}_1 - m\dot{x}_0 = \int_{t_0}^{t_1} X dt, \quad m\dot{y}_1 - m\dot{y}_0 = \int_{t_0}^{t_1} Y dt, \quad m\dot{z}_1 - m\dot{z}_0 = \int_{t_0}^{t_1} Z dt.$$

The quantities in the left-hand members of these equations are the components of a vector, which is the *change of momentum* of the particle during the interval. The quantities in the right-hand members of the same equations are the components of another vector which is called the "impulse of the force" acting on the particle during the interval. The equations can be expressed in words in the statement:—*The change of momentum of a particle in any interval is equal to the impulse of the force acting on the particle during the interval.*

82. Sudden changes of motion. Changes of motion of bodies sometimes take place so rapidly that it is difficult to observe the gradual transition from one state of motion to another. We may allow for the possibility of sudden changes of motion by supposing that the force acting on a particle becomes very great during a very short interval of time, in such a way that the impulse of the force has a finite limit when the interval is diminished indefinitely. Let t' denote the instant at which the sudden change of motion takes place. In the equations of the type

$$m\dot{x}_1 - m\dot{x}_0 = \int_{t_0}^{t_1} X dt,$$

the right-hand members have finite limits when $t_0 = t' - \frac{1}{2}\tau$, and $t_1 = t' + \frac{1}{2}\tau$, and τ is diminished indefinitely. We write

$$\operatorname*{Lt}_{\tau=0} \int_{t'-\frac{1}{2}\tau}^{t'+\frac{1}{2}\tau} X dt = \underset{\cdot}{X}, \quad \operatorname*{Lt}_{\tau=0} \int_{t'-\frac{1}{2}\tau}^{t'+\frac{1}{2}\tau} Y dt = \underset{\cdot}{Y}, \quad \operatorname*{Lt}_{\tau=0} \int_{t'-\frac{1}{2}\tau}^{t'+\frac{1}{2}\tau} Z dt = \underset{\cdot}{Z}.$$

Then the equations are

$$m\dot{x}_1 - m\dot{x}_0 = \underset{\cdot}{X}, \quad m\dot{y}_1 - m\dot{y}_0 = \underset{\cdot}{Y}, \quad m\dot{z}_1 - m\dot{z}_0 = \underset{\cdot}{Z}.$$

We define the vector, localized at the position of the particle, of which the components parallel to the axes are $\underset{\cdot}{X}$, $\underset{\cdot}{Y}$, $\underset{\cdot}{Z}$, to be the "impulse exerted on the particle" at the instant t', at which the sudden change of motion takes place.

83. Constancy of momentum. The equations of motion of the form

$$m\ddot{x} = X$$

may also be written

$$\frac{d}{dt}(m\dot{x}) = X,$$

and this equation may be expressed in words in the statement:— "The rate of increase of the momentum of a particle in any direction is equal to the sum of the resolved parts in that direction of all the forces which act upon the particle."

If the line of action of the resultant force acting on the particle is at right angles to a fixed line, the resolved part of the momentum in the direction of this line is constant.

We had an example of this in the parabolic motion of projectiles (Art. 33).

If the velocity of a particle undergoes a sudden change, the resolved part of the momentum in any direction at right angles to the direction of the resultant impulse is unaltered.

84. Moment of force, momentum and kinetic reaction about an axis. Let the axis be the axis of z, and consider a force applied at the point (x', y', z'). Let F be the force, and X, Y, Z its components parallel to the axes. Let a plane pass through the point (x', y', z') and cut the axis of z at right angles in the point P.

Resolve the force F into components: Z parallel to the axis of z, and F' at right angles to this axis. Then the moment of F about the axis of z is defined to be the same as the moment of F' about P. The rule of signs is that when the axis of z and the direction of F' are related like the directions of translation and rotation in an ordinary right-handed screw the sign is +. Otherwise the sign is −.

The theorem of Art. 22 gives for the moment of F about the axis of z the expression

$$x'Y - y'X.$$

Let (x, y, z) be any point on the line, r its distance from (x', y', z'), l, m, n the cosines of the angles which the line, supposed drawn from the point (x', y', z') to the point (x, y, z), makes with the axes of x, y, z, drawn in positive senses. Then

$$x - x' = lr,\quad y - y' = mr,\quad z - z' = nr.$$

Now the sense from (x', y', z') to (x, y, z) is either the sense of the force F or the opposite sense, and we have therefore

either $$X = lF,\quad Y = mF,\quad Z = nF,$$

or $$X = -lF,\quad Y = -mF,\quad Z = -nF.$$

In both cases we have the equations

$$\frac{x - x'}{X} = \frac{y - y'}{Y} = \frac{z - z'}{Z},$$

and therefore $$xY - yX = x'Y - y'X.$$

It follows that, so long as the magnitude, line of action and sense of the force remain the same, the moment is independent of the point of application.

Now let the force be supposed to be applied at that point in its line of action at which the common perpendicular to the line of action and the axis of z meets the line of action. Then the force F' is at right angles to this common perpendicular. Hence the moment is the product, with a certain sign, of the length of the common perpendicular and the resolved part of the force at right angles to the axis. The rule of signs is, as before, the rule of the right-handed screw.

This result leads to a *general definition* of the moment of a localized vector about an axis:—Let the axis be a line L to which a certain sense is assigned, and let the vector be localized in a line L', or be localized at a point in L' and have for direction the direction of L'. Resolve the vector into components parallel to L and at right angles to L. The moment of the vector about the axis L is the product, with a certain sign, of the resolved part of the vector at right angles to L, and the length of the common perpendicular to L and L'. The rule of signs is the rule of the right-handed screw.

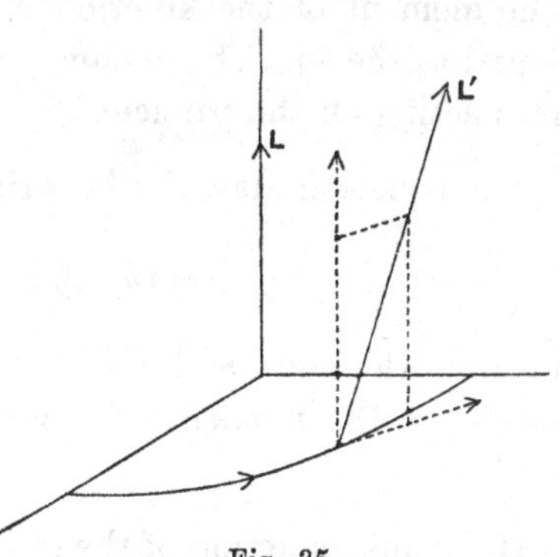

Fig. 35.

From what precedes it is clear that, if the vector is resolved into any components, or is the resultant of given component vectors, the moment of the resultant about any axis is the sum of the moments of the components.

The moments of a force (X, Y, Z), applied at a point (x, y, z), about the axes of x, y, z, are respectively

$$yZ - zY,\ zX - xZ,\ xY - yX.$$

The moments of the momentum of a particle about the axes are

$$m(y\dot{z} - z\dot{y}),\ m(z\dot{x} - x\dot{z}),\ m(x\dot{y} - y\dot{x}),$$

where x, y, z are the coordinates of the position of the particle at time t. The moments of the kinetic reaction of the particle about the axes are

$$m(y\ddot{z} - z\ddot{y}),\ m(z\ddot{x} - x\ddot{z}),\ m(x\ddot{y} - y\ddot{x}).$$

85. Constancy of moment of momentum.

Let x, y, z be the coordinates at time t of a particle which is subject to any forces, and let X, Y, Z be the components of the resultant force parallel to the axes. We have the equations

$$m\ddot{x} = X,\ m\ddot{y} = Y,\ m\ddot{z} = Z.$$

Multiply both members of the second of these equations by x, and both members of the first by y, and subtract the results. We have

$$m(x\ddot{y} - y\ddot{x}) = xY - yX.$$

This equation may be expressed in words in the statement:—"The moment of the kinetic reaction of a particle about an axis is equal to the sum of the moments about the same axis of all the forces acting on the particle."

The equation may also be written

$$\frac{d}{dt}\{m(x\dot{y}-y\dot{x})\}=xY-yX\,;$$

and now the left-hand member may be read as "The rate of increase of the moment of momentum of the particle about the axis."

If the line of action of the resultant force acting on the particle meets a fixed axis, or is parallel to such an axis, the moment of momentum of the particle about the axis is constant.

We have had an example of this in central orbits.

If the velocity of a particle undergoes a sudden change the moment of the momentum about any line which meets, or is parallel to, the line of the resultant impulse is unaltered.

85 A. Note. In Art. 85 the formula

$$\frac{d}{dt}\{m(x\dot{y}-y\dot{x})\}=m(x\ddot{y}-y\ddot{x})$$

is interpreted in the statement:—"The moment of the kinetic reaction of a particle about a *fixed* axis is equal to the rate of increase, per unit of time, of the moment of momentum of the particle about the same axis." It is sometimes convenient to take moments about an axis defined by means of a body which is in motion relatively to the axes of reference, or, as it would usually be described, a "moving axis." Let the moving axis be parallel to the axis of z, and meet the plane of (x, y) in the point (ξ, η). Then the moment of the kinetic reaction of the particle about the instantaneous position of the moving axis is

$$m\{(x-\xi)\ddot{y}-(y-\eta)\ddot{x}\},$$

and the moment of momentum about the same line is

$$m\{(x-\xi)\dot{y}-(y-\eta)\dot{x}\}.$$

Now
$$\frac{d}{dt}[m\{(x-\xi)\dot{y}-(y-\eta)\dot{x}\}]$$
$$=m\{(x-\xi)\ddot{y}-(y-\eta)\ddot{x}\}+m(\dot{x}\dot{\eta}-\dot{y}\dot{\xi}),$$

and thus the moment of the kinetic reaction is not in general equal to the rate of increase of the moment of momentum when the axis is in motion.

WORK AND ENERGY

86. Work done by a variable force. Let a particle move along a curved path, of which the arc measured from a fixed point to a variable point is denoted by s, and let F be a force acting on the particle, θ the angle which the line of action of F at any point of the curve makes with the tangent to the curve at the point. We suppose this tangent to be drawn in the sense in which the curve is described.

Let the arc between any two points A and B of the curve be replaced by a polygon of n sides, $s_1, s_2, \dots s_n$, having all its vertices on the curve. If the force F were the same at all points of any of these sides, and, at any point on the side s_κ $(\kappa = 1, 2, \dots n)$, its magnitude were F_κ and the angle which its line of action makes with the side were θ_κ, the work done by the force, as the particle describes the polygon, would be

$$F_1 . s_1 \cos\theta_1 + F_2 . s_2 \cos\theta_2 + \dots + F_n . s_n \cos\theta_n.$$

When the number of sides of the polygon is increased indefinitely, and the lengths of all of them are diminished indefinitely, this expression tends to a limit, called "the line-integral of the tangential component of F" along the arc of the curve between the points A and B. It is expressed by

$$\int_A^B F\cos\theta\, ds.$$

If X, Y, Z are the components of the force at any point (x, y, z), this expression is the same as the line-integral

$$\int\left(X\frac{dx}{ds} + Y\frac{dy}{ds} + Z\frac{dz}{ds}\right)ds, \text{ or } \int(Xdx + Ydy + Zdz),$$

taken along the curve from the point A to the point B.

This expression represents the work done by the force upon the particle in the displacement from A to B along the curve.

It is clear from the form of the expression that the work done by the resultant of any forces acting on a particle is equal to the sum of the works done by the separate forces.

87. Calculation of work. For the actual calculation of the work it would in general be necessary to know how to express the coordinates of a point of the curve in terms of some parameter, say θ, and also to know the values of the components of the force in terms of the position of the particle. Then at any point on the curve we could express X, Y, Z in terms of x, y, z, and therefore of θ, and we could also express $\frac{dx}{d\theta}, \frac{dy}{d\theta}, \frac{dz}{d\theta}$ in terms of θ, and thus we should have to integrate an expression of the form

$$\int \left(X\frac{dx}{d\theta} + Y\frac{dy}{d\theta} + Z\frac{dz}{d\theta} \right) d\theta$$

between two fixed values of θ, corresponding to the points A and B. In this expression $X, \ldots$ and $\frac{dx}{d\theta}, \ldots$ would be expressed in terms of θ.

It is clear that the result, if it could be obtained, would depend in general upon the curve; that is to say it would be different for different curves joining the same two points.

In the case where the force is a central attractive force, $mf(r)$, which is a function of the distance r from a fixed point, the tangential component of the force is $-mf(r)\frac{dr}{ds}$, and the work done is

$$-\int_{r_0}^{r_1} mf(r)\, dr,$$

where r_0 and r_1 are the distances of A and B from the fixed point. Now let $\phi(r)$ be the indefinite integral of $f(r)$, so that

$$\frac{d\phi(r)}{dr} = f(r),$$

then the work done is $m[\phi(r_0) - \phi(r_1)]$. It depends on r_0 and r_1, but is the same for any two curves joining the points A and B.

Another example in which the work is independent of the curve is afforded by a constant force as we saw in Art. 67.

88. Work function. When the work is independent of the path, we may choose arbitrarily a fixed point A, and take the integral

$$\int (Xdx + Ydy + Zdz)$$

along *any* path drawn from the point A to a point P. The result is a function of the coordinates of P. This function is the *work function.* The value of the work function at any point P is equal to the work done by the forces upon the particle as the particle moves along any path from the chosen fixed point A to the assigned point P.

When the work is independent of the path, so that a work function exists, the forces are said to be "conservative."

89. Potential function. In the case of a particle moving in a field of force, we denote by f the intensity of the field at any point. Let A be an arbitrary fixed point in the field, s the arc of a curve measured from A, θ the angle which the direction of the field at any point makes with the tangent to the curve at the point, the sense of the tangent being that in which the curve would be described by a particle starting from A. The work done by the force of the field in the displacement of a particle of mass m along the curve from the chosen point A to a variable point P is

$$m\int_A^P f \,.\, \cos\theta \,.\, ds.$$

If the force of the field is conservative, this expression is equal to the value of the work function at P; we write it

$$mV(P) \text{ or } mV.$$

Then $V(P)$ is defined to be the value of the *potential function* at the point P, and the function V is called the "potential" at a point. It is the line-integral of the tangential component of the force of the field (estimated by its intensity) taken along any curve joining the chosen point A to the variable point P.

The potential function vanishes at the point A.

If we replace the point A by any other fixed point B, the potential function is increased by a constant, which is the value of the integral

$$\int_A^B f \,.\, \cos\theta \,.\, ds.$$

In the case of a central field, of which the intensity at a distance r from the centre of force is $\frac{\mu}{r^2}$, we take the point A at an infinite distance. The potential function is then given by the equation

$$V = \int_\infty^P \frac{\mu}{r^2}\left(-\frac{dr}{ds}\right) ds = \frac{\mu}{r},$$

or the potential at any point is the product of the constant μ and the reciprocal of the distance of the point from the centre of force.

In the case of a uniform field of intensity g, we may draw the axis z in the direction opposite to that of the field, then the potential at a point is $-gz$.

90. Forces derived from a potential. Let mX, mY, mZ be the components of the force of a field acting on a particle of mass m, so that the direction of the vector (X, Y, Z) is the direction of the field, and the resultant of (X, Y, Z) is the intensity of the field. Let V be the potential of the field, supposed conservative.

Let P be any point (x, y, z), and P' any neighbouring point $(x+\delta x,\ y+\delta y,\ z+\delta z)$. The difference $V(P')-V(P)$ is the value of

$$\int_A^{P'}(Xdx+Ydy+Zdz)-\int_A^{P}(Xdx+Ydy+Zdz),$$

and this is the same as the value of the integral

$$\int_P^{P'}(Xdx+Ydy+Zdz)$$

taken along the straight line drawn from P to P'.

Now there exist some values X', Y', Z', intermediate between the greatest and least values of X, Y, Z that occur on the line PP', which are such that

$$\int_P^{P'}(Xdx+Ydy+Zdz)=X'\delta x+Y'\delta y+Z'\delta z.$$

This is, of course, a fundamental theorem of Integral Calculus.

Hence we have

$$X'\delta x+Y'\delta y+Z'\delta z=V(x+\delta x,\ y+\delta y,\ z+\delta z)-V(x,\ y,\ z).$$

Let δy and δz be zero, so that the line PP' is parallel to the axis of x. Then we have

$$X'=\frac{V(x+\delta x,\ y,\ z)-V(x,\ y,\ z)}{\delta x};$$

and therefore, in the limit, when P' moves up to P,

$$X=\frac{\partial V}{\partial x}.$$

In like manner we should find

$$Y = \frac{\partial V}{\partial y}, \quad Z = \frac{\partial V}{\partial z}.$$

The result may be interpreted in the statement:—*The force of the field (estimated per unit of mass), in any direction, is equal to the rate of increase of the potential per unit of length in that direction.*

If, adopting a different notation, we denote by X, Y, Z the components parallel to the axes of the force acting on a particle, and if a work function U exists, we have

$$X = \frac{\partial U}{\partial x}, \quad Y = \frac{\partial U}{\partial y}, \quad Z = \frac{\partial U}{\partial z}.$$

When the components of force are, as here, the partial differential coefficients of a function of the coordinates, the force is said to be "derived from a potential."

91. Energy equation. Multiply the left-hand and right-hand members of the equations of motion

$$m\ddot{x} = X, \; m\ddot{y} = Y, \; m\ddot{z} = Z$$

by $\dot{x}$, $\dot{y}$, $\dot{z}$ respectively, and add the results. The sum of the left-hand members, viz.

$$m(\dot{x}\ddot{x} + \dot{y}\ddot{y} + \dot{z}\ddot{z}),$$

is

$$\frac{d}{dt}[\tfrac{1}{2}m(\dot{x}^2 + \dot{y}^2 + \dot{z}^2)],$$

where the quantity differentiated is the kinetic energy of the particle at time t. The sum of the right-hand members is

$$X\dot{x} + Y\dot{y} + Z\dot{z},$$

and this expression represents the *rate at which work is done* by the forces.

Hence we have the equation

$$\frac{d}{dt}[\tfrac{1}{2}m(\dot{x}^2 + \dot{y}^2 + \dot{z}^2)] = X\dot{x} + Y\dot{y} + Z\dot{z};$$

and this equation can be expressed in words in the statement:—*The rate of increase of the kinetic energy of a particle is equal to the rate at which work is done by the forces acting on the particle.*

Let s denote the arc of the path measured from a fixed point

A of it to a variable point P of it. We multiply both sides of the equation just written by $\frac{dt}{ds}$. It becomes

$$\frac{d}{ds}\left[\tfrac{1}{2}m(\dot{x}^2+\dot{y}^2+\dot{z}^2)\right]=X\frac{dx}{ds}+Y\frac{dy}{ds}+Z\frac{dz}{ds},$$

and we hence find the equation

$$\tfrac{1}{2}mv^2-\tfrac{1}{2}mv_0^2=\int_A^P(Xdx+Ydy+Zdz),$$

where v and v_0 are the values of the velocity of the particle at P and A, and the integral is a line-integral taken along the path.

The equation can be expressed in words in the statement:—*The increment of kinetic energy in any displacement is equal to the work done by the forces in that displacement.*

When the forces are conservative, and U denotes the work function, the right-hand member of the equation last written is $U(P)-U(A)$, and we have

$$\tfrac{1}{2}mv^2-U(P)=\text{const.}$$

We call this equation the "energy equation."

We have already had several examples of energy equations. In the parabolic motion of projectiles we have the result in Art. 34, Ex. 3, in the case of simple harmonic motion we have a result used in Art. 38, in the case of central orbits we have the result in equation (2) of Art. 50 and the special results in Art. 40, Ex. 4, and Art. 48, Exx. 1 and 2.

92. Potential energy of a particle in a field of force. The work function at a point P, with its sign changed, is the work that would be done by the force of the field upon a particle which moves from the point P by any path to the chosen fixed point A.

This quantity is called the "potential energy of the particle in the field."

The energy equation can be written

"Kinetic Energy + Potential Energy = const."

The potential energy of a body, treated as a particle, in the field of the Earth's gravity is mgz, where z is the height of the particle above some chosen fixed level, and m is the mass of the body.

93. Forces which do no work. When a particle moves on a fixed curve or surface, forming part of the surface of a body, the pressure of the curve or surface does no work; for it is always directed at right angles to the path.

Forces which do no work are frequently called "constraints."

In forming the energy equation we may always omit such forces from the calculation.

94. Conservative and non-conservative fields. All fields of force which are found in nature are conservative.

It is easy to invent analytical expressions for non-conservative fields. For example, let the force at a distance r from a fixed point be always directed at right angles to the radius vector drawn from the point, and be equal to μr; and let a particle be guided by a "constraint" to describe, under the action of the force, a plane closed curve containing the point.

The work done can be shown easily to be equal to the product of 2μ and the area of the curve. Hence every time that the particle moves round the curve it acquires an increment of kinetic energy expressed by this product.

If such a system could be devised it could be used to drive a machine. We should then have a "perpetual motion." The statement that natural fields of force are conservative is included in the statement that there cannot be a perpetual motion.

By a "perpetual motion" is meant a self-acting machine which continually performs work. In the above example the particle, after each circuit of the curve, might yield up its increment of kinetic energy by striking against an external body. It would then start always from the same initial position with the same initial velocity. Its motion would be periodic, and yet it would transfer kinetic energy to an external body. In natural systems, when periodic motions are performed without friction, there can be no increment of kinetic energy available for transfer to an external body. In general there are forces of the nature of friction which have the effect that, when the initial position is recovered, the kinetic energy is diminished. For this reason an ordinary machine, once started, and subject to natural forces, does not go on for ever, but gradually comes to rest.

It is to be observed that a function U may exist which is such that the force (X, Y, Z) satisfies the equations

$$X = \frac{\partial U}{\partial x}, \quad Y = \frac{\partial U}{\partial y}, \quad Z = \frac{\partial U}{\partial z},$$

and yet the field of force may not be conservative. In a conservative field the work done in displacing a particle round *any* closed curve whatever vanishes. Now if U were of the form

$A \tan^{-1}(y/x)$ an amount of work equal to $2\pi A$ would be done in displacing a particle round any curve surrounding the axis of z. We may express the restriction to which this example points by saying that, in a conservative field, not only is the force derived from a potential, but also the potential is a *one-valued* function.

MISCELLANEOUS EXAMPLES

1. Prove that the time of quickest descent along a straight line from a point on one vertical circle to another in the same plane is

$$\sqrt{\{2(c^2-a^2)/g(a+h)\}},$$

where c is the distance between their centres, a is the sum of the radii, and h the vertical height of the centre of the former circle above that of the latter.

2. A parabola of latus rectum $4a$ is placed in a vertical plane with its vertex downwards and its axis inclined to the vertical at an angle β. Prove that the time down the chord of quickest descent from the focus to the curve is $\sqrt{(2ag^{-1}\sec^3\tfrac{1}{3}\beta)}$.

3. A train of mass m runs from rest at one station to stop at the next at a distance l. The full speed is V, and the average speed is v. The resistance of the rails when the brake is not applied is uV/lg of the weight of the train, and when the brake is applied it is $u'V/lg$ of the weight of the train. The pull of the engine has one constant value while the train is getting up speed, and another constant value while it is running at full speed; prove that the average rate at which the engine works in starting the train is

$$\tfrac{1}{2}m\frac{V^2}{l}\left\{u-\frac{1}{2/v-2/V-1/u'}\right\}.$$

4. Two equal bodies, each of mass M, are attached to the chain of an Atwood's machine, and oscillate up and down through two fixed horizontal rings so that each time one of them passes up through a ring it lifts a bar of mass m, while at the same instant the other passes down through its ring and deposits on it a bar of equal mass. Prove, neglecting friction, that the period of an excursion of amplitude a is

$$2\sqrt{\{2a(2M+\mu+m)/mg\}},$$

and that the successive amplitudes form a diminishing geometric progression of which the ratio is

$$\{(2M+\mu)/(2M+\mu+m)\}^2,$$

where μ is a mass which distributed over the circumference of the pulley will produce the same effect on the motion as the inertia of the actual mechanism.

---

5. A particle is projected along the circumference of a smooth vertical circle of radius a. It starts from the lowest point and leaves the circle before reaching the highest point. Prove that, if the coefficient of restitution between the circle and the particle is unity, and if the initial velocity is

$$\sqrt{[ag\{2+\tfrac{3}{2}\sqrt{(3-\sqrt{3})}\}]},$$

the particle after striking the circle will retrace its former path.

6. A particle moves on the outside of a smooth elliptic cylinder whose generators are horizontal, starting from rest on the highest generator, which passes through extremities of major axes of the normal sections. Prove that it will leave the cylinder at a point whose eccentric angle ϕ is given by the equation

$$e^2 \cos^3 \phi = 3 \cos \phi - 2,$$

where e is the eccentricity of the normal sections.

7. Two cycloids are placed in the same vertical plane, with their axes vertical, and their vertices downwards and at the same level. Two particles start to describe the cycloids from points at the same level. Show that they will next be at the same level after a time $2\pi \sqrt{(aa')}/\{(\sqrt{a}+\sqrt{a'})\sqrt{g}\}$, and next after that at time $4\pi \sqrt{(aa')}/\{(\sqrt{a}+\sqrt{a'})\sqrt{g}\}$ or $2\pi \sqrt{(aa')}/\{(\sqrt{a}\sim\sqrt{a'})\sqrt{g}\}$, whichever is less, a and a' being the radii of the generating circles.

---

8. A railway carriage is travelling on a curve of radius r with velocity v, $2a$ is the distance between the rails and h is the height of the centre of gravity of the carriage above the rails. Show that the weight of the carriage is divided between the rails in the ratio $gra - v^2h : gra + v^2h$, and hence that the carriage will upset if

$$v > \sqrt{(gra/h)}.$$

9. A train starts from rest on a level uniform curve, and moves round the curve so that its speed increases at a constant rate f. The outer rail is raised so that the floor of a carriage is inclined at an angle a to the horizon. Show that a body cannot rest on the floor of the carriage unless the coefficient of friction between the body and the floor exceeds

$$\sqrt{(f^2 + g^2 \sin^2 a)}/g \cos a.$$

---

10. Prove that the impulse necessary to make a particle of unit mass, moving in an equiangular spiral of angle a under the action of a force to the pole, describe a circle under the action of the same force, is

$$2\sqrt{(Fr)} \sin\left(\frac{\pi}{4} \pm \frac{a}{2}\right),$$

r being the distance from the pole, and F the force at the moment of impact.

11. A particle is describing an ellipse of eccentricity e about a focus and when its radius vector is half the latus rectum it receives a blow which makes it move towards the other focus with a momentum equal to that of the blow. Find the position of the axis of the new orbit and show that its eccentricity is $\frac{1}{2}(e^{-1} - e)$.

12. A particle is describing an ellipse about a centre of force in one focus S, and when it is at the end E of the further latus rectum it receives a blow in direction SE which makes it move at right angles to SE. Find the momentum generated by the blow, and prove that the particle will proceed to describe an ellipse of eccentricity $\{2e^2/(1+e^2)\}$.

CHAPTER IV†

MOTION OF A PARTICLE UNDER GIVEN FORCES

95. THE application of the principles which have been laid down in previous Chapters to the discussion of the motions of particles in particular circumstances is the part of our subject usually described as "Dynamics of a Particle." We shall devote to it the two following Chapters. This part of our subject divides itself into two main branches, referring respectively to motions under given forces, and to constrained and resisted motions taking place under forces which are not all given. We confine our attention in the present Chapter to motions under given forces.

96. Formation of equations of motion. The method of formation of the equations of motion has been described in Article 64. It consists in equating the product of the mass of the particle and its resolved acceleration in any direction to the resolved part of the force acting upon it in that direction. The equations thus arrived at are differential equations. The left-hand member of any equation contains differential coefficients of geometrical quantities with respect to the time. The right-hand member is, in general, a given function of geometrical quantities; in special cases it may be a given function of the time. Although there are many cases in which equations of this kind can be solved, there exists no general method for solving them.

Diversity can arise, in regard to the formation of the equations, only from the choice of different directions in which to resolve. Thus we may resolve parallel to the axes of reference, or we may resolve along the radius vector from the origin to a particle, and in directions at right angles thereto, or again we may resolve along the tangent to the path of a particle and in directions at right angles thereto. The most suitable directions to choose in particular cases are determined by the circumstances.

† Articles in this Chapter which are marked with an asterisk (*) may be omitted in a first reading.

Methods by which the components of acceleration in chosen directions can be expressed in terms of suitable geometrical quantities have been exemplified in Arts. 36 and 43. Further illustrations are given in the next two Articles.

***97. Acceleration of a point describing a tortuous curve.**

We recall the facts that, if x, y, z are the rectangular coordinates of a point of a curve and s the arc measured from some particular point of the curve to the point (x, y, z), the direction cosines of the tangent, in the sense in which s increases, are $\frac{dx}{ds}, \frac{dy}{ds}, \frac{dz}{ds}$, satisfying the relation $\left(\frac{dx}{ds}\right)^2+\left(\frac{dy}{ds}\right)^2+\left(\frac{dz}{ds}\right)^2=1$; the direction cosines of the principal normal directed towards the centre of curvature are $\rho\frac{d^2x}{ds^2}, \rho\frac{d^2y}{ds^2}, \rho\frac{d^2z}{ds^2}$, satisfying the relation

$$\frac{1}{\rho^2}=\left(\frac{d^2x}{ds^2}\right)^2+\left(\frac{d^2y}{ds^2}\right)^2+\left(\frac{d^2z}{ds^2}\right)^2,$$

where ρ is the radius of circular curvature; and the direction cosines of the binormal are $\rho\left(\frac{d^2y}{ds^2}\frac{dz}{ds}-\frac{d^2z}{ds^2}\frac{dy}{ds}\right), \rho\left(\frac{d^2z}{ds^2}\frac{dx}{ds}-\frac{d^2x}{ds^2}\frac{dz}{ds}\right), \rho\left(\frac{d^2x}{ds^2}\frac{dy}{ds}-\frac{d^2y}{ds^2}\frac{dx}{ds}\right)$. We recall also the relation $\frac{dx}{ds}\frac{d^2x}{ds^2}+\frac{dy}{ds}\frac{d^2y}{ds^2}+\frac{dz}{ds}\frac{d^2z}{ds^2}=0$.

In the expressions $\ddot{x}, \ddot{y}, \ddot{z}$ for the component accelerations parallel to the axes we change the independent variable from t to s.

We have, writing v for the speed, so that v stands for $\dot{s}$,

$$\ddot{x}=\frac{d^2x}{dt^2}=\frac{d}{dt}\left(\frac{dx}{dt}\right)=\frac{ds}{dt}\frac{d}{ds}\left(\frac{ds}{dt}\frac{dx}{ds}\right)=v\frac{d}{ds}\left(v\frac{dx}{ds}\right),$$

so that

$$\ddot{x}=v\frac{dv}{ds}\frac{dx}{ds}+v^2\frac{d^2x}{ds^2}.$$

Similarly

$$\ddot{y}=v\frac{dv}{ds}\frac{dy}{ds}+v^2\frac{d^2y}{ds^2},$$

and

$$\ddot{z}=v\frac{dv}{ds}\frac{dz}{ds}+v^2\frac{d^2z}{ds^2}.$$

If we multiply these component accelerations in order by the direction cosines of the tangent and add, we obtain the component acceleration parallel to the tangent to the curve in the sense in which s increases; we thus find for this component the expression

$$v\frac{dv}{ds}\left[\left(\frac{dx}{ds}\right)^2+\left(\frac{dy}{ds}\right)^2+\left(\frac{dz}{ds}\right)^2\right]+v^2\left(\frac{dx}{ds}\frac{d^2x}{ds^2}+\frac{dy}{ds}\frac{d^2y}{ds^2}+\frac{dz}{ds}\frac{d^2z}{ds^2}\right), \text{ or } v\frac{dv}{ds}.$$

Again, if we multiply by the direction cosines of the principal normal and add, we obtain the component acceleration parallel to the principal normal directed towards the centre of curvature; we thus find for this component the expression

$$v\frac{dv}{ds}\rho\left[\frac{dx}{ds}\frac{d^2x}{ds^2}+\frac{dy}{ds}\frac{d^2y}{ds^2}+\frac{dz}{ds}\frac{d^2z}{ds^2}\right]+v^2\rho\left[\left(\frac{d^2x}{ds^2}\right)^2+\left(\frac{d^2y}{ds^2}\right)^2+\left(\frac{d^2z}{ds^2}\right)^2\right], \text{ or } \frac{v^2}{\rho}.$$

Finally, if we multiply by the direction cosines of the binormal and add, we find no component acceleration parallel to the binormal.

Thus the acceleration of a point describing a tortuous curve is in the osculating plane of the curve, and its resolved parts parallel to the tangent and principal normal are $v\frac{dv}{ds}$ and $\frac{v^2}{\rho}$, exactly as in the case of a point describing a plane curve. As in that case, the expression for the former component may be replaced by $\dot{v}$, or by $\ddot{s}$.

***98. Polar coordinates in three dimensions.** The coordinates are r the distance from the origin, θ the angle between the radius vector and the axis z, ϕ the angle between the plane containing the radius vector and the axis z and a fixed plane passing through the axis z.

The plane containing the radius vector and the axis z will be called the "meridian plane," and the circle in which this plane cuts a sphere $r=$const. the "meridian."

We denote distance from the axis z by ϖ, so that $\varpi=r\sin\theta$.

In a plane parallel to the plane (x, y), ϖ and ϕ are plane polar coordinates; in the meridian plane z and ϖ are Cartesian coordinates, and r and θ are plane polar coordinates.

Hence the velocity $(\dot{x}, \dot{y})$ parallel to the plane (x, y) is equivalent to $\dot{\varpi}$ at right angles to the axis z in the meridian plane, and $\varpi\dot{\phi}$ at right angles to this plane; and the velocity $(\dot{x}, \dot{y}, \dot{z})$ is equivalent to $(\dot{z}, \dot{\varpi})$ in the meridian plane and $\varpi\dot{\phi}$ at right angles to this plane. Also the velocity $(\dot{z}, \dot{\varpi})$ in this plane is equivalent to $\dot{r}$ along the radius vector and $r\dot{\theta}$ along the tangent to the meridian. The components of velocity are therefore

$\dot{r}$ along the radius vector,

$r\dot{\theta}$ along the tangent to the meridian,

$r\sin\theta\dot{\phi}$ at right angles to the meridian plane.

The accelerations $\ddot{x}$, $\ddot{y}$ parallel to the axes x, y are equivalent to $\ddot{\varpi}-\varpi\dot{\phi}^2$ and $\frac{1}{\varpi}\frac{d}{dt}(\varpi^2\dot{\phi})$ in and perpendicular to the meridian plane. Hence the acceleration is equivalent to $\ddot{z}$ parallel to the axis z, $\ddot{\varpi}-\varpi\dot{\phi}^2$ at right angles to the axis z and in the meridian plane, $\frac{1}{\varpi}\frac{d}{dt}(\varpi^2\dot{\phi})$ at right angles to the meridian plane.

Taking the components $\ddot{z}$, $\ddot{\varpi}$, which are in the meridian plane and are parallel and perpendicular to the axis z, we see that these are equivalent to $\ddot{r}-r\dot{\theta}^2$ along the radius vector and $\frac{1}{r}\frac{d}{dt}(r^2\dot{\theta})$ along the tangent to the meridian.

We resolve the acceleration $-\varpi\dot{\phi}^2$, which is in the meridian plane and at right angles to the axis z, into components parallel to the radius vector and to the tangent to the meridian. These components are $-\varpi\dot{\phi}^2 \sin\theta$ and $-\varpi\dot{\phi}^2 \cos\theta$. Hence the components of acceleration are

$$\ddot{r} - r\dot{\theta}^2 - r\sin^2\theta\dot{\phi}^2 \text{ along the radius vector,}$$

$$\frac{1}{r}\frac{d}{dt}(r^2\dot{\theta}) - r\sin\theta\cos\theta\dot{\phi}^2 \text{ along the tangent to the meridian,}$$

$$\frac{1}{r\sin\theta}\frac{d}{dt}(r^2\sin^2\theta\dot{\phi}) \text{ at right angles to the meridian plane.}$$

99. Integration of the equations of motion. Whenever there is an energy equation (Art. 91) it is an integral of the equations of motion.

When the particle moves in a straight line under conservative forces the energy equation expresses the velocity in terms of the position; and the position at any time, or the time of reaching any position, is determined by integration. For an example see Art. 54.

When the particle does not move in a straight line other integrals of the equations are requisite before the position at any time can be determined. If there is an equation of constancy of momentum (Art. 83), or of moment of momentum (Art. 85), these also are integrals of the equations of motion. These, combined with the energy equation, are sometimes sufficient to determine the position at any time. Examples are afforded by the parabolic motion of projectiles and by elliptic motion about a focus.

100. Example.

Deduce the result that the path of a particle moving freely under gravity is a parabola from the equation expressing the constancy of the horizontal component of momentum and the energy equation.

101. Motion of a body attached to a string or spring. Simple examples of Dynamics of a Particle are afforded by problems of the motion of a body attached to an extensible string or spring. We consider cases in which the particle moves in the line of the string or spring (supposed to be a straight line).

When the mass of the string is neglected†, and there is no friction acting upon it, the tension is constant throughout it (Chapter VI).

When the length of a string can change there is a particular length which corresponds to a state of zero tension. This state

† A string of which the mass is neglected is often called a "thread."

is called the "natural state," and the corresponding length the "natural length."

Let l_0 be the natural length, l the length in any state. The quantity $(l - l_0)/l_0$ is called the "extension."

The *law connecting the tension and the extension* is that the *tension is proportional to the extension.* If ϵ is the extension, the tension is equal to the product of ϵ and a certain constant. This constant is called the "modulus of elasticity" of the string.

If, in the course of any motion of an extensible string, the string recovers its natural length, the tension becomes zero, and the string becomes "slack." A particle attached to the string is then free from force exerted by the string until the length again comes to exceed the natural length.

A string which exerts tension, but is never sensibly extended, must be thought of as an ideal limit to which an extensible string approaches when the extension ϵ tends to zero, and the modulus λ tends to become infinite, in such a way that the product $\lambda\epsilon$ is the finite tension of the string. Such a string would be described as "inextensible."

A spring, when extended, exerts tension in the same way as an extensible string; when contracted, it exerts pressure which is the same multiple of the contraction $(l_0 - l)/l_0$ as the tension is of the extension.

A body attached to a spring, of which one end is fixed, and moveable in the line of the spring, is subject to a force equal to μx, where μ is a constant called the "strength of the spring," and x is the displacement of the body from the position in which the spring has its natural length. When the length is increased by x the force is tension; when it is diminished by x the force is pressure. The equation of motion of the body, considered as a particle of mass m, is $m\ddot{x} = -\mu x$.

It follows that the motion of the particle is simple harmonic motion of period $2\pi\sqrt{(m/\mu)}$.

This result may also be obtained by forming the energy equation. For the work done by the force in the displacement x is

$$\int_0^x -\mu x\,dx,$$

or it is $-\frac{1}{2}\mu x^2$; and the kinetic energy of the body, treated as a particle, is $\frac{1}{2}m\dot{x}^2$. Hence the energy equation is

$$\tfrac{1}{2}m\dot{x}^2+\tfrac{1}{2}\mu x^2=\text{const.},$$

and the result that x is of the form $a\cos\{t\sqrt{(\mu/m)}+\alpha\}$ can be obtained by integrating this equation.

102. Examples.

1. A particle of mass m is attached to the middle point of an elastic thread, of natural length a and modulus λ, which is stretched between two fixed points. Prove that, if no forces act on the particle other than the tensions in the parts of the thread, it can oscillate in the line of the thread with a simple harmonic motion of period $\pi\sqrt{(ma/\lambda)}$.

2. A particle of mass m is attached to one end of an elastic thread, of natural length a and modulus λ, the other end of which is fixed. The particle is displaced until the thread is of length $a+b$, and is then let go. Prove that, if no forces act on the particle except the tension of the thread, it will return to the starting point after a time $2\left(\pi+2\dfrac{a}{b}\right)\sqrt{\dfrac{ma}{\lambda}}$.

3. Prove that, if a body is suddenly attached to an unstretched vertical elastic thread and let fall under gravity, the greatest subsequent extension is twice the statical extension of the thread when supporting the body.

4. Prove that, if a spring is held compressed by a given force and the force is suddenly reversed, the greatest subsequent extension is three times the initial contraction.

5. An elastic thread of natural length a has one end fixed, and a particle is attached to the other end, the modulus of elasticity being n times the weight of the particle. The particle is at first held with the thread hanging vertically and of length a', and is then let go from rest. Show that the time until it returns to its initial position is

$$2(\pi-\theta+\theta'+\tan\theta-\tan\theta')\sqrt{(a/ng)},$$

where θ, θ' are acute angles given by

$$\sec\theta=na'/a-n-1,\quad \sec^2\theta'=\sec^2\theta-4n,$$

and a' is so great that real values of these angles exist.

102 A. Force of simple harmonic type. The most important case of forces, which are given in terms of the time, is that where simple harmonic motion is disturbed by a force proportional at time t to a function of the form $\cos(pt+\alpha)$, in which p and α are constants. Such forces have definite periods, and are often loosely described as "periodic forces"; they may be described strictly as "forces of simple harmonic type."

Let a particle of mass m be attached to a spring of such strength that, if free from the action of all forces except that exerted by the spring, it would have a simple harmonic motion of period $2\pi/n$; and let it be subject also to a force of magnitude $mP\cos(pt+\alpha)$ in the line of the spring. The equation of motion is

$$\ddot{x} + n^2 x = P\cos(pt+\alpha).$$

A solution of this equation would be found by putting x equal to $Q\cos(pt+\alpha)$ if $(n^2-p^2)Q = P$. To obtain a more general solution we put

$$x = \{P/(n^2-p^2)\}\cos(pt+\alpha) + \xi,$$

and then find that ξ satisfies the equation $\ddot{\xi} + n^2\xi = 0$, so that ξ must be of the form $A\cos nt + B\sin nt$, where A and B are arbitrary constants.

The motion of the particle is compounded of two simple harmonic motions in the same straight line, one having the period that the motion in the absence of the disturbing force would have, and the other having the period of the disturbing force. These are called the "free oscillation" and the "forced oscillation." The phase of the forced oscillation is the same as that of the force producing it if $n > p$, that is if the period of the force is longer than the period of free oscillation. When the period of the force is shorter than the period of free oscillation, or $p > n$, the phase of the forced oscillation is opposite to that of the force producing it, or the displacement in the forced oscillation always has the opposite sign to the force.

The amplitude of the forced oscillation becomes very great when p is nearly equal to n. When $p = n$ the equation of motion is satisfied by putting $x = (P/2n)t\sin(nt+\alpha)$, as may be verified easily, and the complete primitive of the equation is then

$$x = A\cos nt + B\sin nt + (P/2n)t\sin(nt+\alpha).$$

In this case the forced oscillation may be described as a simple harmonic motion of variable amplitude, which increases continually with the time. The phase of this motion is always one quarter of a period behind that of the force producing it.

The above is an example of a principle of wide application, to the effect that a system, which can oscillate in a definite period, can be thrown into a state of violent oscillation by the action of forces, which are of simple harmonic type and the same period. On account of its importance in the Theory of Sound this result is known as the "principle of resonance."

103. The problem of central orbits. We have already investigated this problem in some detail in Arts. 49—52. We found that a particle moving under a central force directed to a fixed point, moves in a fixed plane which contains the centre of force and the tangent to the path at any chosen instant. We found that the equations of motion could be expressed in the form

$$m(\ddot{r} - r\dot{\theta}^2) = -mf, \quad m\frac{1}{r}\frac{d}{dt}(r^2\dot{\theta}) = 0,$$

where m is the mass of the particle, and f is the intensity of the field of force, taken to be an attraction. We suppose that f is given as a function of r.

The energy equation is

$$\tfrac{1}{2}m(\dot{r}^2 + r^2\dot{\theta}^2) = \text{const.} - m\int f dr,$$

and the equation of constancy of moment of momentum about an axis through the centre of force at right angles to the plane of motion is

$$mr^2\dot{\theta} = mh,$$

where h is a constant which represents twice the rate of description of area by the radius vector.

We found that these equations lead to the equation

$$\left(\frac{du}{d\theta}\right)^2 + u^2 = \frac{2A}{h^2} + \frac{2}{h^2}\int \frac{f}{u^2} du,$$

where A is a constant, u is written for $1/r$, and f is now supposed to be expressed in terms of u. This equation determines the path of the particle.

When f is given, and the particle starts from a point at a distance a from the centre of force, with a velocity V, in a direction making an angle α with the radius vector, the value of h is $Va \sin \alpha$. The initial value of $\left(\frac{du}{d\theta}\right)^2 + u^2$ is $1/a^2 \sin^2 \alpha$, for it is the reciprocal of the square of the perpendicular from the origin of r upon the tangent to the path. Hence the equation of the path takes the form

$$\left(\frac{du}{d\theta}\right)^2 + u^2 - \frac{1}{a^2 \sin^2 \alpha} = \frac{2}{V^2 a^2 \sin^2 \alpha}\int_{\frac{1}{a}}^{u} \frac{f}{u^2} du.$$

When the path is known, so that u becomes a known function of θ, the time of describing any arc of the path is the value of the integral

$$\int \frac{d\theta}{u^2 Va \sin \alpha},$$

taken between limits for θ which correspond to the ends of the arc.

104. Apses. An apse is a point of a central orbit at which the tangent is at right angles to the radius vector.

There is a theory concerning the distribution of the apses when the central acceleration is a single-valued function of the distance, *i.e.* for the case where the acceleration depends only on the distance and is always the same at the same distance.

Let A be an apse on a central orbit described about a point O, f the central acceleration, supposed a single-valued function of distance, TAT' a line through A at right angles to AO. Then a point starting from A at right angles to AO with a certain velocity would describe the orbit. Let V be this velocity.

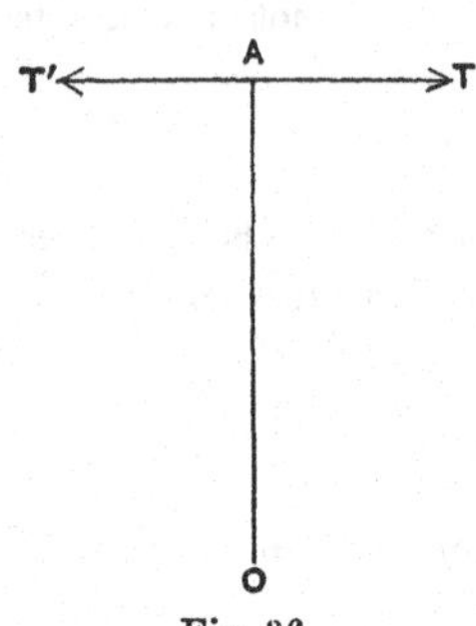

Fig. 36.

If a point starts from A with velocity V in direction AT or AT', and has the acceleration f towards O, it describes the orbit; so that two points starting from A in these two directions with the same velocity V and the same acceleration f describe the same orbit. Since the two points have the same acceleration at the same distance, the curves they describe are clearly equal and similar, and are symmetrically placed with respect to the line AO. Thus the orbit is symmetrical with respect to AO in such a way that chords drawn across it at right angles to AO are bisected by AO. The parts of the orbit on either side of AO are therefore optical images in the line AO.

Now let the point start from A in direction AT, and let B be the next apse of the orbit that it passes through, also let A' be the next apse after B that it passes through. Then the parts AOB, BOA' of the orbit are optical images in the line OB, and the angle AOB is equal to the angle $A'OB$, and the line AO is equal to the line $A'O$. In the same way the next apse the point passes through will be at a distance from O equal to OB, and thus all the apses are at distances from O equal to either OA or OB; these are called the *apsidal distances*, and the angle between consecutive apses in the order in which the moving point passes through them is always equal to AOB; this is called the *apsidal angle*.

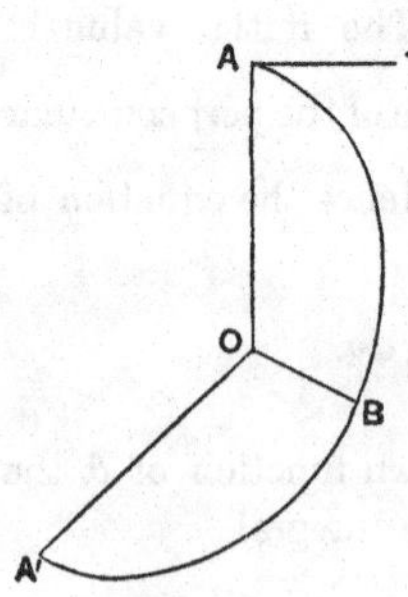

Fig. 37.

The theory just explained is usually stated in the form:— *There are two apsidal distances and one apsidal angle.*

It is clear that the radius vector is a periodic function of the vectorial angle with period twice the apsidal angle.

105. **Examples.**

1. If the apsidal distances are equal the orbit is a circle described about its centre.

2. Write down the lengths of the apsidal distances and the apsidal angle for (1) elliptic motion about the centre, (2) elliptic motion about a focus, (3) all the orbits that can be described with a central acceleration varying inversely as the cube of the distance.

3. Explain the following paradox:—Four real normals can be drawn to an ellipse from a point within its evolute, and in Ex. 6 of Art. 46 we found the central acceleration to any point requisite for the description of an ellipse; there are apparently in this case four apsidal distances and four apsidal angles.

106. Apsidal angle in nearly circular orbit. Let the central acceleration be $f(r)$ at distance r, then a circle of radius c described about its centre is a possible orbit with $\frac{1}{2}h$ for rate of describing area provided that

$$\frac{1}{c}\left(\frac{h}{c}\right)^2 = f(c),$$

or

$$h^2 = c^3 f(c).$$

Let us suppose the point to be at some instant near to the circle, and to be describing an orbit about the origin with moment of momentum specified by this h.

The equation of its path is

$$\frac{d^2u}{d\theta^2} + u = \frac{f(r)}{h^2u^2}.$$

At the instant in question u is nearly equal to $1/c$; if it was precisely $1/c$, and if the point was moving at right angles to the radius vector, the point would describe the circle of radius c. We assume that it is always so near to the circle that the difference $u - 1/c$ is so small that we may neglect its square; the investigation we give will determine under what condition this assumption is justifiable.

Write $\phi(u)$ for $f(r)$ and a for $1/c$, and put $u = a + x$, so that

$$h^2 = \phi(a)/a^3.$$

Then

$$\frac{d^2x}{d\theta^2}+x+a=\frac{a^3\phi(a+x)}{\phi(a)}\frac{1}{(a+x)^2}$$

$$=\frac{a^3}{\phi(a)}\left[\frac{\phi(a)}{a^2}+x\frac{d}{da}\left\{\frac{\phi(a)}{a^2}\right\}+\dots\right],$$

or

$$\frac{d^2x}{d\theta^2}+x\left\{3-\frac{a\phi'(a)}{\phi(a)}\right\}=0,$$

if x^2 is neglected.

Now if $3-a\phi'(a)/\phi(a)$ is positive we may put it equal to κ^2, and then the solution of the above equation is of the form

$$x=A\cos(\kappa\theta+\alpha),$$

so that the greatest value of x is A, and by taking A small enough x will be as small as we please and the neglect of x^2 will be justified.

In this case u, and therefore r, will be a periodic function of θ with period $2\pi/\sqrt{\{3-a\phi'(a)/\phi(a)\}}$, the orbit is nearly circular and its apsidal angle is $\pi/\sqrt{\{3-a\phi'(a)/\phi(a)\}}$.

Again, if $3-a\phi'(a)/\phi(a)$ is negative we may put it equal to $-\kappa^2$, and then the solution of the above equation is of the form

$$x=Ae^{\kappa\theta}+Be^{-\kappa\theta},$$

and it is clear that one of the terms increases in geometrical progression whether θ increases or diminishes, so that x will very soon be so great that its square can no longer be neglected, whatever the number we agree to neglect may be. In this case the orbit tends to depart widely from the circular form.

In the former of these cases the circular motion is said to be *stable*, in the latter *unstable*.

107. Examples.

1. If $f(r)=r^{-n}$ or $\phi(u)=u^n$, prove that the possible circular orbits are stable when $n<3$ and unstable when $n>3$.

2. For $n=3$ prove that the circular orbit is unstable, and find the orbit described by a point moving with the moment of momentum required for circular motion in a circle of radius c through a point near the circle.

3. If $f(r)=r^{-4}$, prove that the curve described with the moment of momentum required for circular motion in a circle of radius c, when the point of projection is near to or on this circle, is either the circle $r=c$ or one of the curves

$$\frac{r}{c}=\frac{\cosh\theta+1}{\cosh\theta-2},\quad \frac{r}{c}=\frac{\cosh\theta-1}{\cosh\theta+2}.$$

108. Examples of equations of motion expressed in terms of polar coordinates.

1. When the radial and transverse components of force acting on a particle which moves in one plane are R, T, the equations of motion are

$$m(\ddot{r}-r\dot{\theta}^2)=R,\quad \frac{m}{r}\frac{d}{dt}(r^2\dot{\theta})=T.$$

2. When the forces are derived from a potential V we have

$$R=m\frac{\partial V}{\partial r},\qquad T=\frac{m}{r}\frac{\partial V}{\partial \theta},$$

and there is an energy equation

$$\tfrac{1}{2}m(\dot{r}^2+r^2\dot{\theta}^2)=mV+\text{const.}$$

3. Put $r^2\dot{\theta}=h$, $u=r^{-1}$; in general h is variable. The equation of the path can be found by eliminating h between the equations

$$\frac{d}{d\theta}(\tfrac{1}{2}h^2)=\frac{T}{mu^3},\quad h^2\left(\frac{d^2u}{d\theta^2}+u\right)=-\frac{1}{mu^2}\left(R+\frac{T}{u}\frac{du}{d\theta}\right).$$

4. When the forces are derived from a potential, as in Ex. 2, the equation of the path can be written in the form

$$\frac{\partial V}{\partial \theta}=u^2\frac{d}{d\theta}\frac{V}{u^2+\left(\dfrac{du}{d\theta}\right)^2},$$

where $\dfrac{d}{d\theta}$ stands for $\dfrac{\partial}{\partial\theta}+\dfrac{du}{d\theta}\dfrac{\partial}{\partial u}$.

109. Examples of motion under several central forces.

1. A particle of mass m moves under the action of forces to two fixed points A, A' of magnitudes $m\mu/r^2$, $m\mu'/r'^2$ respectively, where r and r' are the

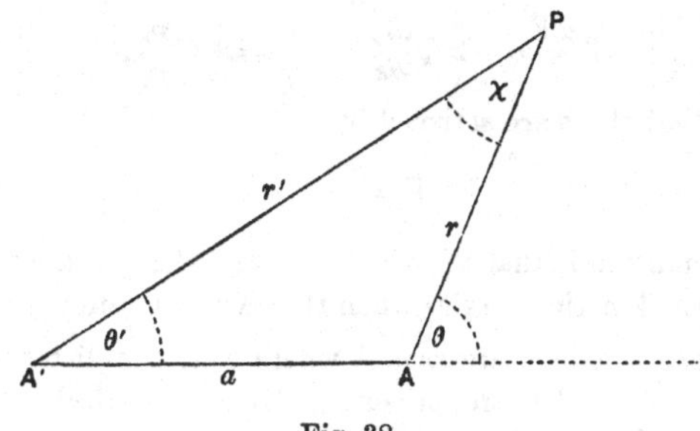

Fig. 38.

distances of the particle from A and A', and μ and μ' are constants. The equations of motion possess an integral of the form

$$r^2r'^2\dot{\theta}\dot{\theta}'=a(\mu\cos\theta-\mu'\cos\theta')+\text{const.},$$

where a is the distance AA'.

Resolving at right angles to the radius vector r, we have

$$m\frac{1}{r}\frac{d}{dt}(r^2\dot{\theta})=m\frac{\mu'}{r'^2}\sin\chi,\ \text{where } \chi \text{ is the angle } APA',$$

so that
$$r'^2\frac{d}{dt}(r^2\dot{\theta})=\mu' r\sin\chi=\mu' a\sin\theta',$$

similarly
$$r^2\frac{d}{dt}(r'^2\dot{\theta}')=-\mu r'\sin\chi=-\mu a\sin\theta.$$

Multiplying by $\dot{\theta}'$ and $\dot{\theta}$, adding, and integrating, we have an equation of the given form.

This equation with the energy equation determines the motion.

2. A particle of mass m moves under the action of forces to two fixed points of magnitudes $m\mu r$, $m\mu' r'$. Prove, with the notation of Ex. 1, that there is an integral equation of the form

$$\mu r^2\dot{\theta}+\mu' r'^2\dot{\theta}'=\text{const.}$$

3. A given plane curve can be described by a particle under central forces to each of n given points, when the forces act separately. Prove that it can be described under the action of all the forces, provided that the particle is properly projected.

Let f_κ be the acceleration produced in the particle by the force to the κth centre O_κ, v_κ the velocity of the particle at any point when the curve is described under this force, r_κ the distance of the point from O_κ, and p_κ the perpendicular from O_κ on the tangent to the curve at the point, ρ the radius of curvature and ds the element of arc of the curve at the point. Then we are given that

$$v_\kappa\frac{dv_\kappa}{ds}=-f_\kappa\frac{dr_\kappa}{ds},\quad \frac{v_\kappa^2}{\rho}=f_\kappa\frac{p_\kappa}{r_\kappa}.$$

Now the curve can be described under all the forces if there exists a velocity V satisfying the two equations

$$V\frac{dV}{ds}=-\sum_1^n f_\kappa\frac{dr_\kappa}{ds},\quad \frac{V^2}{\rho}=\sum_1^n f_\kappa\frac{p_\kappa}{r_\kappa},$$

and it is clear that these are satisfied by

$$V^2=\sum_1^n v_\kappa^2.$$

Thus the condition is that the kinetic energy when all the forces act must be the sum of the kinetic energies when they act separately.

4. Prove that a lemniscate $rr'=c^2$, where $2c$ is the distance between the points from which r and r' are measured, can be described under the action of forces $m\mu/r$ and $m\mu/r'$ directed to those points, and that the velocity is constant and equal to $\frac{2}{3}\sqrt{(3\mu)}$.

5. A particle describes a plane orbit under the action of two central forces each varying inversely as the square of the distance, directed towards two points symmetrically situated in a line perpendicular to the plane of the orbit. Show that the general (p, r) equation of the orbit, referred to the

point where the line joining the centres of force meets the plane as origin, is of the form

$$(1-a^2/p^2)^2 = b^2/(c^2+r^2),$$

where c is the distance of either centre of force from the plane, and a and b are constants.

6. A point describes a semi-ellipse, bounded by the axis minor, and its velocity, at a distance r from the nearer focus, is $a\sqrt{\{f(a-r)/r(2a-r)\}}$, $2a$ being the axis major, and f a constant. Prove that its acceleration is compounded of two, each varying inversely as the square of the distance, one tending to the nearer focus, and the other from the further focus.

110. Disturbed elliptic motion. The motion of the Planets about the Sun does not take place exactly in accordance with Kepler's Laws (Art. 41). Although the Sun's gravitational attraction preponderates very greatly over the attractions between the Planets, these attractions are not entirely negligible. The theory of the motion of the Planets presents us with the problem of determining a motion which, apart from relatively small forces, would be elliptic motion about a focus.

We shall consider here some examples of elliptic motion disturbed by small impulses in lines which lie in the plane of the orbit. The ellipse described after the impulse is a little different from that described before. The ellipses, having a given focus, are determined by the lengths of the major axes, the eccentricities, and the angles which the apse lines make with some fixed line in the plane of the orbit. We denote the major axis by a, the eccentricity by e, and the angle in question by ϖ.

111. Tangential impulse. Let a particle P, describing an elliptic orbit about a focus S, receive a small tangential impulse increasing its velocity by δv. Let R be the distance of the particle from S at the instant, μ/r^2 the acceleration to S when the distance is r, $a+\delta a$ the semi-axis major of the orbit immediately after the impulse.

We have, by Ex. 2 of Art. 48,

$$v^2 = \mu\left(\frac{2}{R}-\frac{1}{a}\right),$$

$$(v+\delta v)^2 = \mu\left(\frac{2}{R}-\frac{1}{a+\delta a}\right),$$

giving

$$\frac{\delta a}{a^2} = \frac{2v\delta v}{\mu} \text{ approximately.}$$

Again, if h is the moment of the velocity about S before the impulse, $h + \delta h$ afterwards, since the tangent to the path is unaltered, we have

$$\frac{h+\delta h}{v+\delta v} = \frac{h}{v},$$

giving $$\delta h = h\frac{\delta v}{v}.$$

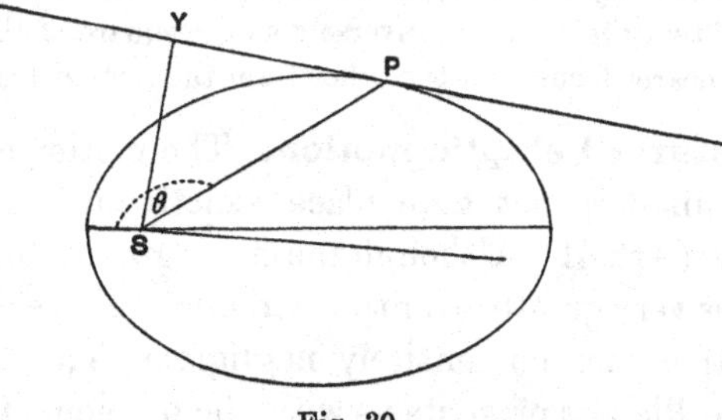

Fig. 39.

Hence if l is the semi-latus rectum before the impulse, $l + \delta l$ afterwards, we have

$$\mu(l+\delta l) = h^2\left(1+\frac{\delta v}{v}\right)^2, \text{ with } h^2 = \mu l,$$

giving $$\delta l = 2l\frac{\delta v}{v} \text{ approximately.}$$

Now $l = a(1-e^2)$, and if e becomes $e + \delta e$,

$$(1-e^2)\,\delta a - 2ea\delta e = 2a(1-e^2)\frac{\delta v}{v},$$

giving $$\delta e = \frac{(1-e^2)}{e}\left[a\frac{v\delta v}{\mu} - \frac{\delta v}{v}\right] = \frac{1-e^2}{e}\frac{\delta v}{v}a\left[\frac{v^2}{\mu} - \frac{1}{a}\right],$$

or $$\delta e = \frac{1-e^2}{e}\frac{2\delta v}{v}a\left(\frac{1}{R} - \frac{1}{a}\right).$$

The equation of the orbit will be changed from $l/r = 1 + e\cos(\theta - \varpi)$ to $(l+\delta l)/r = 1 + (e+\delta e)\cos\{\theta - (\varpi + \delta\varpi)\}$. Taking $\varpi = 0$ we find

$$e\sin\theta\,.\,\delta\varpi = \frac{\delta l}{R} - \frac{\delta e}{e}\left(\frac{l}{R} - 1\right).$$

If the particle is subject to a disturbing force producing a small tangential acceleration f we shall have

$$\dot{a} = \frac{2a^2vf}{\mu},\quad \dot{e} = \frac{2f}{v}\frac{l}{e}\left(\frac{1}{R} - \frac{1}{a}\right),$$

$$e\sin\theta\dot{\varpi} = \frac{2l}{R}\frac{f}{v} - \frac{\dot{e}}{e}\left(\frac{l}{R} - 1\right).$$

112. Normal impulse. Suppose the particle to receive an impulse imparting to it a velocity δv in the direction of the normal inwards. Then the resultant velocity is, to the first order, unaltered, and consequently a is unaltered, or $\delta a = 0$.

If p is the perpendicular from the focus S on the tangent at P, meeting it in Y, then the value of h is increased by $PY\delta v$, or we have

$$\delta h = \sqrt{(R^2 - p^2)}\, \delta v.$$

Hence $$\mu \delta l = 2h\delta h = 2pv\delta v \sqrt{(R^2 - p^2)};$$

also $\delta l = -2ae\delta e$, so that

$$\delta e = -\frac{pv\delta v}{\mu a e} \sqrt{(R^2 - p^2)}.$$

The equation giving $\delta \varpi$ is

$$-2ae\delta e/R = \left(\frac{l}{R} - 1\right) \frac{\delta e}{e} + e \sin\theta \,.\, \delta \varpi.$$

If the particle is subject to a disturbing force producing a small normal acceleration f we have

$$\dot{a} = 0,\ \dot{e} = -\frac{pfv}{\mu a e} \sqrt{(R^2 - p^2)},\ e \sin\theta \,.\, \dot{\varpi} = -\dot{e} \left(\frac{2ae}{R} + \frac{l - R}{eR}\right).$$

113. Examples.

1. For a small tangential impulse prove that

$$\delta e = 2\delta v\,(e + \cos\theta)/v, \quad \delta \varpi = 2\delta v \sin\theta/ev.$$

2. For a small normal impulse prove that

$$\delta e = -r\delta v \sin\theta/av, \quad \delta \varpi = \delta v\,(2ae + r\cos\theta)/aev.$$

3. For a small radial impulse prove that

$$\delta a = 2a^2 e\delta v \sin\theta/h, \quad \delta e = h\delta v \sin\theta/\mu, \quad \delta \varpi = -h\delta v \cos\theta/e\mu.$$

4. For a small transversal impulse prove that

$$\delta a = 2\delta v a^2 (1 + e\cos\theta)/h, \quad \delta e = \delta v \{r(e + \cos\theta) + l\cos\theta\}/h, \quad \delta \varpi = \delta v \sin\theta\,(l + r)/eh.$$

MISCELLANEOUS EXAMPLES

1. Relatively to a certain frame a point O describes a straight line uniformly with velocity V, and a second point P describes a curve in such a way that the line OP describes areas uniformly; prove that the resolved part perpendicular to OP of the acceleration of P is $2Vv \sin\phi/OP$, where v is the velocity of P, and ϕ the angle which the tangent to its path makes with that of O.

2. Relatively to a certain frame, a point A describes a circle (centre O) uniformly, and a point B moves with an acceleration always directed to A. If the area covered by the line AB is described uniformly, prove that the resolved part parallel to OA of the velocity of B is proportional to the perpendicular from B on OA produced.

3. Prove that, if the acceleration of a point describing a tortuous curve makes an angle ψ with the principal normal, then $\tan\psi = \frac{\rho}{v}\frac{dv}{ds}$.

In the case of a plane curve the condition that the acceleration is always directed to the same point is that the equation $\sin\psi + \frac{d}{ds}\frac{\rho\cos\psi}{1-\rho\frac{d\psi}{ds}} = 0$ must be satisfied at every point.

4. The position of a point is given by x, y, r, where x, y, z, r have their usual signification relative to rectangular axes; show that the component accelerations are

$$\dot{u} + \frac{uw}{r}, \quad \dot{v} + \frac{vw}{r}, \quad \dot{w} - (uwx + vwy)/r^2,$$

u, v, w being component velocities in the directions x, y, r.

5. If x, y are the coordinates of a point referred to rectangular axes turning with angular velocity ω, prove that the accelerations in the directions of the axes are

$$\ddot{x} - y\dot{\omega} - 2\dot{y}\omega - \omega^2 x \text{ and } \ddot{y} + x\dot{\omega} + 2\dot{x}\omega - \omega^2 y.$$

6. The radii vectores from two fixed points distant c apart to the position of a particle are r_1, r_2, and the velocities in these directions are u_1, u_2; prove that the accelerations in the same directions are

$$\dot{u}_1 + \frac{u_1 u_2}{2r_1^2 r_2}(r_1^2 - r_2^2 + c^2) \text{ and } \dot{u}_2 + \frac{u_1 u_2}{2r_1 r_2^2}(r_2^2 - r_1^2 + c^2).$$

7. The radii vectores from three fixed points to the position of a particle are r_1, r_2, r_3 and the velocities in these directions are u_1, u_2, u_3; prove that the accelerations in these directions are

$$\dot{u}_1 + u_1\left(\frac{u_2}{r_2} + \frac{u_3}{r_3}\right) - \frac{u_1}{r_1}(u_2\cos\theta_{12} + u_3\cos\theta_{13}),$$

and the two similar expressions, in which θ_{23}, θ_{31}, θ_{12} are the angles contained by the directions of (r_2, r_3), (r_3, r_1) and (r_1, r_2).

---

8. A particle is suspended from a point by an elastic thread and oscillates in the vertical line through the point of suspension. Prove that the period is the same as that of a simple pendulum of length equal to the excess of the length of the thread in the position of equilibrium above its natural length.

9. A particle is attached to a fixed point by means of an elastic thread of natural length $3a$, whose coefficient of elasticity is six times the weight of the particle. When the thread is at its natural length, and the particle

is vertically above the point of attachment, the particle is projected horizontally with a velocity $3\sqrt{(\frac{1}{2}ag)}$; verify that the angular velocity of the thread can be constant, and that the particle can describe the curve

$$r = a(4 - \cos\theta).$$

10. A particle moves in a nearly circular orbit with an acceleration $\mu + \nu(r - a)$, a being the mean radius; show that the apsidal angle is $\pi\omega/\sqrt{(3\omega^2 + \nu)}$, where ω is the mean angular velocity.

11. A particle describes a central orbit with acceleration $\mu/(r-a)^2$ towards the origin, starting with the velocity from infinity at a distance c (which is greater than a and less than $2a$) at an angle $2\cos^{-1}\sqrt{(a/c)}$. Prove that the path is given by the equation

$$\tfrac{1}{2}\theta = \tanh^{-1}\sqrt{\{(r-a)/a\}} - \tan^{-1}\sqrt{\{(r-a)/a\}}.$$

---

12. A particle is projected with velocity less than that from infinity under a force tending to a fixed point and varying inversely as the nth power of the distance. Prove that if n is not < 3 the particle will ultimately fall into the centre of force.

13. A particle moves under a central force varying inversely as the nth power of the distance $(n > 1)$, the velocity of projection is that due to a fall from rest at infinity, and the direction of projection makes an angle β with the radius vector of length R. Prove that the maximum distance is $R(\operatorname{cosec}\beta)^{2/(n-3)}$ when $n > 3$, and that the particle goes to infinity if $n =$ or < 3.

14. Prove that, if a possible orbit under a central force $\phi(r)$ is known, a possible orbit under a central force $\phi(r) + \lambda r^{-3}$ can be found. In particular prove that a particle projected from an apse at distance a with velocity $\sqrt{(\lambda+\mu)}/a$, under an attraction

$$\tfrac{1}{2}\mu(n-1)a^{n-3}r^{-n} + \lambda r^{-3}, \quad (n > 3),$$

will arrive at the centre in time

$$\frac{a^2}{2}\left(\frac{\pi}{\mu}\right)^{\frac{1}{2}} \Gamma\left(\frac{n+1}{2n-6}\right) \Big/ \Gamma\left(\frac{2}{n-3}\right).$$

15. A particle is describing a circular orbit of radius a under a force to the centre producing an acceleration $f(r)$ at distance r, and a small increment of velocity Δu is given to it in the direction of motion. Prove that the apsidal distances of the disturbed orbit are

$$a \text{ and } a + 4\Delta u\sqrt{\{af(a)\}}/\{3f(a) + af'(a)\}.$$

Prove also that, if the increment of velocity imparted to the particle is directed radially, the apsidal distances are approximately

$$a \pm 4\Delta u\sqrt{a}/\sqrt{\{3f(a) + af'(a)\}}.$$

---

16. A particle moves in a plane under a radial force P and a transverse force T, where

$$P = -\mu u^3(3 + 5\cos 2\theta), \qquad T = \mu u^3 \sin 2\theta;$$

prove that a first integral of the differential equation of the path can be expressed in the form

$$h_0^2\left(\sin\theta\frac{du}{d\theta}-u\cos\theta\right)-\frac{\mu}{2}\left[(\sin 3\theta-\sin\theta)\frac{du}{d\theta}-2u\cos 3\theta\right]=C,$$

where h_0^2 and C are constants.

17. A particle moves under the action of a central force P and a transverse disturbing force $\frac{1}{r}f(t)$. Prove that

$$\frac{d^2u}{d\theta^2}+u=\frac{P-f(t)\frac{du}{d\theta}}{u^2\{F(t)\}^2},$$

where $F(t)=\int f(t)\,dt$.

---

18. A particle describes a circle under the action of forces, tending to the extremities of a fixed chord, which are to each other at any point inversely as the distances r, r' from the point to the ends of the chord. Determine the forces, and prove that the product of the component velocities along r and r' varies inversely as the length of the perpendicular from the position of the particle to the chord; also show that the time from one end of the chord to the other is

$$\frac{a}{V}\frac{(\pi-a)\cos a+\sin a}{\cos^2\frac{1}{2}a},$$

where V is the velocity of the particle when moving parallel to the chord, a the radius of the circle, and a the angle between r and r'.

19. A particle is projected from an apse of Bernoulli's Lemniscate ($rr'=c^2$) along the tangent with velocity $\sqrt{\mu/2c}$ and moves under the action of forces

$$\mu r'^2\frac{r'-r}{(3rr'-r^2)^3},\quad \mu r^2\frac{r-r'}{(3rr'-r'^2)^3},$$

to the nearer and further poles respectively, r being the distance from the nearer pole, and r' from the further pole. Show that it describes the lemniscate.

20. A particle P moves under the action of two fixed centres of force S_1, S_2 producing accelerations μ_1/r_1^2 and μ_2/r_2^2 towards S_1 and S_2, where r_1, r_2 are the distances S_1P, S_2P. Prove that, if the motion does not take place in a fixed plane, there is an integral equation of the form

$$(r_1^2\dot\theta_1)(r_2^2\dot\theta_2)+h^2\cot\theta_1\cot\theta_2=c(\mu_1\cos\theta_1+\mu_2\cos\theta_2)+\text{const.},$$

where θ_1, θ_2 are the angles S_2S_1P and S_1S_2P, c is the distance S_1S_2, and h is the moment of the velocity about the line of centres.

---

21. An ellipse of eccentricity e and latus rectum $2l$ is described freely about a focus, with moment of momentum equal to h. When the particle is at the nearer apse it receives a small radial impulse μ. Prove that the apse line is turned through the angle $l\mu/eh$.

22. A particle of mass m describes an ellipse about a focus, μm being the force at unit distance; when the particle is at an extremity of the minor axis it receives a small impulse mV in a direction perpendicular to the plane of the orbit; prove that the eccentricity of the orbit will be diminished by $\frac{1}{2}V^2ae/\mu$, and that the angle which the axis major of the orbit makes with the distance from the focus will be increased by

$$V^2a(2-e^2)/\{2\mu e\sqrt{(1-e^2)}\},$$

where $2a$ is the axis major, and e the eccentricity of the orbit.

23. A comet describes about the Sun an ellipse of eccentricity e nearly equal to unity. At a point where the radius vector makes an angle θ with the apse line, the comet is instantaneously affected by a planet so that its velocity is increased in the ratio $n+1:n$, where n is great, without altering its direction. Show that, if the new orbit is a parabola,

$$e=1-(4/n)\cos^2\tfrac{1}{2}\theta \text{ nearly.}$$

24. A particle is describing an ellipse under a force to a focus S, and, when the particle is at P, the centre of force is suddenly moved a short distance x parallel to the tangent at P. Prove that the axis major is turned through the angle $(x/SG)\sin\phi\sin(\theta-\phi)$, where G is the foot of the normal, θ the angle which the normal makes with SG, and ϕ the angle which the tangent makes with SP.

25. Defining the instantaneous orbit under a central force varying as the distance as that orbit which would be described if the resistance ceased to act, show that, if at any point the resistance produces a retardation f, the rates of variation of the principal semi-axes are given by the equations

$$\frac{\dot{a}}{a(a^2-r^2)}=\frac{\dot{b}}{b(r^2-b^2)}=-\frac{f}{v(a^2-b^2)},$$

where v is the velocity and r the radius vector at the instant.

26. In the last Example there is a disturbance which produces a normal acceleration g instead of the resistance. Show that the maxima of the rates of variation of the principal semi-axes of the instantaneous ellipse are given by the equations

$$\frac{\dot{a}}{b}=-\frac{\dot{b}}{a}=\frac{\pm g}{(a+b)\sqrt{\mu}},$$

where μ is the central force on unit mass at unit distance.

CHAPTER V†

MOTION UNDER CONSTRAINTS AND RESISTANCES

114. THE second main subdivision of "Dynamics of a Particle" relates to motion of a particle in a given field of force when the force of the field is not the only force acting on the particle, but there are other, unknown, forces acting upon it.

Such forces may be *constraints*, that is to say they may do no work. Another class of forces to be included in the discussion are known as *resistances*. We had an example in the friction between an inclined plane and a body placed upon it (Art. 71). The characteristics of a resistance are that its line of action is always the line of the velocity of the particle on which it acts, and its sense is always opposed to the sense of the velocity. It follows that the work done by a resistance is always negative. This work, with its sign changed, is called the "work done against the resistance."

When a particle moves in a given field of force, and is at the same time subject to resistances, the increment of the kinetic energy in any displacement is less than the work done by the force of the field by the work done against the resistances.

115. Motion on a smooth plane curve under any forces. Let a particle of mass m be constrained to move on a given smooth plane curve under the action of given forces in the plane. Let s be the arc of the curve measured from some point of the curve up to the position of the particle at time t. Let S be the tangential component of the forces in the direction in which s increases, and N the component along the normal inwards. Let v be the velocity of the particle in the direction in which s increases, and R the pressure of the curve on the particle. We shall write down the equations for the case where the particle is on the inside of the curve, and R accordingly acts inwards. The equations for the case in which R acts outwards can be obtained by changing the sign of R.

† Articles in this Chapter which are marked with an asterisk (*) may be omitted in a first reading.

By resolving along the tangent and normal we obtain the equations of motion

$$\left.\begin{aligned} mv\frac{dv}{ds} &= S, \\ m\frac{v^2}{\rho} &= N + R \end{aligned}\right\}.$$

When the forces are conservative, the first of these equations has an integral, which is identical with the energy equation. It may be written

$$\tfrac{1}{2}mv^2 = \int S ds + \text{const.}$$

When v is known from this equation, the second of the equations of motion determines the pressure R.

In the case of one-sided constraint (Art. 77) the particle may leave the curve. This happens when R vanishes.

116. Examples.

1. Prove that, when the particle leaves the curve, the velocity is that due to falling under the force kept constant through one quarter of the chord of curvature in the direction of the force.

2. Prove that, when the curve is a free path under the given forces for proper velocity of projection, then, for any other velocity of projection, the pressure varies as the curvature.

117. Motion of two bodies connected by an inextensible string. We shall suppose that the bodies may be treated as particles, that the mass and extension of the string can be neglected, and that the tension of the string is the same throughout. (See Chapter VI.) When this is the case the tension of the string does no work, for the sum of the rates at which it does work on the two particles vanishes. The equations of motion of the bodies can be formed in the manner explained in Art. 73. In forming the equations of motion we take account of the condition that the length of the string is constant. For example, if the string is in two portions, separated by a ring or a peg, the sum of the lengths of the two portions is constant. If there is an energy equation, or an equation of constancy of momentum, or of moment of momentum, it is an integral of the equations of motion.

118. Examples.

1. Two particles of masses M, m are connected by an inextensible thread of negligible mass which passes through a small smooth ring on a smooth

fixed horizontal table. When the thread is just stretched, so that M is at a distance c from the ring, and the particles are at rest, M is projected on the table at right angles to the thread. Prove that until m reaches the ring M describes a curve whose polar equation is of the form

$$r = c \sec [\theta \sqrt{\{M/(M+m)\}}].$$

2. Two particles of masses M, m are connected by an inextensible thread of negligible mass; M describes on a smooth table a curve which is nearly a circle with centre at a point O, and the thread passes through a small smooth hole at O and supports m. Prove that the apsidal angle of M's orbit is

$$\pi \sqrt{\{\tfrac{1}{3}(1 + m/M)\}}.$$

***119. Oscillating pendulum.** The motion of a simple circular pendulum, whether it executes small oscillations (Art. 75) or not, can be determined by the energy equation.

Let θ be the angle which the radius of the circle drawn through the position of the particle at time t makes with the vertical drawn downwards. The kinetic energy is $\frac{1}{2}ml^2\dot{\theta}^2$, where m is the mass of the particle, and l the radius of the circle, or length of the pendulum. The potential energy of the particle in the field of the Earth's gravity (Art. 92) is $mgl(1 - \cos\theta)$, if the chosen fixed level from which it is measured is that of the lowest point. Hence the energy equation can be written

$$\tfrac{1}{2}l\dot{\theta}^2 = g\cos\theta + \text{const.}$$

If the pendulum is displaced initially so that $\theta = \alpha$, and is let go from this position, the energy equation is

$$\tfrac{1}{2}l\dot{\theta}^2 = g(\cos\theta - \cos\alpha),$$

or

$$\tfrac{1}{4}\dot{\theta}^2 = \frac{g}{l}\left(\sin^2\frac{\alpha}{2} - \sin^2\frac{\theta}{2}\right),$$

showing that the pendulum oscillates between two positions in which it is inclined to the vertical at an angle α on the right and left sides of the vertical.

To express the position of the pendulum in terms of the time t, since it was in the equilibrium position, we introduce a new variable ψ defined by the equation

$$\sin\frac{\alpha}{2}\sin\psi = \sin\frac{\theta}{2},$$

with the further conditions that as θ increases from 0 to α, ψ increases from 0 to $\frac{1}{2}\pi$; as θ diminishes from α to 0, ψ increases from $\frac{1}{2}\pi$ to π; as θ diminishes from 0 to $-\alpha$, ψ increases from π

to $\frac{3}{2}\pi$; and as θ increases from $-\alpha$ to 0, ψ increases from $\frac{3}{2}\pi$ to 2π. With these conventions there is one value of ψ corresponding to every instant in a complete period.

Now we have

$$\tfrac{1}{2}\dot{\theta}\cos\frac{\theta}{2} = \dot{\psi}\sin\frac{\alpha}{2}\cos\psi,$$

$$\sin^2\frac{\alpha}{2} - \sin^2\frac{\theta}{2} = \sin^2\frac{\alpha}{2}\cos^2\psi,$$

$$\dot{\psi}^2 = \frac{g}{l}\left(1 - \sin^2\frac{\alpha}{2}\sin^2\psi\right).$$

Hence the time t from the instant when the particle was passing through the lowest point in the direction in which θ increases is given by the equation

$$t = \sqrt{\frac{l}{g}}\int_0^{\psi}\frac{d\psi}{\sqrt{\left(1 - \sin^2\frac{\alpha}{2}\sin^2\psi\right)}},$$

where the square root is always to be taken positively. The complete period is

$$4\sqrt{\frac{l}{g}}\int_0^{\frac{\pi}{2}}\frac{d\psi}{\sqrt{\left(1 - \sin^2\frac{\alpha}{2}\sin^2\psi\right)}}.$$

With the above relation between t and ψ, $\sin\psi$ is said to be an Elliptic Function of $t\sqrt{\frac{g}{l}}$, and the relation is written

$$\sin\psi = \operatorname{sn}\left(t\sqrt{\frac{g}{l}}\right) \qquad \left(\operatorname{mod}\sin\frac{\alpha}{2}\right).$$

The function has a real period, and the integral

$$\int_0^{\frac{\pi}{2}}\frac{d\psi}{\sqrt{\left(1 - \sin^2\frac{\alpha}{2}\sin^2\psi\right)}}$$

is one quarter of this period.

The position of the pendulum at any time t is determined by the equation

$$\sin\frac{\theta}{2} = \sin\frac{\alpha}{2}\operatorname{sn}\left(t\sqrt{\frac{g}{l}}\right) \qquad \left(\operatorname{mod}\sin\frac{\alpha}{2}\right).$$

***120. Complete Revolution.** If the constant in the energy equation of Art. 119 is such that $\dot{\theta}$ never vanishes, it must be greater than g, and the velocity at the lowest point is greater than that due to falling from the highest point. Hence there will be some velocity at the highest point. Let us suppose the velocity at the highest point to be that due to falling through a height h; then, when $\theta = \pi$

$$l^2\dot{\theta}^2 = 2gh,$$

and for any other value of θ

$$\tfrac{1}{2}l\dot{\theta}^2 = g\left(\cos\theta + 1 + \frac{h}{l}\right),$$

or

$$\tfrac{1}{4}\dot{\theta}^2 = \frac{g(h+2l)}{2l^2}\left(1 - \frac{2l}{h+2l}\sin^2\frac{\theta}{2}\right),$$

giving $\sin\frac{\theta}{2} = \operatorname{sn}\left(\frac{t}{k}\sqrt{\frac{g}{l}}\right)$ (mod k), where $k^2 = 2l/(h+2l)$.

The period of a complete revolution is

$$2k\sqrt{\frac{l}{g}}\int_0^{\frac{\pi}{2}} \frac{d\phi}{\sqrt{(1-k^2\sin^2\phi)}}.$$

***121. Limiting case.** In the case where the pendulum is projected from the position of equilibrium with velocity equal to that due to falling from the highest point the equation can be integrated by logarithms.

The constant in the energy equation of Art. 119 must then be chosen so that $\dot{\theta}$ vanishes when $\theta = \pi$, and the equation therefore is

$$\tfrac{1}{2}l\dot{\theta}^2 = g(1+\cos\theta),$$

which may be written

$$\tfrac{1}{4}\dot{\theta}^2 = \frac{g}{l}\cos^2\frac{\theta}{2}.$$

The time of describing an angle θ is therefore t, where

$$t = \sqrt{\frac{l}{g}}\int_0^{\frac{\theta}{2}} \frac{dx}{\cos x} = \sqrt{\frac{l}{g}}\log\left(\sec\frac{\theta}{2} + \tan\frac{\theta}{2}\right).$$

It is to be noted that the particle approaches the highest point indefinitely, but does not reach it in any finite time.

The same equations may be used to describe the motion of a

particle which starts from a position indefinitely close to the unstable position of equilibrium at the highest point of the circle.

*122. Examples.

1. Prove that the time of a finite oscillation when the fourth power of a, the angle of oscillation, is neglected, is $2\pi(1+\frac{1}{16}a^2)\sqrt{(l/g)}$.

2. Prove that, in the limiting case of Art. 121,

$$\theta = 2\tan^{-1}\sinh\{t\sqrt{(g/l)}\}.$$

3. Prove that, if a seconds' pendulum makes a complete finite oscillation in four seconds, the angle a is about 160°.

***123. Smooth plane tube rotating in its plane.** Let a particle of mass m move in a smooth plane tube, and let the tube rotate in its plane about a point O rigidly connected with it. Let OA be any particular radius vector of the tube, and ϕ the angle which OA makes with a fixed line in the plane of the tube. Then $\dot{\phi}$ is the angular velocity of the tube. We shall write ω for $\dot{\phi}$.

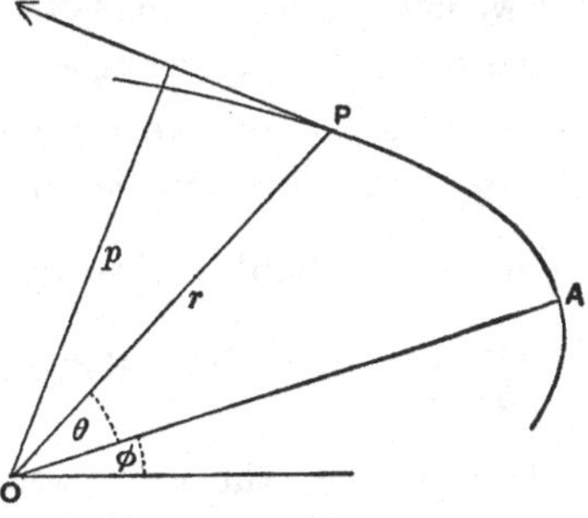

Fig. 40.

Let P be the position of the particle in the tube at time t. Let $OP = r$, and $\angle AOP = \theta$. Then r and θ are polar coordinates of P referred to OA as initial line, and r and $\theta + \phi$ are polar coordinates of P referred to a fixed initial line. Let ρ be the radius of curvature of the tube at P.

Let v be the velocity of the particle relative to the tube. Then, if arc $AP = s$, v is $\dot{s}$, the direction of v is that of the tangent to the tube, and the resolved parts of v along OP and at right angles to OP are $\dot{r}$ and $r\dot{\theta}$.

Now the resolved accelerations of the particle along OP and at right angles to OP are

$$\ddot{r} - r(\dot{\theta} + \dot{\phi})^2,$$

and

$$\frac{1}{r}\frac{d}{dt}\{r^2(\dot{\theta} + \dot{\phi})\}.$$

These may be written

$$\left.\begin{aligned} &\ddot{r} - r\dot{\theta}^2 - 2r\dot{\theta}\omega - r\omega^2, \\ &\frac{1}{r}\frac{d}{dt}(r^2\dot{\theta}) + 2\dot{r}\omega + r\dot{\omega} \end{aligned}\right\}.$$

Of these the terms independent of ω are equivalent to $v\dfrac{dv}{ds}$ along the tangent to the tube at P and v^2/ρ inwards along the normal to the tube.

The terms containing 2ω as a factor are equivalent to $2\omega v$ inwards along the normal to the tube. This can be seen by considering that $\dot{r}$ along OP and $r\dot{\theta}$ transverse to OP are equivalent to v along the tangent in the direction in which s increases, and that we have, as multipliers of 2ω, the components of this resultant turned through a right angle.

Now we can resolve a vector in the direction OP into components along the tangent at P to the tube and inwards along the normal by multiplying by $\dfrac{dr}{ds}$ and $\dfrac{p}{r}$, where p is the perpendicular from O on the tangent; similarly for a vector transverse to OP.

Hence finally the accelerations resolved along the tangent and normal to the tube are

$$\left.\begin{aligned} &v\frac{dv}{ds} - \omega^2 r\frac{dr}{ds} + \dot{\omega}p, \\ &\frac{v^2}{\rho} + 2\omega v + \omega^2 p + \dot{\omega}r\frac{dr}{ds} \end{aligned}\right\}.$$

Now let the particle move in the tube under the action of forces in the plane of the tube whose resolved parts along the tangent and normal to the tube are S and N, and let R be the pressure of the tube on the particle. Then the equations of motion are

$$\left.\begin{aligned} m\left[v\frac{dv}{ds} - \omega^2 r\frac{dr}{ds} + \dot{\omega}p\right] &= S, \\ m\left[\frac{v^2}{\rho} + 2\omega v + \omega^2 p + \dot{\omega}r\frac{dr}{ds}\right] &= N + R \end{aligned}\right\}.$$

***124. Newton's Revolving Orbit.** Suppose that the form of the tube in Art. 123 is a free path under a central force to O.

Let the tube turn about O with an angular velocity ϕ which is always equal to $n\dot{\theta}$, where n is constant, and $\dot{\theta}$ is the angular velocity of the radius vector in the free path when the particle is at (r, θ). Then the path traced out by the particle is a free path under the original central force and an additional central force which varies inversely as the cube of the distance.

Let f be the central acceleration in the free path, and $\frac{1}{2}h$ the rate of description of areas. Then we are given

$$\left.\begin{aligned} \ddot{r} - r\dot{\theta}^2 &= -f, \\ r^2\dot{\theta} &= h \end{aligned}\right\}.$$

Now, in the tube $\dot{\phi} = n\dot{\theta}$, so that

$$r^2(\dot{\theta} + \dot{\phi}) = h(1+n),$$

and

$$\begin{aligned} \ddot{r} - r(\dot{\theta} + \dot{\phi})^2 &= -f - r\dot{\theta}^2(2n + n^2) \\ &= -f - \frac{h^2}{r^3}(2n + n^2). \end{aligned}$$

Hence the path traced out by the particle in the revolving tube is a free path with a central acceleration to O made up of two terms, one of them being f, and the other being inversely proportional to r^3.

This result may be stated in another form as follows:—Relatively to a certain frame a particle describes a central orbit about the origin with central acceleration f; if a second frame with the same origin rotates about the origin relatively to the first frame, with an angular velocity always the same multiple of that of the radius vector in the said central orbit, the path of the particle relatively to the second frame is again a central orbit with the central acceleration increased by an amount inversely proportional to the cube of the distance.

***125. Examples.**

1. A particle moves in a tube in the form of an equiangular spiral which rotates uniformly about the pole, and is under the action of a central force to the pole of the spiral. Prove that, if there is no pressure on the tube, the central force at distance r must be of the form $Ar + Br^{-3}$, where A and B are constants.

2. Prove that motion which, relatively to any frame, can be described as motion in a central orbit with acceleration $\mu/(\text{distance})^3$ towards the origin and moment of velocity h may be described, relatively to a different frame with the same origin, as uniform motion in a straight line, provided $h^2 > \mu$.

3. A particle moves in a smooth plane tube, and is under a central force to a fixed point about which the tube rotates uniformly. Prove that, if the pressure is always zero, the central force is

$$m\left[r\omega^2+2r\omega\,(h-r^2\omega)/p^2+(h-r^2\omega)^2p^{-3}\,dp/dr\right],$$

where m is the mass of the particle, mh is its moment of momentum about the fixed point, ω is the angular velocity of the tube, r is the radius vector, and p the perpendicular from the fixed point on the tangent to the tube at the position of the particle.

***126. Motion on a rough plane curve under gravity.** When a particle is constrained to describe a plane curve in a vertical plane under gravity, but there is frictional resistance to the motion as well as pressure on the curve, we assume that the friction is μ times the pressure, where μ is the coefficient of friction. The friction acts along the tangent to the curve in the sense opposite to that of the velocity.

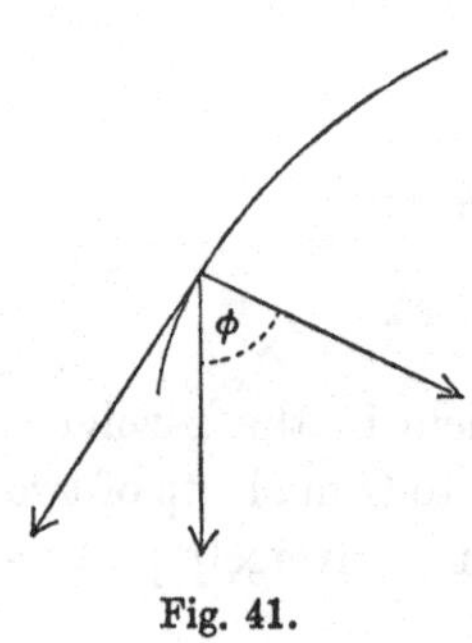

Fig. 41.

The equations of motion take different forms in different circumstances. We shall choose for investigation the case where the particle is on the outside of the curve, and is descending.

Let the arc s of the curve be measured from some point of the curve so that it increases in the sense of the velocity, and let ϕ be the angle contained between the inwards normal and the downwards vertical. Then ϕ increases with s, and $ds/d\phi\,(=\rho)$ is the length of the radius of curvature.

Let v be the velocity of the particle, m its mass, R the pressure of the curve on the particle. The equations of motion are

$$\left.\begin{aligned} mv\frac{dv}{ds} &= mg\sin\phi-\mu R,\\ m\frac{v^2}{\rho} &= mg\cos\phi-R \end{aligned}\right\}.$$

Eliminating R we obtain the equation

$$v\frac{dv}{ds}-\mu\frac{v^2}{\rho}=g\,(\sin\phi-\mu\cos\phi),$$

or
$$v\frac{dv}{d\phi}-\mu v^2=g\rho\,(\sin\phi-\mu\cos\phi).$$

This equation can be integrated after multiplication by the factor $e^{-2\mu\phi}$, in fact it becomes

$$\frac{d}{d\phi}(\tfrac{1}{2}v^2e^{-2\mu\phi}) = g\rho e^{-2\mu\phi}(\sin\phi - \mu\cos\phi),$$

so that $$v^2e^{-2\mu\phi} = 2g\int \rho e^{-2\mu\phi}(\sin\phi - \mu\cos\phi)\,d\phi + \text{const.},$$

an equation which determines v as a function of ϕ, and therefore gives the velocity at any point of the curve. The velocity being determined, the second of the equations of motion gives the pressure, and, just as in the case of a smooth curve, if R vanishes the particle leaves the curve.

The equations of motion take different forms according as the particle is inside or outside the curve, and according as it is ascending or descending. But in each case the equations can be integrated by the above method. There is accordingly no definite expression for the velocity at any point of the curve in terms of the position, but the expressions obtained are different in the different cases.

*127. Examples.

1. Write down the equations of motion in the three cases not investigated in Art. 126 and the integrating factor in each case.

2. A particle is projected horizontally from the lowest point of a rough sphere of radius a, and returns to this point after describing an arc αa, ($\alpha < \frac{1}{2}\pi$), coming to rest at the lowest point. Prove that the initial velocity is $\sin\alpha\sqrt{\{2ga(1+\mu^2)/(1-2\mu^2)\}}$, where μ is the coefficient of friction.

3. A particle slides down a rough cycloid, whose base is horizontal and vertex downwards, starting from rest at a cusp and coming to rest at the vertex. Prove that, if μ is the coefficient of friction, $\mu^2 e^{\mu\pi} = 1$.

4. A ring moves on a rough cycloidal wire whose base is horizontal and vertex downwards; prove that during the ascent the direction of motion at time t makes with the horizontal an angle ϕ, given by the equation

$$\frac{d^2}{dt^2}\{e^{\phi\tan\epsilon}\sin(\phi+\epsilon)\} = -\frac{g}{4a}\sec^2\epsilon\, e^{\phi\tan\epsilon}\sin(\phi+\epsilon),$$

where ϵ is the angle of friction.

***128. Motion on a curve in general.** When a particle moves on a given curve under any forces, we take m for the mass of the particle, S for the tangential component of the resultant force of the field, N for the component along the principal normal, and B for the component along the binormal. Also we take R_1 for the component of the pressure along the principal normal towards the centre of curvature, and R_2 for the component of the pressure along the binormal in the same sense as B. Further if the curve is rough we take F for the friction.

We take s to be the arc of the curve from some point to the position of the particle at time t, ρ to be the radius of curvature, and v to be the velocity, and we suppose the sense in which s increases to be that of v. Then the equations of motion are

$$\left.\begin{aligned} mv\frac{dv}{ds} &= S - F, \\ m\frac{v^2}{\rho} &= N + R_1, \\ 0 &= B + R_2 \end{aligned}\right\}.$$

When the curve is smooth F is zero, and we can integrate the first equation, in the same way as in Art. 115, in the form

$$\tfrac{1}{2}mv^2 = \int S ds + \text{const.},$$

and this result can be expressed in the form

change of kinetic energy = work done,

so that the velocity is determined in terms of the position. The other two equations then determine the pressure.

When the curve is rough we have to eliminate F, R_1, R_2 by means of the equation

$$F^2 = \mu^2 (R_1^2 + R_2^2),$$

which expresses that the friction is proportional to the resultant pressure. There results a differential equation for v^2, and, if we can integrate this equation, we shall obtain an equation giving the velocity in terms of the position. As in Art. 126 the velocity in any position depends partly on the way in which that position has been reached.

*129. Motion on a smooth surface of revolution with a vertical axis.

Let the axis of revolution be the axis x (x being measured upwards), and let the particle at time t be at distance y from the axis, and be on a meridian curve of the surface in an axial plane making an angle ϕ with a given axial plane, and let σ be the arc of the meridian from some particular circular section to the position of the particle.

Then it is clear that the velocity along the tangent to the meridian is $\dot{\sigma}$, and the velocity along the tangent to the circular section is $y\dot{\phi}$. Thus the energy equation is

$$\tfrac{1}{2}(\dot{\sigma}^2 + y^2\dot{\phi}^2) + gx = \text{const.}$$

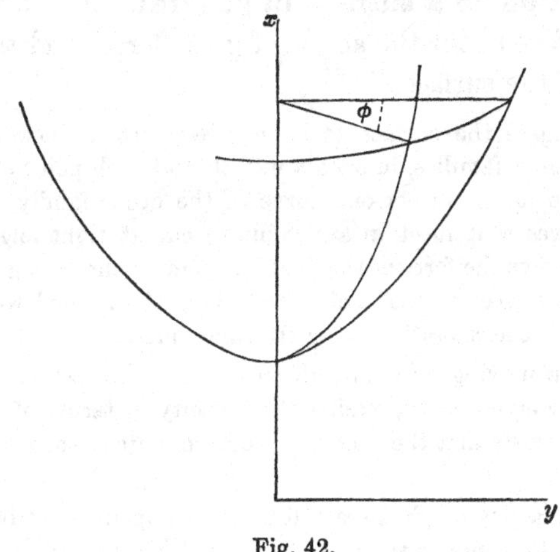

Fig. 42.

Again, since the pressure of the surface on the particle acts along the normal to the surface, and the normal meets the axis of revolution, while the force of gravity acts in a line parallel to this axis, the forces acting on the particle have no moment about this axis. Hence the moment of the momentum about the axis is constant, or we have

$$y^2\dot{\phi} = \text{const.}$$

The equations which have been written down determine $\dot{\sigma}$ and $\dot{\phi}$, that is they determine the two components of velocity ($\dot{\sigma}$ and $y\dot{\phi}$) in two directions, at right angles to each other, which lie in the tangent plane to the surface.

***130. Examples.**

1. If the particle is projected properly it can describe a circle. If y is the radius of the circle, and β the angle which the normal to the surface at any point on the circle makes with the vertical, the required velocity of projection is $(gy \tan\beta)^{\frac{1}{2}}$.

In this case the pressure of the surface is equal to $mg \sec\beta$, where m is the mass of the particle.

2. Prove that, if $1/u$ is put for y, and $x=f(u)$ is the equation of the meridian curve of the surface, the projection of the path of the particle on a horizontal plane is given by an equation of the form

$$\left(\frac{du}{d\theta}\right)^2[1+\{u^2f'(u)\}^2]+u^2+\frac{2g}{h^2}f(u)=\text{const.},$$

where h is a constant.

***131. Motion on a surface in general.** Let a particle move on a fixed surface under the action of given forces and the pressure and friction of the surface.

We may imagine the surface to be covered with a network of curves belonging to distinct families, in such a way that at each point of the surface one curve of one family meets one curve of the other family, and we may suppose the curves that meet in any point to cut at right angles. At any point we may resolve the force of the field into components along the tangents to the curves that meet in that point, and along the normal to the surface. We may resolve the acceleration along the same lines.

For a particle moving on a smooth surface in a conservative field there will be an energy equation expressing the velocity in terms of the position. We shall see presently that the pressure is determinate as soon as the velocity is known.

When the surface is rough there will be two components of friction in the directions of the tangents to the two curves that meet at any point, and the resultant friction has the same direction as the velocity but the opposite sense. Also the resultant friction is equal in magnitude to the product of the coefficient of friction and the pressure.

We have thus the means of writing down equations of motion of the particle, but the process can in general be simplified by using methods of Kinematics and Analytical Dynamics which are beyond the scope of the present work. We shall therefore confine ourselves to the simplest cases.

We proceed to investigate a general expression for the resolved part of the acceleration along the normal to the surface.

Let v be the velocity of the particle, ρ the radius of curvature of its path. The tangent to the path touches the surface, and we suppose a normal section of the surface drawn through it. This section is not, in general, the osculating plane of the path; we suppose that it makes an angle ϕ with this osculating plane. We take ρ' to be the radius of curvature of the normal section of the surface through the tangent to the path.

Since the normal to the surface is at right angles to the tangent to the path the resolved part of the acceleration along the normal to the surface is the resolved part in that direction of the acceleration along the principal normal to the path, it is therefore

$$\frac{v^2}{\rho}\cos\phi.$$

Also by a well-known theorem we have $\rho = \rho' \cos\phi$.

Hence the acceleration along the normal to the surface is v^2/ρ', and the pressure is determined by resolving along the normal..

***132. Osculating plane of path.** In Ex. 1 of Art. 130 it is stated that a particle may be projected along a horizontal tangent of a smooth surface of revolution whose axis is vertical with such velocity that it describes the circular section under the action of

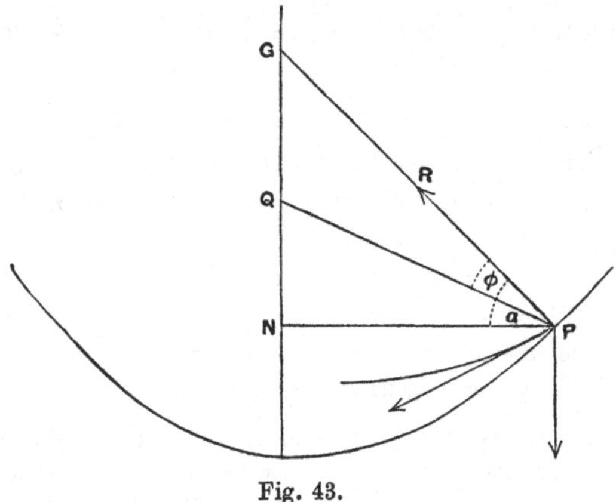

Fig. 43.

gravity and the pressure of the surface. It is almost obvious that if the velocity exceeds that requisite for description of the circle the path of the particle rises above the circle, otherwise it falls below the circle. We may use the result of Art. 131 to find the position of the osculating plane of the path for any velocity of projection.

Let P be the point of projection, PG the normal to the surface at P, $PN = y$ the ordinate of P at right angles to the axis of revolution, Q the point where the osculating plane of the path meets the axis. Let $\angle GPN = \alpha$, and $\angle GPQ = \phi$.

When the particle is projected along the tangent to the circular section with velocity V there is initially no acceleration along a line in the meridian plane at right angles to PQ.

Hence resolving along this line we have

$$R \sin \phi - mg \cos (\alpha - \phi) = 0,$$

where m is the mass of the particle, and R is the pressure.

Again, resolving along PN, we have

$$m \frac{V^2}{\rho} \cos (\alpha - \phi) = R \cos \alpha,$$

where ρ is the radius of curvature of the path.

Now, with the notation of Art. 131,

$$\rho' = PG, \quad \rho = PG \cos \phi.$$

Also $$y = PN = PG \cos \alpha.$$

Hence $$\tan \phi = gy/V^2.$$

This equation determines the position of the osculating plane of the path.

Now if $\tan \phi > \tan \alpha$, or $V^2 < gy \cot \alpha$, the osculating plane of the path initially lies below the horizontal plane through the point of projection, and if $\tan \phi < \tan \alpha$, or $V^2 > gy \cot \alpha$, it lies above that plane.

***133. Examples.**

1. A particle moving on a surface (smooth or rough) under no forces but the reaction of the surface describes a geodesic.

2. A particle moves on a rough cylinder of radius a under no forces but the reaction of the surface, starting with velocity V in a direction making an angle a with the generators; prove that in time t it moves over an arc

$$a\mu^{-1} \operatorname{cosec}^2 a \log (1 + \mu V t a^{-1} \sin^2 a),$$

μ being the coefficient of friction.

3. A hollow circular cylinder of radius a is rough on the inside, and is made to rotate uniformly with angular velocity ω about its axis which makes an angle a with the vertical. Show that a particle can slide down a fixed line parallel to the axis with uniform velocity

$$a\omega \sqrt{\{(\mu^2 + 1)/(\mu^2 \tan^2 a - 1)\}},$$

where μ is the coefficient of friction, and $\mu > \cot a$.

4. An ellipsoidal shell whose principal semi-axes are a, b, c $(a > b > c)$ is placed with the greatest axis vertical, and a particle is projected from one of the lower umbilics with velocity v along the tangent to the horizontal section within the ellipsoid. Show that the osculating plane of the path is initially above or below this section according as

$$v^2 > \text{ or } < gab^2 (b^2/c^2 - 1)/\sqrt{\{(a^2 - c^2)(a^2 - b^2)\}}.$$

134. Motion in Resisting Medium. We consider cases of the motion of a particle in a known field of force when, in addition to the force of the field, there is exerted on the particle a force proportional to a power of its velocity having the same direction as the velocity and the opposite sense.

Problems of this kind are related to facts of observation in regard to the motions of bodies in the air and in other fluid media. In many cases it is found that the observed facts can be approximately represented by the supposition that the resistance is proportional to the velocity, this is true for instance for the motion of a pendulum swinging in air.

135. Resistance proportional to the Velocity. Since the velocity of a particle is a vector whose direction and sense are determined by the resolved parts $\dot{x}$, $\dot{y}$, $\dot{z}$, the resistance has resolved parts $-\kappa\dot{x}$, $-\kappa\dot{y}$, $-\kappa\dot{z}$, where κ is a constant.

Let the motion take place under gravity parallel to the negative direction of the axis y, and first suppose the particle to move vertically. The equation of motion is

$$m\ddot{y} = -mg - \kappa\dot{y},$$

or

$$\ddot{y} + \lambda\dot{y} + g = 0,$$

where λ is written for κ/m. Multiplying by e^{λ} and integrating, we have

$$\dot{y}e^{\lambda t} + \frac{g}{\lambda}e^{\lambda t} = C,$$

where C is a constant of integration. Hence

$$\dot{y} = Ce^{-\kappa t/m} - mg/\kappa.$$

If the particle continues to fall for a sufficiently long time the value of $\dot{y}$ will ultimately differ very little from $-gm/\kappa$, or the particle falls with a practically constant velocity when it has been falling for some seconds.

This velocity is called the *terminal velocity* in the medium.

The equation last written can be integrated again so as to express y as a function of t.

Again suppose that the particle is projected in any other than a vertical direction; then the vertical motion is the same as before, but for the horizontal motion we have an equation

$$m\ddot{x} = -\kappa\dot{x},$$

giving $$\dot{x} = Ae^{-\kappa t/m},$$
where A is a constant of integration. This equation can be integrated again so as to express x as a function of t.

Since x and y are known, as functions of t, the path can be determined.

136. Damped Harmonic Motion. Consider the case where, apart from the resistance, the motion would be simple harmonic in period $2\pi/n$, and the resistance is proportional to the velocity.

We have the equation
$$m\ddot{x} = -mn^2x - \kappa\dot{x},$$
or
$$\ddot{x} + \lambda\dot{x} + n^2x = 0,$$
where λ is written for κ/m. The complete primitive of this equation takes different forms according as $n^2 >$ or $< \frac{1}{4}\lambda^2$. In the former case, which is practically the more important, it is
$$x = e^{-\frac{1}{2}\lambda t}\left[A\cos\{t\sqrt{(n^2 - \tfrac{1}{4}\lambda^2)}\} + B\sin\{t\sqrt{(n^2 - \tfrac{1}{4}\lambda^2)}\}\right].$$

The motion may be described roughly as simple harmonic motion with period $2\pi/\sqrt{(n^2 - \frac{1}{4}\lambda^2)}$, and with amplitude diminishing according to the exponential function $e^{-\frac{1}{2}\lambda t}$. It will be observed that the period is lengthened by the resistance, and that the amplitude falls off in geometric progression as the time increases in arithmetic progression. Thus the motion rapidly dies away.

136 A. Effect of damping on forced oscillation. Let a particle, which can move with a damped harmonic motion, be disturbed by a force, which at time t is proportional to $\cos(pt + \alpha)$, where p and α are constants. The equation of motion is of the form
$$\ddot{x} + \lambda\dot{x} + n^2x = P\cos(pt + \alpha).$$
A solution of this equation can be found by putting x equal to a function of the form $Q\cos(pt + \alpha - \epsilon)$ provided
$$(n^2 - p^2)\,Q\cos\epsilon + \lambda p\,Q\sin\epsilon = P,$$
$$-\lambda pQ\cos\epsilon + (n^2 - p^2)\,Q\sin\epsilon = 0,$$
whence $$Q\{(n^2 - p^2)^2 + \lambda^2p^2\}^{\frac{1}{2}} = P, \quad \tan\epsilon = \lambda p/(n^2 - p^2).$$

When $n > \frac{1}{2}\lambda$ the complete primitive is
$$x = e^{-\frac{1}{2}\lambda t}(A\cos n't + B\sin n't) + Q\cos(pt + \alpha - \epsilon),$$
where $n'^2 = n^2 - \frac{1}{4}\lambda^2$, and the motion is compounded of a damped harmonic motion, the free oscillation, and a simple harmonic motion,

the forced oscillation. The amplitude of the free oscillation diminishes in geometric progression as the time increases in arithmetic progression, and thus after a time the motion is practically simple harmonic. The period of the forced oscillation is the same as the period of the force producing it; its amplitude never becomes infinite, even when the period of the force is the same as that of the free oscillation; its phase is behind or in advance of that of the force according as the period of the free oscillation, when there is no resistance, is less or greater than that of the force.

When $n < \frac{1}{2}\lambda$ the complete primitive is

$$x = e^{-\frac{1}{2}\lambda t}\left(Ae^{n''t} + Be^{-n''t}\right) + Q\cos(pt + \alpha - \epsilon),$$

where $n''^2 = \frac{1}{4}\lambda^2 - n^2$. The term with coefficient Q represents, as before, a forced oscillation.

137. Examples.

1. A particle is projected vertically upwards with velocity v in a medium in which the resistance is proportional to the velocity. It rises to a height h and returns to the point of projection with velocity w. Prove that

$$gh/v^2 = \tfrac{1}{2} - \tfrac{1}{3}(v/V) + \tfrac{1}{4}(v/V)^2 - \tfrac{1}{5}(v/V)^3 + \ldots,$$
$$gh/w^2 = \tfrac{1}{2} + \tfrac{1}{3}(w/V) + \tfrac{1}{4}(w/V)^2 + \tfrac{1}{5}(w/V)^3 + \ldots,$$

where V is the terminal velocity in the medium.

2. A particle moves under gravity in a medium whose resistance varies as the velocity, starting with horizontal and vertical component velocities u_0, v_0, and returning to the horizontal plane through the point of projection with component velocities u_1, v_1; show that the range R and time of flight t are given by the equations

$$v_0 - v_1 = gt, \quad R = t(u_0 - u_1)/(\log u_0 - \log u_1).$$

Prove also that $R = u_0 Vt/(V + v_0)$, where V is the terminal velocity in the medium.

3. A body performs rectilinear vibrations under an attractive force to a fixed centre proportional to the distance in a medium whose resistance is proportional to the velocity. Prove that, if T is the period, and a, b, c are the coordinates of the extremities of three consecutive semi-vibrations, then the coordinate of the position of equilibrium and the time of vibration if there were no resistance are respectively

$$\frac{ac - b^2}{a + c - 2b} \text{ and } T\left[1 + \frac{1}{\pi^2}\left(\log\frac{a-b}{c-b}\right)^2\right]^{-\frac{1}{2}}.$$

4. If in the problem considered in Art. 136, $\lambda > 2n$, and the particle starts from rest in any displaced position, it creeps asymptotically towards its position of equilibrium, according to the formula

$$x = a\left(\alpha e^{-\beta t} - \beta e^{-\alpha t}\right)/(\alpha - \beta),$$

where α and β are the roots of the quadratic $\xi^2 - \lambda\xi + n^2 = 0$.

5. A particle of unit mass is fastened to one end of an elastic thread of natural length a and modulus an^2, in a medium the resistance of which to the motion of the particle is 2κ (velocity). The other end of the thread is fixed and the particle is held at a distance $b\ (>a)$ below the fixed point. Prove that, when set free, (i) it will begin to rise or fall according as $n^2(b-a)>$ or $<g$, (ii) in its subsequent motion it will oscillate about a point O which is at a distance $a+g/n^2$ below the fixed point, (iii) the distances from O of successive positions of rest form a geometric series of ratio $e^{-\pi\kappa/m}$, (iv) the interval between any two positions of rest is π/m, where $m^2=n^2-\kappa^2$.

6. A particle moves on a smooth cycloid whose axis is vertical and vertex downwards under gravity and a resistance varying as the velocity. Prove that the time of falling from any point to the vertex is independent of the starting point.

7. A particle moves under a central force $\phi(r)$ in a medium of which the resistance varies as the velocity. Investigate the equations

$$\ddot{r}+\mu\dot{r}-\frac{h^2}{r^3}e^{-2\mu t}+\phi(r)=0,\quad \frac{h^2}{p^3}\frac{dp}{dr}=\phi(r)\,e^{2\mu t},$$

where h and μ are constants.

*138. Motion in a vertical plane under gravity.

For any law of resistance we can make some progress with the equations of motion of a particle moving in a vertical plane under gravity.

Let $mf(v)$ be the magnitude of the resistance when the velocity is v, m being the mass of the particle, then resolving horizontally we have

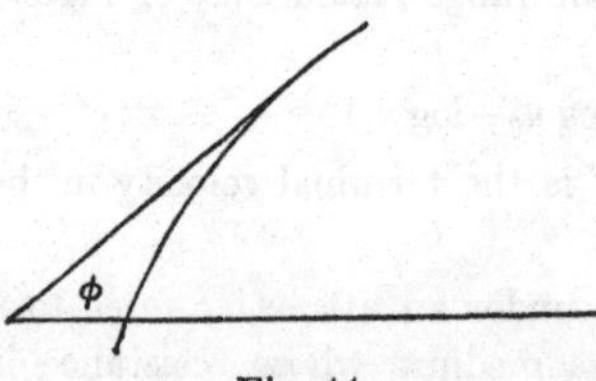

Fig. 44.

$$\dot{u}=-f(v)\cos\phi,$$

where ϕ is the angle which the direction of motion at time t makes with the horizontal and u is the horizontal velocity, so that $u=v\cos\phi$.

Again resolving along the normal to the path, since the resistance is directed along the tangent, we have

$$\frac{v^2}{\rho}=g\cos\phi,$$

where ρ is the radius of curvature. Since ϕ diminishes as s increases, ρ is $-ds/d\phi$, and the above equation may be written

$$v\dot{\phi}=-g\cos\phi,$$

and thus, eliminating t, we get

$$\frac{du}{d\phi} = \frac{vf(v)}{g}, \text{ where } v = u \sec \phi.$$

This equation can be integrated when $f(v) = \kappa v^n$, and we have

$$\frac{1}{u^n} + \frac{n\kappa}{g}\int \frac{d\phi}{\cos^{n+1}\phi} = \text{const.},$$

an equation giving u, and therefore also v, in terms of ϕ.

Now the equation

$$v\frac{d\phi}{dt} = -g\cos\phi$$

gives $$t = -\int \frac{v}{g}\sec\phi\, d\phi + \text{const.},$$

so that t is found in terms of ϕ. Also the equations

$$\frac{dx}{ds} = \cos\phi, \quad \frac{dy}{ds} = \sin\phi, \quad \frac{ds}{dt} = v$$

give us $x = -\int \frac{v^2}{g} d\phi + \text{const.}, \quad y = -\int \frac{v^2}{g}\tan\phi\, d\phi + \text{const.},$

and thus the time and the position of the particle are determined in terms of a single parameter ϕ.

It is not generally possible to integrate the equation for *vertical rectilinear motion* even for the case here described where $f(v) = \kappa v^n$. In the special case, however, where the resistance is proportional to the square of the velocity the velocity can be found in any position. We have, when the particle is ascending,

$$\ddot{y} = -g - \kappa\dot{y}^2,$$

y being measured upwards. Now

$$\ddot{y} = \dot{y}\frac{d\dot{y}}{dy} = \frac{d}{dy}(\tfrac{1}{2}\dot{y}^2);$$

hence $$\frac{d}{dy}(\tfrac{1}{2}\dot{y}^2) + \kappa\dot{y}^2 = -g.$$

Multiplying by $e^{2\kappa y}$ and integrating, we have

$$\tfrac{1}{2}\dot{y}^2 e^{2\kappa y} = -\frac{g}{2\kappa}e^{2\kappa y} + \text{const.},$$

giving $$\dot{y}^2 = Ce^{-2\kappa y} - g/\kappa.$$

Again, when the particle is descending we have, measuring y downwards,

$$\ddot{y} = g - \kappa\dot{y}^2,$$

or
$$\frac{d}{dy}(\tfrac{1}{2}\dot{y}^2) + \kappa\dot{y}^2 = g,$$
giving
$$\dot{y}^2 = \frac{g}{\kappa} - Ce^{-2\kappa y}.$$

As in the case of resistance proportional to the velocity, there is a terminal velocity, $\surd(g/\kappa)$, which is practically attained when the particle has fallen through a considerable height.

***139. Examples.**

1. A particle is projected vertically upwards in a medium whose resistance varies as the square of the velocity. Prove that the interval that elapses before it returns to the point of projection is less than it would be if there were no resistance.

Prove also that, if the particle is let fall from rest, then in time t it acquires a velocity $U \tanh(gt/U)$ and falls a distance $U^2 g^{-1} \log\cosh(gt/U)$, where U is the terminal velocity in the medium.

2. A particle of weight W moves in a medium whose resistance varies as the nth power of the velocity. Prove that, if F is the resistance when the direction of motion makes an angle ϕ with the horizon, then
$$\frac{W}{F} = n\cos^n\phi \int \sec^{n+1}\phi\, d\phi.$$

3. If v is the velocity of a projectile when the inclination of its path to the horizontal is ϕ, a point whose polar coordinates are v and ϕ traces out a curve called the "hodograph" of the trajectory. Prove (i) that, when the resistance is proportional to the velocity, the hodograph is a straight line; (ii) that, whatever the law of resistance may be, the sectorial area bounded by an arc of the hodograph and two of its radii vectores is the product of $\tfrac{1}{2}g$ and the difference of the values of x in the corresponding arc of the trajectory.

4. A particle of unit mass moves in a straight line under an attraction μ (distance) to a point in the line, and a resistance κ (velocity)^2. Prove that, if it starts from rest at a distance a from the centre of force, it will first come to rest at a distance b, where
$$(1 + 2a\kappa)\, e^{-2a\kappa} = (1 - 2\kappa b)\, e^{2\kappa b}.$$

5. The bob of a simple pendulum moves under gravity in a medium of which the resistance per unit of mass is κ (velocity)^2, and starts from the lowest point with such velocity that if it were unresisted the angle of oscillation would be a. Prove that it comes to rest after describing an angle θ which satisfies the equation
$$(1 + 4\kappa^2 l^2)\cos a = 4\kappa^2 l^2 - 2\kappa l \sin\theta e^{2\kappa l\theta} + \cos\theta e^{2\kappa l\theta},$$
where l is the length of the pendulum.

MISCELLANEOUS EXAMPLES

1. A particle is projected horizontally from the lowest point of a smooth elliptic arc, whose major axis $2a$ is vertical, and moves under gravity along the concave side. Prove that it will leave the curve if the velocity of projection lies between $\sqrt{(2ga)}$ and $\sqrt{\{ga(5-e^2)\}}$.

2. A particle moves on a smooth curve in a vertical plane, the form of the curve being such that the pressure on the curve is always m times the weight of the particle. Prove that the time of a complete revolution is $2\pi m\sqrt{a}/\{\sqrt{g}\,.\,(m^2-1)^{\frac{3}{2}}\}$, and that the length of the vertical axis of the curve is $2ma/(m^2-1)^2$, the whole length of the curve being $\pi a(2m^2+1)/(m^2-1)^{\frac{5}{2}}$.

3. A bead moves on a smooth circular wire in a vertical plane its velocity being that due to falling from a horizontal line HK above the circle. Prove that, if I is the internal limiting point of the co-axal system of which the circle and the line HK are members, then any chord through I divides the wire into two parts which are described in equal times.

4. Prove that the time of a beat of a circular pendulum of length a oscillating through an angle 2α is equal to the time of complete revolution of a pendulum of length $a\operatorname{cosec}^2\frac{1}{2}\alpha$, the height of the line of zero velocity above the lowest point being $2a\operatorname{cosec}^4\frac{1}{2}\alpha$.

5. The point of support of a simple pendulum of length l and weight w is attached to a massless spring so that it can move to and fro in a horizontal line; prove that the time of a small oscillation is

$$2\pi\sqrt{\{(W+w)l/Wg\}},$$

where W is the weight required to stretch the spring a length l.

6. A platform is sliding down a smooth spherical hill from rest at the summit. From a point fixed on it a plumb-line is suspended in a tube which is always held perpendicular to the surface of the hill at the point of contact of the platform. Prove that the tension of the cord, when the platform has descended a distance x measured vertically, is $w(a-3x)/a$, where a is the radius of the sphere, and w is the weight of the lead.

7. Prove that, if the suspending fibre of a simple pendulum is slightly extensible, the period of small oscillation is that due to the stretched length of the fibre in the position of equilibrium.

8. A particle moves in a smooth tube in the form of a catenary being attracted to the directrix with a force proportional to the distance from the directrix. Prove that the period of oscillation is independent of the amplitude.

---

9. Two particles of masses P and Q lie near to each other on a smooth horizontal table, being connected by a thread on which is a ring of mass R hanging just over the edge of the table. Prove that it falls with acceleration

$$g(1/P+1/Q)\div(1/P+1/Q+4/R).$$

10. Two particles of masses m, m' are attached to the ends of a thread passing over a pulley, and are held on two inclined planes each of angle α placed back to back with their highest points beneath the centre of the pulley. Prove that, if each portion of the thread makes an angle β with the corresponding plane, the particle of greater mass m will at once pull the other off the plane if

$$m'/m < 2 \tan \alpha \tan \beta - 1.$$

11. A straight smooth groove is cut in a horizontal table, and a straight slit is cut in the bottom of the groove. A thread of length l, attached at one end to a shot of mass m resting in the groove, passes through the slit and supports a particle of mass κm. The suspended particle is held displaced in the vertical plane containing the slit with the thread straight, and is let go. Prove that its path is part of an ellipse of semi-axes l, and $l/(1+\kappa)$, the major axis being vertical.

12. Two particles of masses M, m are connected by a cord passing over a small smooth pulley; the smaller (m) hangs vertically and the other (M) moves in a smooth circular groove on a fixed plane of inclination α to the vertical, the highest point of the groove being the foot of the normal from the pulley to the plane. M starts from a point close to the highest point of the groove without initial velocity. Prove that, if it makes complete revolutions, the radius of the groove must not exceed

$$hmM \cos \alpha/(m^2 - M^2 \cos^2 \alpha),$$

where h is the distance of the pulley from the plane.

13. A particle of weight W moves in a smooth elliptic groove on a horizontal table, and is attached to two threads which pass through holes at the foci, and each thread supports a body of weight W. One of the bodies is pulled downwards with velocity Ve when the particle is at an end of the minor axis. Prove that, if $V^2 < ab^2g/\{e(3a^2 - 2b^2)\}$, the threads do not become slack, and that in this case the horizontal pressures, R and R', on the groove when the particle is at the ends of the axes are connected by the equation

$$Rb^3 \sim R'a(3a^2 - 2b^2) = 6 Wa^2be^2,$$

where $2a$ and $2b$ are the principal axes, and e is the eccentricity of the ellipse.

14. A smooth parabolic wire, on which is a bead of weight w, is fixed in a horizontal plane. To the bead is attached a thread, which passes through a smooth ring fixed at the focus of the parabola and carries, at its other end, a weight $w/(e-1)$. Prove that the tension T of the thread at any stage of the motion is given by an equation of the form

$$(eT - w)(er - a)^2 = \text{const.},$$

where r is the focal distance of the bead and $4a$ the latus rectum of the parabola.

15. Two smooth straight horizontal non-intersecting wires are fixed at right angles to each other at a distance d apart. Two small rings of equal mass, connected by an inextensible thread of length l, slide on the wires, and they are projected with velocities u and v from points at distances a and b

from the shortest distance between the wires. Prove that after the thread becomes tight the motion is oscillatory and of period $2\pi\,(l^2-d^2)/(av \sim bu)$.

16. One end of a thread of length l is attached to the highest point of a fixed horizontal circular cylinder of radius a. A particle attached to the other end is dropped from a position in which the thread is straight and horizontal and at right angles to the axis of the cylinder. Prove that, if $l \nless 2\pi a$, the thread will become slack before the particle comes to rest, and that it will then have turned through an angle whose circular measure is

$$\pi+\tfrac{4}{3}a/l+\tfrac{4}{3}\pi\,(a/l)^2+\tfrac{4}{3}\,(\tfrac{32}{27}+\pi^2)\,(a/l)^3+\ldots.$$

17. Two particles, masses m, m', on a smooth horizontal table are connected by a thread passing through a small smooth ring fixed in the table. Initially the thread is just extended and in two straight pieces meeting at the ring, the lengths of the pieces being a and a'. The particles are projected at right angles to the thread with velocities v and v'. Prove that, if T is the tension at any time and r, r' the distances from the ring, then

$$T\left(\frac{1}{m}+\frac{1}{m'}\right)=\frac{a^2v^2}{r^3}+\frac{a'^2v'^2}{r'^3}.$$

Prove also that the other apsidal distances will be equal if

$$mv^2 : m'v'^2 = 3a'+a : 3a+a'.$$

---

18. A particle slides down a rough cycloid whose axis is vertical and vertex downwards. Prove that the time of reaching a certain point on the cycloid is independent of the starting point.

Prove also that, if λ is the angle of friction, and if the tangent at the starting point makes with the horizontal an angle greater than α, where α is the least positive angle which satisfies the equation

$$\sin(\alpha-\lambda)=e^{(\alpha+\lambda)\tan\lambda}\sin 2\lambda,$$

the particle will oscillate.

19. A ring moves on a rough cycloidal wire with its axis vertical and vertex downwards. Prove that, if it starts from the lowest point with velocity u_0, its velocity u when its direction of motion is inclined at an angle ϕ to the horizontal is given by

$$u^2=(u_0{}^2+4ag\sin^2\epsilon)\,e^{-2\phi\tan\epsilon}-4ag\sin^2(\phi+\epsilon),$$

where a is the radius of the generating circle and ϵ is the angle of friction.

Prove also that, if it starts from a cusp with velocity v_0, its velocity v during its descent is given by

$$v^2=(v_0{}^2+4ag\cos^2\epsilon)\,e^{-(\pi-2\phi)\tan\epsilon}-4ag\sin^2(\phi-\epsilon).$$

20. A particle is projected from a point on the lowest generator of a rough horizontal cylinder of radius a with velocity V at right angles to the generator, and moves under no forces except the pressure and friction of the surface. Prove that it returns to the point of projection after a time $a\,(e^{2\mu\pi}-1)/(\mu V)$, where μ is the coefficient of friction.

---

21. A point P moves along a plane curve which rotates in its plane about a point O with uniform angular velocity ω. Prove that the curvature of its path is

$$\frac{V(\sigma V+2\omega)(V+r\omega\sin\psi)+r\omega(V\omega\sin\psi-f\cos\psi+r\omega^2)}{(V^2+r^2\omega^2+2Vr\omega\sin\psi)^{\frac{3}{2}}},$$

where r is the length OP, σ is the curvature of the curve at P, ψ the angle between OP and the tangent, V the velocity of P relative to the curve, and f the rate of increase of V.

22. A bead is initially at rest on a smooth circular wire of radius a in a horizontal plane; the wire is made to rotate with uniform angular velocity ω about an axis perpendicular to its plane and passing through a point on the diameter through the bead at a distance c from the centre. When the bead has moved a distance $a\theta$ on the wire, the wire is suddenly stopped. Prove that the bead will subsequently move with velocity

$$\omega\{\sqrt{(a^2+c^2+2ac\cos\theta)}-(a+c\cos\theta)\}.$$

23. Two small beads of masses m_1, m_2 slide along two smooth straight rods which intersect at an angle α, and the beads are connected by an elastic thread of natural length c and modulus λ. The rods are made to revolve uniformly in their plane, about their point of intersection, with angular velocity ω. Prove that throughout the motion

$$m_1(\dot{r}_1^2-r_1^2\omega^2)+m_2(\dot{r}_2^2-r_2^2\omega^2)+\lambda\epsilon^2c=\text{const.},$$

where ϵ is the extension of the thread, and r_1, r_2 are the distances of the beads from the intersection of the wires at any time.

24. A smooth elliptic tube rotates about a vertical axis through its centre perpendicular to its plane with uniform angular velocity ω. Prove that a particle can remain at an extremity of the axis major, and, if slightly disturbed, will oscillate in a period $2\pi\sqrt{(1-e^2)}/e\omega$, where e is the eccentricity.

25. A body is describing an ellipse of semi-axes a, b about a centre of gravitation, and when it is at a distance r from this centre it comes under the influence of a small disturbing force directed to the same point and varying inversely as the cube of the distance. Prove that the effect is the same as if the body described under the original force an orbit which at the same time rotated (with the body) round the centre of force with angular velocity n times the angular velocity of the body, where n is a small constant such that the semi-axes of this new free orbit are equal to those of the original one reduced by fractions $2nb^2/r^2$ and $n(1+b^2/r^2)$ of themselves.

---

26. A particle moves on a helical wire whose axis is vertical. Prove that the velocity v after describing an arc s is given by the equations

$$v^2=ag\sec\alpha\sinh\phi,\quad \frac{ds}{d\phi}=\tfrac{1}{2}a\frac{\sec^2\alpha\cosh\phi}{\tan\alpha-\mu\cosh\phi},$$

where a is the radius of the cylinder on which the helix lies, α the inclination of the helix to the horizon, and μ the coefficient of friction.

27. A small smooth groove is cut on the surface of a right circular cone whose axis is vertical and vertex upwards in such a manner that the tangent is always inclined to the vertical at the same angle β. A particle slides down the groove from rest at the vertex; show that the time of descending through a vertical height h is equal to the time of falling freely through a height $h \sec^2 \beta$. Show also that the pressure is constant and makes with the principal normal to the path a constant angle

$$\tan^{-1} \{\tfrac{1}{2} \sin a/\sqrt{(\cos^2 a - \cos^2 \beta)}\},$$

where $2a$ is the angle of the cone.

28. A smooth helical tube of pitch a has its axis inclined at an angle $\beta\,(> a)$ to the vertical, and a particle rests in the tube. The tube is made to turn about its axis with uniform angular velocity ω. Prove that the particle makes at least one complete revolution round the axis if

$$\tfrac{1}{2}a\omega^2/g > [(\pi + 2\gamma) \sin \gamma + 2 \cos \gamma] \sin \beta \cot a \operatorname{cosec}^3 a,$$

where $\sin \gamma = \tan a \cot \beta$, and a is the radius of the helix.

29. A small ring can slide on a smooth plane curved wire which rotates with angular velocity ω about a vertical axis in its plane. Find the form of the curve in order that the ring may be in relative equilibrium at any point.

Prove that, if the angular velocity is increased to ω', the ring will still be in relative equilibrium if the wire is rough and the coefficient of friction between it and the ring is not less than $\frac{1}{2}(\omega'/\omega - \omega/\omega')$.

30. A particle moves in a smooth circular tube of radius a which rotates about a fixed vertical diameter with angular velocity ω. Prove that, if θ is the angular distance of the particle from the lowest point, and if initially it is at rest relative to the tube with the value a for θ where $\omega \cos \frac{1}{2}a = \sqrt{(g/a)}$, then at any subsequent time t

$$\cot \tfrac{1}{2}\theta = \cot \tfrac{1}{2}a \cosh (\omega t \sin \tfrac{1}{2}a).$$

---

31. A particle moves under gravity on a right circular cone with a vertical axis. Show that, if the equations of motion can be integrated without elliptic functions, the particle must be below the vertex, and that its distance r from the vertex at time t is given by an equation of the form

$$(r\dot{r})^2 = 2g \cos a\,(r - r_0)\,(r + 2r_0)^2,$$

where $2a$ is the vertical angle of the cone.

32. A right circular cone of vertical angle $2a$ is placed with one generator vertical and vertex upwards. From a point on the generator o least slope a particle is projected horizontally and at right angles to the generator with velocity v. Prove that it will just skim the surface of the cone without pressure if the distance of the point of projection from the vertex is

$$\tfrac{1}{2}v^2 \operatorname{cosec}^2 a/g.$$

33. A particle is projected horizontally from a fixed point on the interior surface of a smooth paraboloid of revolution whose axis is vertical and vertex downwards. Prove that when it is again moving horizontally its velocity is independent of the velocity of projection.

34. Prove that, if the path of a particle moving on a right circular cone cuts the generators at an angle χ, the acceleration in the tangent plane to the surface and normal to the path is

$$v^2 (d\chi/ds + r^{-1} \sin \chi),$$

where v is the velocity, and r the distance from the vertex.

If the axis of the cone is vertical, and the vertex upwards, and if the velocity is that due to falling from the vertex, prove that, when the particle leaves the cone (supposed smooth),

$$2 \sin^2 \chi = \tan^2 a,$$

$2a$ being the vertical angle of the cone. What happens when $\tan^2 a > 2$?

35. A particle moves on a surface of revolution. The velocity is v at a point where the normal terminated by the axis of revolution is of length ν, and this normal makes an angle θ with the axis; prove that, if ds is the element of arc of the path, and χ the angle at which it cuts the meridian, the acceleration in the tangent plane to the surface and normal to the path is

$$v^2 \left(\frac{d\chi}{ds} + \frac{\sin \chi \cot \theta}{\nu} \right).$$

36. A particle is placed at rest on the smooth inner surface of a vertical circular cylinder, which rotates with uniform angular velocity ω about the generator which is initially furthest from the particle. Prove that the pressure vanishes when the particle has descended a distance

$$\frac{1}{2} \frac{g}{\omega^2} \left(\log \frac{2}{3 - \sqrt{5}} \right)^2.$$

37. A particle is attached by a thread of length a to a point of a rough fixed plane inclined to the horizon at an angle equal to the angle of friction between the particle and the plane. The particle is projected down the plane at right angles to the thread, which is initially straight and horizontal. Prove that it comes to rest at the lowest point of its path if the square of the initial velocity is $(\pi - 2) \mu g a / \sqrt{(1 + \mu^2)}$, where μ is the coefficient of friction.

38. A particle is projected horizontally with velocity V along the interior surface of a rough vertical circular cylinder. Prove that, at a point where the path cuts the generator at an angle ϕ, the velocity v is given by the equation

$$ag/v^2 = \sin^2 \phi \, \{ag/V^2 + 2\mu \log (\cot \phi + \operatorname{cosec} \phi)\},$$

and the azimuthal angle and the vertical descent are respectively

$$\int_\phi^{\frac{1}{2}\pi} \frac{v^2}{ag} \, d\phi \text{ and } \int_\phi^{\frac{1}{2}\pi} \frac{v^2}{g} \cot \phi \, d\phi.$$

39. A particle falls from rest under gravity through a distance x in a medium whose resistance varies as the square of the velocity; v is the velocity acquired by the particle, V the terminal velocity, and v_0 the velocity that would be acquired by falling through a distance x in vacuo; prove that

$$v^2/v_0^2 = 1 - \tfrac{1}{2}v_0^2/V^2 + \frac{1}{2\,.\,3}v_0^4/V^4 - \frac{1}{2\,.\,3\,.\,4}v_0^6/V^6 + \ldots.$$

40. A particle is projected vertically upwards from the surface of the Earth with velocity u, and when its velocity is v and its height above the surface is z the resistance is $\kappa v^2/(a+z)$, where a is the Earth's radius. Prove that, if z is always small compared with a, the velocity V with which it returns to the point of projection is given approximately by the equation

$$V^2/u^2 = 1 - \kappa u^2/ga + (4\kappa^2 - \tfrac{2}{3}\kappa)(u^2/2ga)^2,$$

variations of gravity with height being taken into account.

41. Prove that, if the resistance is proportional to the square of the velocity, the angle θ between the asymptotes of the complete trajectory of a projectile is given by the equation

$$U^2/w^2 = \cot\theta \operatorname{cosec}\theta + \sinh^{-1}\cot\theta,$$

where U is the terminal velocity and w the velocity when the projectile moves horizontally.

42. A particle moves under gravity in a medium whose resistance is proportional to the velocity. Prove that the range on a horizontal plane is a maximum, for given velocity of projection, when the angle of elevation at first and the angle of descent at last are complementary.

43. A particle is projected up a plane of inclination α under gravity and a resistance proportional to the velocity. The direction of projection makes an angle β with the vertical, the range R is a maximum and t is the time of flight. Prove that, if U is the terminal velocity and V the velocity of projection, then

(i) $1 + (V/U)\sec\beta = \exp.\,(gt/U)$,

(ii) $UV(U + V\cos\beta)/(V + U\cos\beta) = g(R\sin\alpha + Ut)$,

(iii) $UV^2\sin\beta/(V + U\cos\beta) = gR\cos\alpha$.

44. A pendulum oscillates in a medium of which the resistance per unit of mass is κ (velocity)^2. Prove that, when powers of the arc above the first are neglected, the period is the same as in the absence of resistance, but the time of descent exceeds that of ascent by $\tfrac{2}{3}\kappa\alpha\sqrt{(l^3/g)}$, where α is the angular amplitude of the descent, and l is the length of the pendulum.

45. A particle of mass m moves in a field of force having a potential V in a medium in which the resistance is k times the velocity. Prove that, if D is the quantity of energy dissipated in time t,

$$\frac{dD}{dt} + \frac{2k}{m}(D - mV) = \text{const.}$$

If the resistance is k (velocity)^2, and if ds is the element of arc of the path of the particle, then

$$\frac{dD}{ds}+\frac{2k}{m}(D-mV)=\text{const.}$$

46. A smooth straight tube rotates in one plane with uniform angular velocity ω about a fixed end, and a particle moves within it under a resistance equal to κ times the square of the relative velocity. Prove that, if the particle is projected so as to come to rest at the fixed end, the relative velocity at a distance r from that end is

$$\tfrac{1}{2}\sqrt{2}\,.\,\omega\kappa^{-1}\sqrt{(e^{2\kappa r}-2\kappa r-1)}.$$

47. A particle moves on a smooth cycloid whose axis is vertical and vertex upwards in a medium whose resistance is $(2c)^{-1}$ (velocity)^2 per unit of mass, and the distance of the starting point from the vertex measured along the curve is c; prove that the time to the cusp is $\sqrt{\{8a(4a-c)/gc\}}$, $2a$ being the length of the axis.

CHAPTER VI

THE LAW OF REACTION

140. Direct impact of spheres. Let the centres of two spheres move in the same line. This line must be that joining the centres. The spheres will come into contact if their centres are moving in opposite senses, or if one of them is at rest, and the other is moving towards it, or if they are moving in the same sense, and one overtakes the other. Let m, m' be the masses of the spheres, determined by weighing them in a common balance. Let U be the velocity of the centre of the sphere m before impact, in the sense from m towards m', U' the velocity of the centre of m' before impact, in the same sense; and let u and u' be the velocities of m and m' in the same sense after impact. When proper arrangements are made for measuring the velocities, it is found that

$$m(u-U)=m'(U'-u').$$

141. Ballistic balance. An instrument by which experiments of the kind just considered may be made is called a "ballistic balance." In principle it comes to this*:—The two spheres are suspended from two fixed points at the same level by cords, and, when the cords are vertical, the spheres are in contact and the line of centres is horizontal (see Fig. 45). The distance between the fixed points is equal to the sum of the radii. One sphere is then raised, the cord attached to it being kept taut, until its centre is at a known height H above the equilibrium position. It is then let fall. At the instant of impact its velocity is $\sqrt{(2gH)}$. The velocities of the spheres immediately after the impact are measured by observing the heights to which the centres rise.

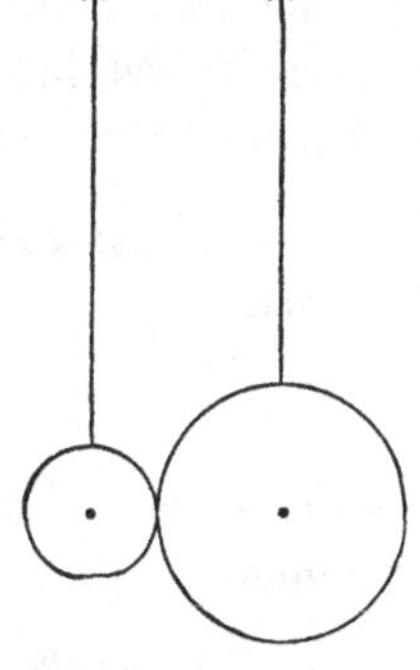

Fig. 45.

* The actual construction and method of using the instrument are described by W. M. Hicks, *Elementary Dynamics of Particles and Solids*, London, 1890. Experimental investigations of the kind referred to in the text were made by Newton. See *Principia*, Lib. I. "Axiomata sive leges motus."

142. Statement of the Law of Reaction. The result stated in Art. 140 may be written

$$m'u' - m'U' = -(mu - mU).$$

The left-hand member is the measure of the "change of momentum" of the sphere m'; the right-hand member, with its sign changed, is the measure of the change of momentum of the sphere m. These changes of momentum are produced, during the very short time of the impact, by forces which the spheres exert one on the other. The result can be stated in the form :—The impulses of these forces are equal and opposite. This result leads us to conclude that the forces also are equal and opposite. The result is generalized in the statement :—In any action between bodies, by which the motion of either is set up, altered or stopped, each body exerts force on the other, and these forces are equal and opposite. The statement may be made more precise when the bodies are replaced by particles, and then it takes the form :—

The magnitude of the force exerted by one particle on another is equal to the magnitude of the force exerted by the second particle on the first, the lines of action of both the forces coincide with the line joining the particles, and the forces have opposite senses.

This abstract statement may be regarded as an induction from experience. The proof of its truth is found in the agreement of results deduced from it with results of experiment.

The statement is frequently called the "Law of Reaction" because it was briefly expressed by Newton in the phrase "action and reaction are equal and opposite."

143. Mass-ratio. The result of Art. 140 may be expressed in the form

$$\frac{-(u-U)}{u'-U'} = \frac{m'}{m},$$

and this result may be generalized, and made precise, in the statement :—

In any action between particles the changes of velocity are inversely proportional to the masses.

This result enables us to assign for any two particles, or for any two bodies treated as particles, a perfectly definite ratio, which may be called the "mass-ratio." If the force between the particles produces in them accelerations f and f' respectively, the mass-ratio is $f' : f$.

The mass-ratio of any two particles is the inverse ratio of the accelerations which, by their mutual action, either produces in the other.

144. Mass. Whenever two bodies can be treated as particles, the mass-ratio of the particles is the ratio of the masses of the bodies.

This statement enables us to assign masses to bodies without weighing them in a common balance.

Whenever the bodies can be so weighed, the ratio of the masses that is determined by the mutual action is, as a matter of fact, the same as the ratio that is determined by the operation of weighing.

It is clear that the definition of mass by means of mutual action is more general and more fundamental than that by means of weighing. We shall show in Chapter X that the determination of masses by weighing is a particular case of the determination by means of mutual action.

Since we are accustomed to estimate the *quantity of matter* in a body by weighing the body, it is customary to state that the quantity of matter in a body is equal to the mass of the body.

To produce any alteration in the velocity of a moving body, to set the body in motion, or to bring it to rest, applications of force are required. This result leads us to recognize a tendency in bodies to maintain an established state of motion when there are no forces which produce changes of motion. This tendency is called "inertia." The impulse of the force required to produce any assigned change of motion in a body is proportional to the mass of the body. Thus the mass of the body provides a measure of its *inertia*.

145. Density. The fraction

$$\frac{\text{number of units of mass in the mass of a body}}{\text{number of units of volume in the volume of the body}}$$

is the "mean density" of the body. In the same way we may define the mean density of any portion of a body.

When the mean density of all parts of the body is the same, the body is said to be "homogeneous," or "uniform," otherwise it is "heterogeneous."

In the case of a heterogeneous body, we may define the *density at a point* as the limit to which the mean density of a volume containing the point tends when the volume is diminished indefinitely.

The densities of sensibly homogeneous substances in assigned circumstances are physical constants. For example, the density of pure water (at a temperature of 4° Centigrade and a barometric pressure represented by 76 centimetres of mercury) is unity, the centimetre and the gramme being the units of length and mass.

Density is a physical quantity of dimensions 1 in mass and -3 in length.

146. Gravitation. The periodic time of a particle describing an elliptic orbit about a focus is $2\pi a^{\frac{3}{2}} \mu^{-\frac{1}{2}}$, where $2a$ is the major axis of the orbit, and μ is the intensity of the field of force at unit distance from the focus (Art. 48, Ex. 5). The result that the squares of the periodic times of the Planets, describing orbits about the Sun, are proportional to the cubes of the major axes of the orbits, was noted by Kepler*. If the *intensity* of the field of the Sun's gravitation is denoted by $\mu/(\text{distance})^2$, the quantity μ is the same for all the Planets.

Let E be the mass of the Earth, P that of any Planet, r, r' the distances from the Sun to the Earth and the Planet respectively. The forces of the Sun's gravitation, acting on the Earth and the Planet respectively, are $\mu E/r^2$ and $\mu P/r'^2$. These therefore are the magnitudes of the forces which the two bodies exert on the Sun, and they are proportional to the masses of the bodies. Thus the force of the Earth's gravitation, and the force of the Planet's gravitation, are proportional to the masses of the Earth and the Planet respectively. We should accordingly expect the force of the Sun's gravitation to be proportional to the mass of the Sun, that is to say, we are led to take for μ the form γS, where S denotes the mass of the Sun and γ is a constant independent of the masses. The force exerted by the Sun on the Earth, or by the Earth on the Sun, is then expressed by the formula

$$\frac{\gamma ES}{r^2}.$$

* *Harmonice Mundi*, 1619. The result is sometimes called Kepler's "third law of planetary motion."

Such forces would arise if bodies were made up of small parts, each of which may be treated as a particle, if these particles acted upon each other with forces in the lines joining their positions, and if the force between two particles of masses m and m' were an attraction of amount

$$\frac{\gamma m m'}{r^2}.$$

The *law of gravitation* states that this formula expresses the law of force between particles (taken to be small parts of bodies) at all distances which can be measured by ordinary means (*e.g.* by a divided scale), and at all greater distances.

The law can be verified by actual observation of the gravitational force between bodies at the Earth's surface. By these observations also the value of γ can be determined. The best determination gives for γ the value $(6{\cdot}65)\,10^{-8}$ in C.G.S. units*.

The quantity γ is a physical constant; it is called the "constant of gravitation." It is of dimensions, 3 in length, -1 in mass, -2 in time.

Since the intensity of the field of the Sun's gravitation is $\gamma S/(\text{distance})^2$, a knowledge of the period of the Earth's revolution about the Sun ($365\frac{1}{4}$ days) enables us to determine the mass of the Sun.

147. Theory of Attractions. When a body is regarded as made up of particles, and the particles of a body, and those of other bodies, act upon each other with forces according to the law of gravitation, the resultant force acting on a particle of any one of the bodies may be calculated. The theory by means of which the calculation is effected is the Theory of Attractions, and accounts of it will be found in books on Statics. From our present point of view, the most important result of the theory is that homogeneous spheres, or spheres of which the material is arranged in concentric spherical strata of constant density, attract an external particle as if their masses were condensed at their centres†.

148. Mean density of the Earth. In consequence of the result last stated, we are led to take the intensity of the field of the Earth's gravitation, even at a moderate distance, to be $\gamma E/R^2$, where E is the mass of the Earth, and R denotes distance from its centre. Now if we take R to be the radius of the Earth, this quantity is the acceleration of a free body at the surface. Apart from the correction on account of the rotation of the Earth, it is the same as g. We denote it by g'. Then we find that the mean density ρ of the Earth is given by the equation

$$\rho = \frac{3g'}{4\pi\gamma R}.$$

* C. V. Boys, *Proc. R. Soc. London*, vol. 56 (1894).

† The result is due to Newton, *Principia*, Lib. I. Sect. XII.

If we ignore the distinction between g' and g, or if we determine g' (cf. Chapter X), this equation gives us ρ when γ is known. Thus the law of gravitation avails for the determination of the mass and the mean density of the Earth. The mean density (in grammes per cubic centimetre) has been determined* to be 5·527, or about $5\frac{1}{2}$ times the density of water.

149. Attraction within gravitating sphere. It is a known result in the Theory of Attractions that a homogeneous shell bounded by concentric spherical surfaces exerts no attraction at any point within its inner surface.

It follows that the attraction at a point within a homogeneous gravitating sphere is that of the concentric sphere which passes through the point.

If the Earth were a homogeneous sphere of radius a, the attraction of the Earth upon an internal particle at a distance r from its centre would be $g'r/a$, where g' is the attraction at the surface.

150. Examples.

1. Consider the motion of a particle under the action of a uniform fixed gravitating sphere, of density ρ and radius a, and suppose the particle to start from rest at a distance $b\,(>a)$ from the centre. It will move directly towards the centre with an acceleration $\frac{4}{3}\pi\gamma\rho a^3/x^2$ at a distance x from the centre, so long as $x>a$, and when $x=a$, it will have a velocity given by

$$\tfrac{1}{2}\dot{x}^2=\tfrac{4}{3}\pi\gamma\rho a^3\left(\frac{1}{a}-\frac{1}{b}\right).$$

Now suppose a fine tunnel to be bored through the centre of the sphere in the direction of motion of the particle. When the particle passes into the tunnel its acceleration becomes $\frac{4}{3}\pi\gamma\rho x$ at a distance x from the centre, and it moves with a simple harmonic motion. The velocity at a distance x from the centre is given by the equation

$$\tfrac{1}{2}\dot{x}^2+\tfrac{2}{3}\pi\gamma\rho x^2=\text{const.},$$

and the constant is determined from the expression given above for the velocity at the instant of entering the tube.

Prove that the velocity at the centre is

$$\sqrt{\{\tfrac{4}{3}\pi\gamma\rho a^2(3-2a/b)\}};$$

and, taking $b=a$, find the time of passing through the tunnel.

2. Prove that, on taking a pendulum down a mine, the time of vibration is increased or diminished according as the mean density of the surface rock is greater or less than two-thirds of the Earth's mean density. [Neglect the distinction between g' and g.]

* C. V. Boys, *loc. cit.*

Theory of a system of particles

151. The Sun and the Planets with their Satellites afford an example of a system of bodies, which can be treated as particles moving under their mutual attractions. The law of gravitation avails for the determination of the masses of the system as well as for the determination of the motions. Much of theoretical Mechanics has been developed from the theory of the motion of such a system of particles. In general we shall suppose that each particle of the system has an assigned mass, and moves under forces, some of which are taken to arise from the mutual actions of particles within the system, and others from the actions exerted upon particles within the system by particles outside the system.

152. Centre of mass. Let x, y, z be the coordinates at time t of a particle of the system, m the mass of the particle; and let a point $(\bar{x}, \bar{y}, \bar{z})$ be determined by the equations

$$\bar{x} = \frac{\Sigma(mx)}{\Sigma m}, \quad \bar{y} = \frac{\Sigma(my)}{\Sigma m}, \quad \bar{z} = \frac{\Sigma(mz)}{\Sigma m},$$

where the summations extend to all the particles. This point is defined to be the "centre of mass" of the system of particles.

The centre of mass coincides with the "centre of gravity" defined in books on Statics. On account of the relation between mass and inertia (Art. 144) it is sometimes called the "centre of inertia." We shall denote it by the letter G.

153. Resultant momentum. The momentum of a particle of mass m, which is at the point (x, y, z) at time t, has been defined to be a vector, localized in a line through the point, of which the resolved parts in the directions of the axes are $m\dot{x}$, $m\dot{y}$, $m\dot{z}$. The momenta of the particles of a system are a system of vectors localized in lines.

The general theory of the reduction of a system of localized vectors (see Appendix to this Chapter) shows that the momenta of the particles of a system are equivalent to a "resultant momentum," localized in a line through any chosen point, together with a vector couple, which is a "moment of momentum." The resolved parts in the directions of the axes of the resultant momentum are

$$\Sigma(m\dot{x}), \quad \Sigma(m\dot{y}), \quad \Sigma(m\dot{z}),$$

where the summations extend to all the particles.

Now we have

$$\dot{\bar{x}}\Sigma m = \Sigma(m\dot{x}), \quad \dot{\bar{y}}\Sigma m = \Sigma(m\dot{y}), \quad \dot{\bar{z}}\Sigma m = \Sigma(m\dot{z}).$$

The left-hand members of these equations are the resolved parts parallel to the axes of the momentum of a fictitious particle, of mass equal to the sum of the masses of the particles, and moving so as to be always at the centre of mass of the system of particles. We call this fictitious particle the "particle G." Then we have the result that the resultant momentum of the system of particles is equal to the momentum of the particle G.

154. Resultant kinetic reaction. The kinetic reaction of a particle of mass m, which is at the point (x, y, z) at time t, has been defined as a vector, localized in a line through the point, of which the resolved parts in the directions of the axes are $m\ddot{x}$, $m\ddot{y}$, $m\ddot{z}$.

The kinetic reactions of a system of particles are equivalent to a "resultant kinetic reaction," localized in a line through any chosen point, and a vector couple, which is a "moment of kinetic reaction."

The components parallel to the axes of the resultant kinetic reaction of a system of particles are

$$\Sigma(m\ddot{x}), \quad \Sigma(m\ddot{y}), \quad \Sigma(m\ddot{z}).$$

Now by differentiating the equations such as $\dot{\bar{x}}\Sigma(m) = \Sigma(m\dot{x})$, we find such equations as $\ddot{\bar{x}}\Sigma(m) = \Sigma(m\ddot{x})$.

Hence the resultant kinetic reaction is the same as the kinetic reaction of the particle G (*i.e.* of a particle of mass equal to the mass of the system, placed at the centre of mass of the system, and moving with it).

155. Relative coordinates. The resultant momentum and resultant kinetic reaction are independent of the chosen point which is used in reducing the system of momenta, or kinetic reactions, to a resultant and a vector couple; but the vector couples depend upon the position of the point. For most purposes it is simplest to take the point either at the origin of coordinates, which is an arbitrary fixed point, or at the centre of mass. We shall take $\bar{x}$, $\bar{y}$, $\bar{z}$ to be the coordinates of the centre of mass, and put

$$x = \bar{x} + x', \quad y = \bar{y} + y', \quad z = \bar{z} + z'.$$

Then x', y', z' are the coordinates of a point relative to the centre of mass.

From the definition of $\bar{x}$, $\bar{y}$, $\bar{z}$ we have

$$\Sigma(mx') = 0, \quad \Sigma(my') = 0, \quad \Sigma(mz') = 0,$$

and it follows that

$$\Sigma(m\dot{x}') = 0, \ldots \quad \Sigma(m\ddot{x}') = 0, \ldots.$$

156. Moment of Momentum. The sum of the moments of the momenta of the particles of the system about any axis is the moment of momentum of the system about the axis.

The moment of momentum of the system about the axis x is

$$\Sigma[m(y\dot{z} - z\dot{y})].$$

See Appendix to this Chapter. This expression is equal to

$$\Sigma[m\{(\bar{y} + y')(\dot{\bar{z}} + \dot{z}') - (\bar{z} + z')(\dot{\bar{y}} + \dot{y}')\}],$$

and this reduces to

$$(\bar{y}\dot{\bar{z}} - \bar{z}\dot{\bar{y}})\Sigma(m) + \Sigma[m(y'\dot{z}' - z'\dot{y}')].$$

The first term of this expression is the moment about the axis x of the momentum of the particle G, and the second term is the moment about an axis drawn through G parallel to the axis x of the system of momenta $m\dot{x}'$, $m\dot{y}'$, $m\dot{z}'$. These are the momenta relative to parallel axes through G, or the momenta in the "motion relative to G." We may therefore state our result in the words:— The moment of momentum of a system about any axis is equal to the moment of momentum of the particle G, together with the moment of momentum in the motion relative to G about a parallel axis through G.

When the momenta of a system of particles are reduced to a resultant momentum at the centre of mass and a vector couple, the couple is the moment of momentum in the motion relative to the centre of mass. It may be called the "resultant moment of momentum at the centre of mass" and its axis "the axis of resultant moment of momentum." Its components are

$$\Sigma[m(y'\dot{z}' - z'\dot{y}')], \ldots.$$

"Moment of momentum" is often called "angular momentum."

157. Moment of kinetic reaction. The sum of the moments of the kinetic reactions about the axis x is

$$\Sigma [m (y\ddot{z} - z\ddot{y})], \text{ or } \frac{d}{dt} \Sigma [m (y\dot{z} - z\dot{y})],$$

and this can be expressed in the form

$$(\bar{y}\ddot{\bar{z}} - \bar{z}\ddot{\bar{y}}) \Sigma (m) + \Sigma [m (y'\ddot{z}' - z'\ddot{y}')],$$

or

$$\frac{d}{dt} [(\bar{y}\dot{\bar{z}} - \bar{z}\dot{\bar{y}}) \Sigma (m)] + \frac{d}{dt} \Sigma [m (y'\dot{z}' - z'\dot{y}')].$$

Hence the sum of the moments of the kinetic reactions about any fixed axis is equal to the rate of increase (per unit of time) of the moment of momentum about the same axis, and this is equal to the moment of the kinetic reaction of the particle G about the axis together with the moment of kinetic reaction in the motion relative to G about a parallel axis through G.

When the kinetic reactions of a system of particles are reduced to a resultant kinetic reaction at the centre of mass and a vector couple, the couple is the rate of increase (per unit of time) of the resultant moment of momentum at the centre of mass.

The sum of the moments of the kinetic reactions about a moving axis is not in general equal to the rate of increase (per unit of time) of the moment of momentum about the same axis. For example, let the axis be parallel to the axis of z and be specified as passing through all the points whose x and y are equal to ξ and η, some functions of t. Then the sum of the moments of the kinetic reactions about the axis is

$$\Sigma [m \{(x - \xi) \ddot{y} - (y - \eta) \ddot{x}\}],$$

and the rate of increase of the moment of momentum about this axis is

$$\frac{d}{dt} [\Sigma m \{(x - \xi) \dot{y} - (y - \eta) \dot{x}\}],$$

which differs from the above by the addition of the terms

$$\dot{\eta} \Sigma (m\dot{x}) - \dot{\xi} \Sigma (m\dot{y}).$$

If the moving axis always passes through the centre of mass, the two expressions are equal, as we saw before.

158. Kinetic energy. The kinetic energy of a particle is half the product of its mass and the square of its velocity.

For a particle of mass m at (x, y, z) it is

$$\tfrac{1}{2} m (\dot{x}^2 + \dot{y}^2 + \dot{z}^2).$$

The kinetic energy of a system of particles is the sum of the kinetic energies of the particles. It is the quantity

$$\tfrac{1}{2}\Sigma\left[m\left(\dot{x}^2+\dot{y}^2+\dot{z}^2\right)\right].$$

This expression is equal to

$$\tfrac{1}{2}\left(\dot{\bar{x}}^2+\dot{\bar{y}}^2+\dot{\bar{z}}^2\right)\Sigma m+\tfrac{1}{2}\Sigma\left[m\left(\dot{x}'^2+\dot{y}'^2+\dot{z}'^2\right)\right].$$

We may state this result in words:—The kinetic energy of a system of particles is the kinetic energy of the particle G together with the kinetic energy in the motion relative to G.

159. Examples.

1. Two particles of masses m, m' move in any manner. V is the velocity of the centre of mass, and v the velocity of one particle relative to the other. The kinetic energy is

$$\tfrac{1}{2}(m+m')V^2+\frac{1}{2}\frac{mm'}{m+m'}v^2.$$

2. In the same case, if p is the perpendicular from the position of one particle to the line drawn through the other in the direction of the relative velocity, the resultant moment of momentum at the centre of mass is

$$\frac{mm'}{m+m'}pv,$$

and the axis of resultant moment of momentum is at right angles to the plane containing the particles and the line of the relative velocity.

160. Equations of motion of a system of particles. Let m_1 be the mass of one particle of the system, x_1, y_1, z_1 its coordinates at time t, X_1, Y_1, Z_1 the sums of the resolved parts parallel to the axes of the forces exerted on this particle by particles not forming part of the system, X_1', Y_1', Z_1' the sums of the resolved parts parallel to the axes of the forces exerted on the same particle by the remaining particles of the system.

The equations of motion of this particle are

$$m_1\ddot{x}_1=X_1+X_1',\quad m_1\ddot{y}_1=Y_1+Y_1',\quad m_1\ddot{z}_1=Z_1+Z_1'.$$

Similarly the equations of motion of a second particle of mass m_2 at (x_2, y_2, z_2) may be written

$$m_2\ddot{x}_2=X_2+X_2',\quad m_2\ddot{y}_2=Y_2+Y_2',\quad m_2\ddot{z}_2=Z_2+Z_2'.$$

We shall write as the type of such equations

$$m\ddot{x}=X+X',\quad m\ddot{y}=Y+Y',\quad m\ddot{z}=Z+Z'.$$

Then (X, Y, Z) is the type of the external forces, and (X', Y', Z') is the type of the internal forces.

161. Law of internal action. *The sum of the resolved parts parallel to any axis, and the sum of the moments about any axis, of all the internal forces between the particles of a system are identically zero.*

The mutual action between any two particles of the system consists of two equal and opposite forces acting upon the two particles in the line joining their positions. The sum of the resolved parts of these two forces parallel to any axis vanishes.

The moment of a force about an axis is the same at whatever point in its line of action the force may be applied. Hence the sum of the moments about any axis of two equal and opposite forces acting in the same line vanishes.

In the notation of Art. 160 the result may be written

$$\Sigma(X')=0,\quad \Sigma(Y')=0,\quad \Sigma(Z')=0,$$

$$\Sigma(yZ'-zY')=0,\quad \Sigma(zX'-xZ')=0,\quad \Sigma(xY'-yX')=0.$$

162. Simplified forms of the equations of motion. Adding the left-hand members of all the x-equations of motion, and remembering that $\Sigma X'=0$, we obtain the equation $\Sigma(m\ddot{x})=\Sigma X$.

In like manner we have

$$\Sigma(m\ddot{y})=\Sigma Y, \text{ and } \Sigma(m\ddot{z})=\Sigma Z.$$

Again multiplying the z-equations by the y's and the y-equations by the z's, and remembering that $\Sigma(yZ'-zY')=0$, we form the equation

$$\Sigma[m(y\ddot{z}-z\ddot{y})]=\Sigma(yZ-zY).$$

In like manner we have

$$\Sigma[m(z\ddot{x}-x\ddot{z})]=\Sigma(zX-xZ), \text{ and } \Sigma[m(x\ddot{y}-y\ddot{x})]=\Sigma(xY-yX).$$

Our equations may be stated in words:—

(1) The sum of the resolved parts in any direction of the kinetic reactions of a system of particles is equal to the sum of the resolved parts of the external forces in the same direction.

(2) The sum of the moments about any axis of the kinetic reactions of a system of particles is equal to the sum of the moments of the external forces about the same axis.

The result may also be briefly stated in the form:—When the external forces are regarded as localized in their lines of action, the kinetic reactions and the external forces are two equivalent systems of localized vectors.

This result, in a slightly different form, was first stated by D'Alembert in his *Traité de Dynamique*, 1743. It is known as *D'Alembert's Principle.*

By integrating both members of the equations such as

$$\Sigma (m\ddot{x}) = \Sigma X$$

with respect to the time, between limits which correspond to the initial and final instants of any interval, we find such results as

$$\Sigma (m\dot{x})_{t=t_1} - \Sigma (m\dot{x})_{t=t_0} = \Sigma \int_{t_0}^{t_1} X\, dt,$$

or, in words:—The change of momentum of the system in any direction is equal to the sum of the impulses of the external forces resolved in that direction.

163. Motion of the centre of mass. Since the resultant kinetic reaction of a system is the kinetic reaction of a particle of mass equal to the mass of the system placed at the centre of mass and moving with it, we see that

$$\ddot{\bar{x}}\Sigma m = \Sigma X, \quad \ddot{\bar{y}}\Sigma m = \Sigma Y, \quad \ddot{\bar{z}}\Sigma m = \Sigma Z,$$

so that the centre of mass moves like a particle, of mass equal to the mass of the system, under the action of the vector resultant of all the external forces applied to the system.

164. Motion relative to the centre of mass. In the equations such as $\Sigma [m(y\ddot{z} - z\ddot{y})] = \Sigma (yZ - zY)$ put $x = \bar{x} + x'$, ... The left-hand member of the equation just written becomes

$$[(\bar{y}\ddot{\bar{z}} - \bar{z}\ddot{\bar{y}})\, \Sigma m] + \Sigma \{m(y'\ddot{z}' - z'\ddot{y}')\},$$

and the right-hand member becomes

$$[\bar{y}\Sigma Z - \bar{z}\Sigma Y] + \Sigma (y'Z - z'Y).$$

The terms in square brackets in the two members are equal, and we thus have such equations as

$$\Sigma \{m(y'\ddot{z}' - z'\ddot{y}')\} = \Sigma (y'Z - z'Y).$$

These can be stated in words:—The rate of increase (per unit of time) of the moment of momentum in the motion relative to G, about any line through G, is equal to the sum of the moments of the external forces about the same line.

165. Independence of translation and rotation. From the results of the last two Articles we see that the motion of the centre of mass is determined by the external forces independently of any motion relative to the centre of mass, and the motion relative to the centre of mass is determined independently of the motion of the centre of mass.

166. Conservation of Momentum. When the resultant external force on a system has no resolved part parallel to a particular line, the sum of the resolved parts of the kinetic reactions of the particles parallel to that line is zero. Hence the rate of increase (per unit of time) of the resolved part of the resultant momentum of the system parallel to that line is zero, or the resolved part of the resultant momentum parallel to the line is constant.

In such a case the resolved part, parallel to the line, of the velocity of the centre of mass is constant.

167. Conservation of moment of momentum. When the sum of the moments of the external forces about any fixed axis vanishes, the sum of the moments of the kinetic reactions about that axis vanishes, and the moment of momentum of the system about the axis is constant.

When the sum of the moments of the external forces about an axis, drawn in a fixed direction through the centre of mass, vanishes, the moment of momentum about that axis in the motion relative to the centre of mass is constant.

168. Sudden changes of motion. As in Art. 160, let $X + X'$ be the sum of the resolved parts parallel to the axis x of all the forces, external and internal, that act on a particle m; and, as in Art. 82, suppose that X and X' do not remain finite at time t, but that the impulses of X and X' are finite, or that $\underset{\cdot}{X}$ and $\underset{\cdot}{X}'$, defined by the equations

$$\mathrm{Lt}_{\tau=0} \int_{t-\frac{1}{2}\tau}^{t+\frac{1}{2}\tau} X\,dt = \underset{\cdot}{X}, \quad \mathrm{Lt}_{\tau=0} \int_{t-\frac{1}{2}\tau}^{t+\frac{1}{2}\tau} X'dt = \underset{\cdot}{X}',$$

are finite. Let $\dot{x}$ and $\dot{\xi}$ be the resolved parts parallel to the axis x of the velocity of m just after the instant t and just before this instant respectively. Then we have the equation

$$m(\dot{x} - \dot{\xi}) = \underset{\cdot}{X} + \underset{\cdot}{X}'.$$

In like manner the impulsive changes of velocity parallel to the axes y and z will be determined by equations which may be written

$$m(\dot{y} - \dot{\eta}) = \underset{.}{Y} + \underset{.}{Y}',$$
$$m(\dot{z} - \dot{\zeta}) = \underset{.}{Z} + \underset{.}{Z}'.$$

Now it follows from the law of internal action (Art. 161) that $\Sigma \underset{.}{X}', \ldots$ and $\Sigma (y\underset{.}{Z}' - z\underset{.}{Y}'), \ldots$ vanish. Hence we have the equations

$$\Sigma [m(\dot{x} - \dot{\xi})] = \Sigma \underset{.}{X}, \ldots,$$
$$\Sigma [m\{y(\dot{z} - \dot{\zeta}) - z(\dot{y} - \dot{\eta})\}] = \Sigma (y\underset{.}{Z} - z\underset{.}{Y}), \ldots.$$

These equations can be expressed in words in the statements:—

(1) The change of momentum of the particle G in any direction is equal to the sum of the resolved parts of the external impulses in that direction.

(2) The change of the moment of momentum of the system about any axis is equal to the sum of the moments of the external impulses about that axis.

169. Work done by the force between two particles. Let x_1, y_1, z_1 and x_2, y_2, z_2 denote the coordinates of the two particles at time t, and r the distance between them, so that

$$r^2 = (x_1 - x_2)^2 + (y_1 - y_2)^2 + (z_1 - z_2)^2.$$

Also let F denote the magnitude of the force between them, and, for definiteness, take this force to be repulsive. The components parallel to the axes of the forces exerted on the particles 1 and 2 respectively are

$$F\frac{x_1 - x_2}{r}, \quad F\frac{y_1 - y_2}{r}, \quad F\frac{z_1 - z_2}{r},$$

and
$$F\frac{x_2 - x_1}{r}, \quad F\frac{y_2 - y_1}{r}, \quad F\frac{z_2 - z_1}{r}.$$

The rate (per unit of time) at which the first force does work is

$$F\frac{x_1 - x_2}{r}\dot{x}_1 + F\frac{y_1 - y_2}{r}y_1 + F\frac{z_1 - z_2}{r}\dot{z}_1,$$

and the rate at which the second force does work is

$$F\frac{x_2 - x_1}{r}\dot{x}_2 + F\frac{y_2 - y_1}{r}\dot{y}_2 + F\frac{z_2 - z_1}{r}\dot{z}_2.$$

Hence the sum of the rates at which the two forces do work is

$$\frac{F}{r}[(x_1 - x_2)(\dot{x}_1 - \dot{x}_2) + (y_1 - y_2)(\dot{y}_1 - \dot{y}_2) + (z_1 - z_2)(\dot{z}_1 - \dot{z}_2)],$$

or $\quad F\dot{r}$.

The work done in any displacement is the value of the integral

$$\int F\dot{r}\,dt \text{ or } \int F\,dr,$$

taken between limits which correspond to the positions of the particles before and after the displacement.

If the distance between the particles remains unaltered throughout the motion, no work is done by the force between them; but if the distance varies, the internal force does work.

170. Work function. We form as in Art. 86 the work done by all the forces acting on any particle of a system as the particles move from their positions at time t_0 to their positions at time t. The expression for the sum of the works of all the forces acting on all the particles may be written

$$\Sigma \int_{t_0}^{t} \{(X+X')\dot{x} + (Y+Y')\dot{y} + (Z+Z')\dot{z}\}\,dt,$$

where the summation extends to all the particles.

When this expression has the same value for all paths joining the initial and final positions of the particles, it is a function of the coordinates of the final positions, the initial positions being prescribed. This function is the "work function."

We refer to the prescribed initial positions as constituting the "standard position."

It is important to observe that the work done by the internal forces may not in general be omitted from the sum.

When a work function exists the system is said to be "conservative."

171. Potential Energy. The work function in any position A with its sign changed is the work that would be done by the forces if the system passed from the position A to the standard position. It is defined to be the *Potential Energy* of the system in the position A.

For the sake of precision we present our previous statements in the following form:—A system in which the work done by all the forces on all the particles, as they pass from one set of positions to another, is independent of the paths of the particles, is said to be a conservative system; and the work done by the forces of such a system, as its particles pass from any set of positions to a prescribed standard set of positions, is called the potential energy of the system in the former set of positions.

172. Potential energy of gravitating system. When the force between two particles of masses m, m' is an attraction $\gamma mm'/r^2$, the work done in a displacement by which the distance r between them changes from r_0 to r_1 is

$$\int_{r_0}^{r_1} -\gamma \frac{mm'}{r^2}\, dr,$$

and this is

$$\gamma mm' \left(\frac{1}{r_1} - \frac{1}{r_0}\right).$$

Hence in a gravitating system the work done in any displacement is

$$\gamma \Sigma \left(\frac{mm'}{r_1} - \frac{mm'}{r_0}\right),$$

where the summation extends to all the pairs of particles.

If we choose the standard position to be that in which all the distances are infinite, the value of the work function in any other position is

$$\gamma \Sigma \frac{mm'}{r},$$

and the potential energy in this position is

$$-\gamma \Sigma \frac{mm'}{r}.$$

The negative sign indicates that there is less potential energy in any other state than there is in the state of infinite diffusion.

173. Energy equation. From the equations of the type

$$m\ddot{x} = X + X'$$

we form the equation

$$\Sigma \left[m\left(\dot{x}\ddot{x} + \dot{y}\ddot{y} + \dot{z}\ddot{z}\right)\right] = \Sigma \left[(X + X')\,\dot{x} + (Y + Y')\,\dot{y} + (Z + Z')\,\dot{z}\right],$$

of which the left-hand member may be written

$$\frac{d}{dt}\left\{\tfrac{1}{2}\Sigma \left[m\left(\dot{x}^2 + \dot{y}^2 + \dot{z}^2\right)\right]\right\}.$$

We deduce the result that the rate of increase (per unit of time) of the kinetic energy of the system is equal to the rate at which work is done by all the forces internal and external; and consequently we deduce the result that the increment of kinetic energy in any displacement is equal to the sum of the works done by all the forces.

When a work function exists this result gives us an integral of the equations of motion, and this integral can be written in the form

kinetic energy + potential energy = const.

This integral of the equations of motion is called the "energy equation."

The work done by the internal forces may not, in general, be omitted. Examples are furnished by a system of gravitating particles, or by the tension in an elastic string.

In some cases the motion of some part of a system is assigned. For example the system may contain a rigid body which is made to rotate uniformly about an axis. In such cases force must usually be applied to the system in order to maintain the assigned motion, and the required force usually does work. It follows that in such cases there is not, in general, an energy equation.

174. Kinetic Energy produced by Impulses. As in Art. 168 let $\dot{x}$, $\dot{y}$, $\dot{z}$ be the resolved parts parallel to the axes of the velocity of the particle of mass m just after an impulse, $\dot{\xi}$, $\dot{\eta}$, $\dot{\zeta}$ the similar resolved parts of the velocity just before the impulse, $\underset{.}{X}$, $\underset{.}{Y}$, $\underset{.}{Z}$ the sums of the resolved parts parallel to the axes of the external impulses applied to m, $\underset{.}{X}'$, $\underset{.}{Y}'$, $\underset{.}{Z}'$ the sums of the similar resolved parts of the internal impulses, T and T_0 the kinetic energies of the system just after and just before the impulses.

We have such equations as

$$m(\dot{x} - \dot{\xi}) = \underset{.}{X} + \underset{.}{X}'.$$

$$\begin{aligned} \text{Also } T - T_0 &= \tfrac{1}{2}\Sigma\,[m(\dot{x}^2 + \dot{y}^2 + \dot{z}^2)] - \tfrac{1}{2}\Sigma\,[m(\dot{\xi}^2 + \dot{\eta}^2 + \dot{\zeta}^2)] \\ &= \tfrac{1}{2}\Sigma\,[m(\dot{x} - \dot{\xi})(\dot{x} + \dot{\xi}) + \text{two similar terms}] \\ &= \Sigma\,[(\underset{.}{X} + \underset{.}{X}')\,\tfrac{1}{2}(\dot{x} + \dot{\xi}) + \text{two similar terms}]. \end{aligned}$$

Thus, *the change of kinetic energy produced by impulses is the sum of the products of all the impulses and the arithmetic means of the velocities, in their directions, of the particles to which they are applied, just before and just after the impulsive actions.*

It is very important to notice that the internal impulses may not be omitted from the equation here obtained, just as the internal forces may not be omitted from the energy equation of Art. 173.

The Problem of the Solar System

175. The problem of n bodies. As we have already explained, the bodies of the Solar system can be treated as a system of particles moving under their mutual gravitation. The mathematical problem of integrating the equations of motion of such a system of particles, supposed to be n in number, is known as the "problem of n bodies." The particular cases of two and three bodies are known as the "problem of two bodies" and the "problem of three bodies." The only one of these problems which has been solved completely is the problem of two bodies.

176. The Problem of Two Bodies*. *Two particles which attract each other according to the law of gravitation are projected in any manner. It is required to show that the relative motion is parallel to a fixed plane, and that the relative orbits are conics, and to determine the periodic time when the orbits are elliptic.*

The principle of the conservation of momentum shows that the centre of mass of the two particles moves uniformly in a straight line. The accelerations of the particles, and the velocity of either relative to the other, are unaltered, if we refer them to a frame whose axes are parallel to those of the original frame of reference, and whose origin is at the centre of mass. We shall suppose this to be done.

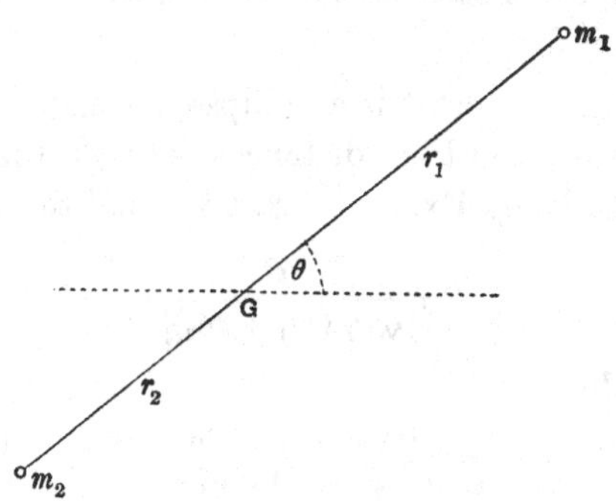

Fig. 46.

Then the acceleration of each particle is in the line joining it to the origin, and the velocities of the particles are localized in lines which lie in a plane containing the origin; the motion of each particle therefore takes place in this plane.

Now let G be the centre of mass, m_1, m_2 the masses of the particles, r_1, r_2 their distances from G at time t, θ the angle which

* The Problem of Two Bodies was solved by Newton, *Principia*, Lib I, Sect. XI, Props. 57—63.

the line joining them makes with any fixed line in the plane of motion, also let $r, = r_1 + r_2$, be the distance between the particles at time t. The force between them is $\gamma m_1 m_2/r^2$.

Then the equations of motion of m_1 are

$$\left.\begin{aligned} m_1(\ddot{r}_1 - r_1\dot{\theta}^2) &= -\gamma m_1 m_2/r^2, \\ m_1 \frac{1}{r_1}\frac{d}{dt}(r_1{}^2\dot{\theta}) &= 0 \end{aligned}\right\}.$$

Since $r_1 = m_2 r/(m_1 + m_2)$, these equations become

$$\left.\begin{aligned} \ddot{r} - r\dot{\theta}^2 &= -\gamma(m_1 + m_2)/r^2, \\ \frac{d}{dt}(r^2\dot{\theta}) &= 0 \end{aligned}\right\},$$

and it is clear that the equations of motion of m_2 would lead us to the same two equations.

The equations last written show that the acceleration of m_1 relative to m_2, or of m_2 relative to m_1, is $\gamma(m_1 + m_2)/r^2$, and that there is no transverse acceleration. Thus either particle describes a central orbit about the other with acceleration varying inversely as the square of the distance, and, by Art. 51, this orbit is a conic described about a focus.

Further, when the orbit is an ellipse, its major axis, $2a$, is the sum of the greatest and least distances between the particles, and the periodic time is, by Ex. 5 of Art. 48, equal to

$$2\pi \frac{a^{\frac{3}{2}}}{\sqrt{\{\gamma(m_1 + m_2)\}}}.$$

177. Examples.

1. If the particles are projected with velocities v, v' in directions containing an angle a from points whose distance apart is R, prove that the relative orbit is an ellipse, parabola, or hyperbola according as

$$v^2 + v'^2 - 2vv' \cos a < = \text{ or } > 2\gamma(m_1 + m_2)/R.$$

2. S, P, and E denote the masses of the Sun, a Planet, and the Earth; the major axis of the Planet's orbit is k times that of the Earth's orbit, and its periodic time is n years; prove, neglecting the mutual attractions of the Planets, that

$$n^2 = k^3(S+E)/(S+P).$$

[Kepler's Third Law of Planetary motion quoted in Art. 146 states that $n^2 = k^3$ approximately. Kepler's law is approximately correct because S is great compared with P or E.]

3. Two gravitating spheres of masses m, m', and radii a, a', are allowed to fall together from a position in which their centres are at a distance c, it is required to find the time until they are in contact.

We may suppose the centre of mass to be at rest, and take x for the distance between the centres of the spheres at time t. Then their velocities are equal in magnitude to

$$\frac{m'\dot{x}}{m+m'}, \text{ and } \frac{m\dot{x}}{m+m'}.$$

Hence the kinetic energy of the system is

$$\tfrac{1}{2}m\left(\frac{m'\dot{x}}{m+m'}\right)^2+\tfrac{1}{2}m'\left(\frac{m\dot{x}}{m+m'}\right)^2=\tfrac{1}{2}\frac{mm'}{m+m'}\dot{x}^2.$$

The potential energy, measured from the position in which the distance was c as standard position, is (see Art. 172)

$$\gamma mm'\left(\frac{1}{c}-\frac{1}{x}\right).$$

Hence the energy equation is

$$\dot{x}^2=2\gamma(m+m')\left(\frac{1}{x}-\frac{1}{c}\right),$$

and the time required is

$$\frac{1}{\sqrt{\{2\gamma(m+m')\}}}\int_{a+a'}^{c}\sqrt{\frac{cx}{c-x}}\,dx.$$

If then we find an angle θ such that $a+a'=c\cos^2\theta$, we shall have for the required time

$$\frac{c^{\frac{3}{2}}(\theta+\sin\theta\cos\theta)}{\sqrt{\{2\gamma(m+m')\}}}.$$

4. Two gravitating spheres, masses m, m', moving freely with relative velocity V when at a great distance apart, would, in the absence of gravitation, pass each other at a minimum distance d. Prove that the relative orbits are hyperbolic, and that the direction of the relative velocity will ultimately be turned through an angle

$$2\tan^{-1}\{V^2d/\gamma(m+m')\}.$$

5. Prove that, if two bodies of masses E and M move under their mutual gravitation and that of a fixed body of mass S, so that the three are always in a fixed plane, then

$$(E+M)^2H+EMh=\text{const.},$$

where h is the rate at which M describes area about E, and H is the rate at which the centre of mass of E and M describes area about S.

Prove that, if all three bodies are free, the equation becomes

$$S(E+M)^2H+(S+E+M)EMh=\text{const.}$$

178. General problem of Planetary motion. In the general case of a system of particles moving under their mutual gravitation we know seven first integrals of the equations of motion. The principle of the Conservation of Momentum gives us three integrals representing the result that the velocity of the centre of mass in any direction is constant. The principle of the Conservation of Moment of Momentum gives us three integrals representing the result that the moment of momentum of the system about any axis drawn in a fixed direction through the centre of mass is constant. The energy equation also is an integral of the equations of motion.

Even in the case of three particles these integrals do not suffice for a complete description of the motion. Except in particular circumstances of projection, no other first integral has, so far, been obtained.

Thus we cannot deduce from the law of gravitation an exact account of the motions of the bodies forming the Solar system. But there are a number of circumstances which conduce to the possibility of deducing from this law such an approximate account of the motions in question as shall be sufficiently exact to agree with observation over a long period of time. Among these we may mention (1) that the mass of the Sun is great compared with that of the other bodies, even the mass of Jupiter being less than $\frac{1}{1000}$th part of that of the Sun, (2) that all the orbits are nearly circular, and all but those of a few Satellites lie nearly in one plane.

It would be outside the scope of this book to explain how these special circumstances can be utilized for the purpose of integrating approximately the equations of motion of the bodies of the Solar system. For this we must refer to books on gravitational Astronomy. A comprehensive treatise is F. Tisserand's *Traité de Mécanique céleste*, tt. 1—4, Paris, 1889–1896.

BODIES OF FINITE SIZE

179. Theory of the motion of a body. We deal with the motion of a body in the same way as with the motion of a system of particles. If the body is divided in imagination into a very large number of very small compartments, and a particle is supposed to be placed in each compartment, the motion of the body is determined when the motions of all the particles are known.

We suppose that the particles move under the actions of forces obeying the law of reaction.

We adjust the masses of the particles so that the sum of the masses of those particles which are in any part of the body shall be equal to the mass of that part of the body. This comes to the same thing as taking the mass of a particle, in any compartment, to be equal to the product of the volume of the compartment and the density of the body in the neighbourhood.

In general we do not attempt to determine the forces between the particles, but we assume that they are adjusted so as to secure the satisfaction of certain conditions. For example, when the body is regarded as *rigid*, we assume that they are adjusted so that the distance between any two particles is invariable. When the body is a *string* or *chain*, we assume that the forces between particles situated on the two sides of a plane, drawn at right angles to the line of the chain, are equivalent to a single force directed along this line. This force is the *tension* of the chain. A more general discussion will be given in Chapter XI.

The *centre of mass* of a body is found by a limiting process from the formulæ of Art. 152. It coincides with the centre of gravity of the body, as determined in books on Statics.

The *momentum* of a body is equivalent to a certain resultant momentum and a certain moment of momentum. The resultant momentum is that of a particle, of mass equal to the mass of the body, placed at the centre of mass and moving with it. The moment of momentum about any axis through the centre of mass is the sum of the moments about that axis of the momenta of the particles relative to the centre of mass.

Like statements hold for the *kinetic reaction*.

The *kinetic energy* of the body is equal to the kinetic energy of a particle, of mass equal to the mass of the body, placed at the centre of mass and moving with it, together with the kinetic energy of the motion relative to the centre of mass.

The *equations of motion* of the body express the statements that the resolved part of the resultant kinetic reaction in any direction is equal to the sum of the resolved parts of the external forces in the same direction, and the moment of kinetic reaction about any axis is equal to the sum of the moments of the external forces about the same axis.

The *equations of motion of any part* of the body are formed in the same way. The forces exerted upon this part of the body across the surface which separates it from the rest of the body are now "external" forces acting on the part in question. The gravitational attractions between particles within the surface and particles outside it are also "external" forces acting on the part within the surface.

The rate (per unit of time) at which the kinetic energy of a body increases is equal to the sum of the rates at which work is done by all the forces external and internal. If the work done can be specified by a "work function" there is an *energy equation*, which is an integral of the equations of motion.

180. Motion of a rigid body. Solid bodies often move in such a way that no apparent change of size or shape takes place in any part of them. To represent the motions of such bodies by those of systems of particles we subject the internal forces between the hypothetical particles to the condition that the distance between any two of the particles is to be maintained invariable.

The system of particles subjected to this condition is said to represent a "rigid body."

The motion of a rigid body is determined when the motion of three of its particles is determined. For the three particles determine a frame of reference relatively to which all the particles of the body have invariable positions.

To determine the positions of all the particles of a rigid body relative to a frame is therefore the same thing as determining the position of one frame, F, relative to another. This requires the determination of the positions of the origin of the frame F, of one of its lines of reference, and of a plane through that line. The position of a point depends on three quantities, the coordinates of the point. The position of a line through a point depends on two quantities, since the line may make any angle with one of the axes, and the plane through it parallel to that axis may make any angle with a coordinate plane, but these two angles determine the line. The position of a plane through a line depends on one quantity, which may be taken to be the angle it makes with the plane passing through the line and parallel to one of the axes of reference. Thus the positions of all the particles of a rigid body relative to a frame are determined when six quantities such as those specified are given.

When a rigid body moves without rotation, the motion of the body is determined by that of a fictitious particle, of mass equal to the mass of the body, placed at the centre of mass and moving with it. The equations of motion of this particle are the same as if all the external forces acting on the body were applied at the centre of mass, their magnitudes, directions and senses being unaltered.

181. Transmissibility of force. The motion of every part of a rigid body is known when the motion of any part of it is known.

Now the equations of motion of the body involve the external forces by containing the sums of the resolved parts of these forces in assigned directions and the sums of the moments of these forces about assigned axes. The forces do not enter into the equations in any other way.

The resolved parts and moments in question depend upon the lines of action of the forces, but not upon their points of application.

Hence the forces may be supposed to act at any points in their lines of action without altering the motion of the body, or of any part of the body.

In the cases of a deformable body and a system of isolated particles, it is manifest that the internal relative motion of the parts of the body or system would be altered by transferring the point of application of a force from one particle to another in the line of action of the force.

We conclude that a force *acting on a rigid body* may be regarded as a vector localized in a line instead of a vector localized at a point. This result is sometimes called the *Principle of the transmissibility of force.*

182. Forces between rigid bodies in contact. The surfaces of two rigid bodies may be regarded as touching at a single point, and the action between the two bodies (apart from their mutual gravitation) may be regarded as consisting of a pair of equal and opposite forces applied at the point of contact.

The force which one of the bodies A exerts upon the other B at the point of contact can be resolved into components along and perpendicular to the common normal. The normal component is the "pressure" of A on B, and the tangential component is the "friction" of A on B. The resultant of the pressure and friction is often called the "total reaction."

In the system of two bodies in contact the pressure does no work; for, so long as the bodies remain in contact, the parts in contact have the same velocity in the direction of the normal, and the pressures acting upon the two bodies are equal and opposite. In general, the pressure does (positive or negative) work on both bodies, and the sum of the rates (per unit of time) at which it does work on the two is zero.

183. Friction. Let P be the point of contact of two bodies A, B, and let R denote the pressure and F the friction.

Each of the bodies is regarded as having a particle at P.

The particle of A at P will have a certain velocity, and similarly for the particle of B at P. The velocity of the particle of A at P, relative to axes parallel to the axes of reference drawn through the particle of B at P, is the velocity of the point of contact, considered as a point of A, relative to B. In like manner there is an equal and opposite velocity of the point of contact, considered as a point of B, relative to A.

The condition of continued contact is that the relative velocity just described is localized in a line in the tangent plane at P, or that the resolved part of this velocity in the direction of the common normal vanishes.

The *first law of Friction* is that the friction acting upon $\begin{Bmatrix} A \\ B \end{Bmatrix}$ at P is opposite in sense to the velocity of the point of contact, considered as a point of $\begin{Bmatrix} A \\ B \end{Bmatrix}$, relative to $\begin{Bmatrix} B \\ A \end{Bmatrix}$.

The *second law of Friction* is that the friction F and the pressure R are connected by a relation of inequality $F \leqslant \mu R$, where μ is a constant depending only on the materials of which the bodies are composed. The constant μ is called the *coefficient of friction.*

When the relative velocity above described is zero, the motion is described as *rolling.* In order that rolling may take place it is generally necessary that the coefficient of friction should exceed a certain number depending on the circumstances of the case. A motion of two bodies in contact which is not one of pure rolling is described as a motion of *sliding* or *slipping.* The rule for the direction of friction may be stated in the form:—Friction tends to prevent slipping. When slipping takes place $F = \mu R$. When the bodies are sufficiently rough to prevent slipping throughout the motion they are sometimes said to be *perfectly rough.*

When the motion is one of rolling, the friction does no work on the system of two bodies, but it may do (positive or negative) work on each of the bodies; and then the sum of the rates (per unit of time) at which it does work on the two is zero.

When the motion is one of sliding, the friction does work on the system, and this work is always negative.

184. Potential energy of a body. For a body under the gravitational attractions of other bodies, and regarded as made up of particles, the external forces X, Y, Z of Art. 160 do work in any displacement; and this work is specified by means of a work function.

Further the work done by those components of the internal forces, which represent the mutual gravitation of the parts of the body, is also specified by means of a work function. The other internal forces may also do work, and this work may also be specified by a work function. When this is the case the portion of the potential energy, corresponding to this work function, represents what may be called "internal potential energy."

In such a case the potential energy is divisible into three parts: potential energy of the body in the field of the external attraction, potential energy of the mutual gravitation of the parts of the body, and internal potential energy.

The potential energy of a body in the field of the Earth's gravity is represented by the expression

$$\Sigma mgz,$$

where m denotes the mass of any of the hypothetical particles, and z is the height of that particle above a fixed level. This expression is equal to

$$Mg\bar{z},$$

where M is the mass of the body, and $\bar{z}$ is the height of its centre of mass above the fixed level.

185. Energy of a rigid body. It follows from the result ot Art. 169 that the internal forces between the particles of a rigid body never do any work.

The potential energy of the mutual gravitation of the parts of a rigid body and the internal potential energy of the body can both be taken to be zero by choosing the actual state of aggregation of the body as the "standard" state.

The kinetic energy of the body and the potential energy of the body in the field of external force are variable quantities.

The equations of motion of a rigid body do not always possess an integral in the form of an energy equation. For the body may be in contact with other rigid bodies, or with deformable bodies such as elastic strings, or with resisting media such as the air; and the forces exerted upon the rigid body by bodies with which it is in contact may do work which is not specified by a work function.

186. Potential energy of a stretched string. Consider a portion of the string of natural length l_0, and let its extension be ϵ, so that its length is $l_0(1+\epsilon)$. Its tension is $\lambda\epsilon$, where λ is the modulus of elasticity. For the purpose of calculating the potential energy we may regard this portion as having one end fixed, and the other attached to a body, which exerts upon it a tension $\lambda\epsilon$, and we may also regard the portion as free from the action of any other external forces. Now let the string be extended further. The rate at which the terminal tension does work (per unit of time) is $\lambda\epsilon\,.\,l_0\dot{\epsilon}$, for $l_0\dot{\epsilon}$ is the velocity of the moving end. Hence the work done in the extension of the string from its natural length to the length $l_0(1+\epsilon)$ is

$$\int \lambda\epsilon\,.\,l_0\dot{\epsilon}\,dt.$$

The integral is taken between limits which correspond to the values 0 and ϵ of the extension, and its value is $\frac{1}{2}\lambda l_0\epsilon^2$.

We may regard the string as being extended so slowly that no sensible kinetic energy is imparted to it. Then the work done by the internal forces together with that done by the external forces vanishes. It follows that the work done by the internal forces is $-\frac{1}{2}\lambda l_0\epsilon^2$.

Since this amount depends only on the initial and final states we can regard it, with changed sign, as an amount of internal potential energy (Art. 184). Hence the potential energy of a portion of a stretched string, which is of natural length l_0, is $\frac{1}{2}\lambda l_0\epsilon^2$, when its extension is ϵ.

A similar result holds for a spring, whether extended or contracted (cf. Art. 101).

When the string is not stretched uniformly, let s_0 be the natural length of any portion measured from one end, $s_0+\Delta s_0$ that of a slightly longer portion, and let s, $s+\Delta s$ be what these lengths become when the string is stretched. Then we define the *extension at the point* corresponding to s_0 to be

$$\text{Lt}_{\Delta s_0=0}\frac{\Delta s-\Delta s_0}{\Delta s_0}.$$

If this is denoted by ϵ, the potential energy of any portion between $s_0=a$ and $s_0=b$ is

$$\int_a^b \tfrac{1}{2}\lambda\epsilon^2 ds_0.$$

187. Localization of Potential Energy. The potential energy of a gravitating system and the potential energy of a stretched string are two examples of the potential energy that arises from internal forces between the parts of a system.

But the two cases present a marked difference. In the case of the string we are able to assign a certain amount of the potential energy to each piece of the string, in such a way that the amount so assigned corresponds to the state of that piece. We may therefore say that the energy is *located* in the string, so much being located in each piece. The amount located in any piece can be expressed as $\frac{1}{2}\lambda\epsilon^2$ per unit of length (in the natural state), ϵ denoting the extension at any point of the piece. We can think of this energy as *possessed by the piece* of string, in the same way as kinetic energy is possessed by a moving body.

In the case of the gravitating system we are not able to assign a certain amount of the potential energy to any part of the system in such a way that changes of the energy so assigned correspond to changes in the state of that part, independently of changes in the position of the part relative to other parts. We cannot, in any way that shall be completely satisfactory, locate some portion of the energy in one part of the system, another portion in another part of the system, and so on. For instance, in the case of a heavy body near the Earth's surface we cannot locate the energy in the body, or in the Earth, or in any definite proportion some of it in the body and some in the Earth. We have to think of it as *possessed by the system*, not by the bodies composing the system.

188. Power. When work is done by the action of a system S upon a system S' the forces exerted by the particles of S upon the particles of S' do work in the displacements of the particles of S'. In cases where the energy can be localized, the energy of the system S' is increased, and that of S diminished, by a quantity equal to the amount of work so done. The number of units of work done in any interval bears a definite ratio to the number of units of time in the interval; and, when the interval is indefinitely short, this ratio has a limit, which is the rate at which work is being done per unit of time.

The *power* of a system acting on another system is the rate per unit of time at which the first system does work upon the second.

Corresponding to each force between particles of the two systems there is a certain power measured by the product of the magnitude of the force and the resolved part, in its direction, of the velocity of the particle on which it acts, or by the product of the magnitude of the velocity of the particle and the resolved part, in its direction, of the force exerted upon it; either of these products measures the rate at which the force does work. The sum of all these powers is the power of the first system acting on the second.

The power can be measured equally by the rate at which work is done upon the second system or by the rate at which the first system does work.

Thus, in any machine performing mechanical work, a certain amount of energy is expended, and an equal amount of work done, per unit of time; and the machine is said to be "working up to a power" measured by the rate at which the work is done. In general much of the work is done against friction.

189. Motion of a string or chain. In general we neglect the thickness of the chain, but suppose that the mass of any finite length of it is finite. When the mass of any portion is proportional to the length of the portion, the chain is *uniform*. When the chain is not uniform, the limit of the ratio of the number of units of mass in the mass of any portion to the number of units of length in the length of the portion, when the length is diminished indefinitely, is the mass per unit of length, or the *line-density*.

If a (geometrical) plane cuts the line of the chain at right angles at any point, the two parts of the chain which are separated by this plane act one on the other with a force directed along the line of the chain at the point. This force is the *tension* of the chain.

Let the chain be divided in imagination into a very large number of very short lengths. In each length let a particle be supposed to be placed, and let the mass of the particle be the mass of that length of the chain. Let each of the hypothetical particles act upon its next neighbours with a force adjusted in accordance with the law of reaction. The force between two neighbouring particles is taken to be equal to the tension of the chain at the corresponding point. The motion of the chain is determined by forming the equations of motion of any particle, and then passing to a limit by increasing the number of particles, and diminishing the lengths of the small portions of the chain, indefinitely.

If any of the short lengths is Δs, and if m is the line-density of the chain in the neighbourhood, $m\Delta s$ is the mass of the corresponding particle.

The tensions in the two directions from the particle to its two next neighbours are in general different, but the difference tends to zero with Δs.

The other forces acting on the hypothetical particles are the forces of the field, when the chain is in a field of force, and the pressure and friction of any curve or surface with which the chain is in contact.

190. String or chain of negligible mass in contact with a smooth surface. The chain lies in a curve drawn on the surface. We resolve the acceleration of any hypothetical particle of the chain in the direction of the tangent to this curve at the point occupied by the particle. We denote the resolved part of the acceleration by f. We resolve the force of the field in the same direction, and denote by F the force of the field per unit of mass in that direction. The pressure of the surface on the hypothetical particle is directed at right angles to the tangent to the curve at the point.

Let T be the tension of the chain at the point; and let T_1 and T_2 be the forces acting between the hypothetical particle and its two next neighbours, ϕ_1 and ϕ_2 the angles which their lines of action make with the tangent to the curve. In the limit

$$T_1 = T_2 = T \text{ and } \phi_2 = 0, \quad \phi_1 = \pi.$$

Resolve along the tangent to the curve for the motion of the hypothetical particle. Denoting the mass per unit of length by m, we have

$$m\Delta s \,.\, f = m\Delta s \,.\, F + T_2 \cos\phi_2 + T_1 \cos\phi_1,$$

or

$$mf = mF + \frac{T_2\cos\phi_2 - T_1}{\Delta s} + T_1\frac{1+\cos\phi_1}{\Delta s}.$$

The limiting form of this equation is

$$mf = mF + \frac{dT}{ds}.$$

If m is very small this equation is nearly the same as $\dfrac{dT}{ds} = 0$. Hence we conclude that, if the mass of the chain may be neglected, the tension is constant.

The result is proved for any portion of the chain which is in contact with a smooth surface. The form of the argument shows that it holds also for any portion which is free.

MISCELLANEOUS EXAMPLES

1. A thin spherical shell of small radius, moving without rotation, describes a circle of radius R with velocity V about a gravitating centre of force O; and, when its centre is at a point A, bursts with an explosion which generates velocity v in each fragment directly outwards from the centre of the shell. Prove that the fragments all pass through the line AO within a length

$$8V^3vR/(V^4 - 6V^2v^2 + v^4),$$

and that, if v is small, the stream of fragments will form a complete ring after a time approximately equal to $\frac{1}{3}\pi R/v$.

2. Two equal particles are under the action of forces tending to a fixed point and varying as the distance from that point, the force being the same at the same distance in each case; the particles also attract each other with a

different force varying as the distance between them; prove that the orbit of either particle relative to the other is an ellipse and the periodic time is $2\pi/\sqrt{(\mu+2\mu')}$, μ and μ' denoting the forces on unit mass respectively at unit distance.

3. A body, of mass km, describes an ellipse of eccentricity e and axis major $2a$ under the action of a fixed gravitating body of mass m. Prove that, if m is let go when the distance between the bodies is R, the eccentricity e' of the subsequent relative orbit is given by the equation

$$e'^2-e^2=\frac{k(1-e^2)}{(1+k)^2}\left\{k+2\left(1-\frac{a}{R}\right)\right\}.$$

4. Two gravitating particles of masses m, m' are describing relatively to each other elliptic orbits of eccentricity e and axis major $2a$, their centre of mass being at rest. Prove that, if m is suddenly fixed when the particles are at a distance R, the eccentricity e' of the orbit subsequently described by m' is given by the equation

$$(m+m')\left(\frac{2}{R}-\frac{m+m'}{am}\frac{1-e'^2}{1-e^2}\right)=m\left(\frac{2}{R}-\frac{1}{a}\right).$$

5. A body of mass M is moving in a straight line with velocity U, and is followed, at a distance r, by a smaller body of mass m moving in the same line with velocity u. The bodies attract each other according to the law of gravitation. Prove that the smaller body will overtake the other after a time

$$\left(\frac{r}{1+w}\right)^{\frac{3}{2}}\frac{\pi-\sqrt{(1-w^2)}-\cos^{-1}w}{\sqrt{\{\gamma(M+m)\}}},$$

where $$1-w=\frac{r(U-u)^2}{\gamma(M+m)}.$$

6. Two bodies, masses m, m', are describing relatively to each other circular orbits under their mutual gravitation, a and a' being their distances from the centre of mass. If V is the relative velocity, and m receives an impulse mV towards m', prove that the two bodies proceed to describe, relatively to the centre of mass, parabolas whose latera recta are $2a$ and $2a'$.

7. In a system of two gravitating bodies, M and m, initially M is at rest, and m is projected with velocity $\sqrt{\{\gamma(M+m)/d\}}$ at right angles to the line joining the bodies, d being the distance between the bodies. Prove that the path of M is a succession of cycloids and that M comes to rest at successive cusps after equal intervals of time.

8. In a system of two gravitating bodies of masses M and m the relative orbit is an ellipse of semi-axes a and b. Prove that, if the mass of the second body could be suddenly doubled, the eccentricity of the new orbit would be

$$\frac{1}{M+2m}\sqrt{\left[(M+2m)^2-\frac{b^2}{a^2}(M+m)(M+2m)-\frac{b^2}{a}\frac{m}{\gamma}v^2\right]},$$

where v is the relative velocity at the instant of the change.

9. Two gravitating particles, whose distance is r, are describing circles uniformly about their common centre of gravity with angular velocity ω, and a small general disturbance in the plane of motion is communicated to the system, so that after any time t the distance is $r+u$, and the line joining the particles is in advance of the position it would have occupied if the steady motion had not been disturbed by the angle ϕ; obtain the equation

$$2\dot{u} - r\omega\phi = 3\omega t(r\dot{\phi} + 2\omega u) + \text{const.},$$

squares of u and $\dot{\phi}$ being neglected.

10. Two equal particles P, Q are projected from points equidistant on opposite sides of a third particle S, with a velocity due to their distance under the attraction of S only. All three particles are gravitating, and the directions of projection are at right angles to PQ. If b is the conjugate axis of the orbit described by either P or Q, e its eccentricity, and b', e' those of the relative orbit of P and S (in the absence of Q), P being projected in the same manner as before, then $b = 2b'$, and

$$(1-e)/(1+e) = \tfrac{1}{4}(1-e')/(1+e').$$

11. If three bodies of masses m_1, m_2, m_3, subject only to their mutual attractions P_{23}, P_{31}, P_{12}, remain at constant distances from one another, those distances are in the ratios

$$m_1 P_{23} : m_2 P_{31} : m_3 P_{12}.$$

12. Three equal particles A, B, C, attracting each other with a force proportional to the distance and equal to μ per unit mass at unit distance, are placed at the corners of an equilateral triangle of side $2a$. The particle A is projected towards the centre of the triangle with velocity $c\sqrt{\mu}$, the other particles being set free at the instant of projection. Prove that the three particles will first be in a straight line after a time

$$\frac{1}{\sqrt{(3\mu)}} \sin^{-1} \frac{a}{\sqrt{(a^2 + \frac{1}{9}c^2)}}.$$

13. Two particles, each of unit mass, attracting each other with a force μ (distance), are placed in two rough straight intersecting tubes at right angles to each other and the friction is equal to the pressure on each tube: prove that, if they are initially at unequal distances from the point of intersection, one moves for a time $\frac{1}{2}\pi/\sqrt{\mu}$ before the other starts, and that, while they are approaching the point of intersection of the tubes, they move in the same manner as the projections of the two extremities of a diameter of a circle upon a straight line on which the circle rolls uniformly.

14. Two particles move in a medium, the resistance of which is proportional to the mass and the velocity, under the action of their mutua attraction, which is any function of their distance. Prove that their centre of mass either remains at rest or moves in a straight line with a velocity which diminishes in geometric progression as the time increases in arithmetic progression.

15. A particle placed at an end of the major axis of a normal section of a uniform gravitating elliptic cylinder of infinite length is slightly disturbed in the plane of the section. Prove that it can move round in contact with the cylinder, and that its velocity v when at a distance y from the major axis of the section is given by the equation

$$v^2=4\pi\gamma\rho y^2 a\,(a-b)/\{b\,(a+b)\},$$

where ρ is the density of the cylinder, and $2a$, $2b$ are the principal axes of a normal section.

16. A particle is projected along a circular section of the surface of a smooth uniform oblate spheroid given by the equation $(x^2+y^2)/a^2+z^2/c^2=1$. Prove that, if it describes the circle with uniform angular velocity ω under the attraction of the spheroid, then

$$\omega^2=(Aa^2-Cc^2)/a^2,$$

where Ax, Ay, Cz are the components of attraction of the spheroid at a point (x, y, z).

17. A ring moves on a rough elliptic wire, of semi-axes a, b, under the attraction of a thin uniform gravitating rod of mass M in the line of foci. Prove that, if it is projected from an end of the minor axis and comes to rest at the end of the major axis through which it first passes, the velocity x of projection is given by the equation

$$v^2=\frac{4\gamma M\mu a}{(a+b)^2}\int_0^{\pi}\frac{e^{-\mu\theta}\,d\theta}{1-2\alpha\cos\theta+\alpha^2},$$

where μ is the coefficient of friction, and $\alpha=(a-b)/(a+b)$.

APPENDIX TO CHAPTER VI

REDUCTION OF A SYSTEM OF LOCALIZED VECTORS

(*a*) **Vector couple.** Two equal vectors, localized in parallel lines, and having opposite senses, are said to form a "vector couple," or, briefly, a "couple."

Draw any line L at right angles to the plane of the couple, and choose a sense for this line. The sum of the moments (with their proper signs) of the two vectors about this line L is always the same, both in magnitude and in sign, whatever line L we take, so long as the chosen sense of the line L remains the same. This sum of moments is the *moment of the couple.* Its magnitude is the product of the measure of either vector of the couple and the measure of the perpendicular distance between the lines in which the vectors are localized. Its sign is determined when the sense of the line L is chosen. The rule of signs is the rule of the right-handed screw, and may be stated as follows:—If the line L meets one of the vectors, and the sense of the line L and that of the other vector are related like the senses of translation and rotation of a right-handed screw, the sign is $+$; otherwise, it is $-$.

When the sense of the line L is such that the moment is positive, a vector (unlocalized), of which the magnitude is the magnitude of the moment of the couple, and the direction and sense are those of the line L, is called *the axis of the couple.*

We shall obtain the result that a couple can be represented in all respects by this unlocalized vector.

(*b*) **Equivalence of couples in the same plane.** We shall prove that two couples in the same plane, of equal moments, in opposite senses, are equivalent to zero.

The lines in which the vectors are localized, being two pairs of parallel lines, form a parallelogram. Let this be $ABCD$ (Fig. 47).

Let the vectors of one couple be of magnitude P, and be localized in the lines AB, CD; and let the vectors of the other couple be of magnitude Q, and be localized in the lines AD, CB.

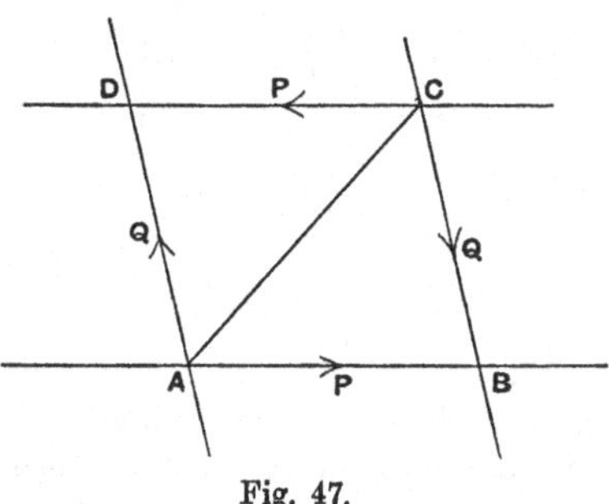

Fig. 47.

Let the unit of length be so chosen that AB represents P in magnitude

Then the area of the parallelogram is of magnitude equal to the moment of the couple.

Hence AD represents Q in magnitude.

Now the vectors P and Q localized in the lines AB, AD, and proportional to those lines, are equivalent to a vector localized in the line AC, and proportional to that line. The sense of this vector is AC.

Also the vectors P and Q localized in the lines CD, CB, and proportional to those lines, are equivalent to a vector localized in the line CA, and proportional to that line. The sense of this vector is CA.

It follows that the set of four vectors P, P and Q, Q are equivalent to zero.

This theorem shows that a couple may be replaced by any other couple in the same plane having the same moment and sense.

(c) **Parallel vectors.** Let P, Q be the magnitudes of two vectors localized in parallel lines, A, B any points on these lines, d the distance between the lines.

When P and Q are in *like senses*, let two vectors each of magnitude Q be introduced in the line of the vector P and in opposite senses. Then the vectors P and Q are equivalent to a vector of magnitude $P+Q$, localized in the line of P, and having the sense of P, and a couple of moment Qd. See Fig. 48. Replace the couple of moment Qd by two vectors, each of magnitude $P+Q$, localized in parallel lines, one of which is the line of P, and let the sense of the vector in this line be opposite to that of P. The line of the other vector is at a distance from the line of P which is equal to $Qd/(P+Q)$, it lies between the lines of P and Q, and the sense of the vector $P+Q$ in it is that of P or Q. See Fig. 49. The two vectors P and Q are equivalent to a single vector $P+Q$ in this line.

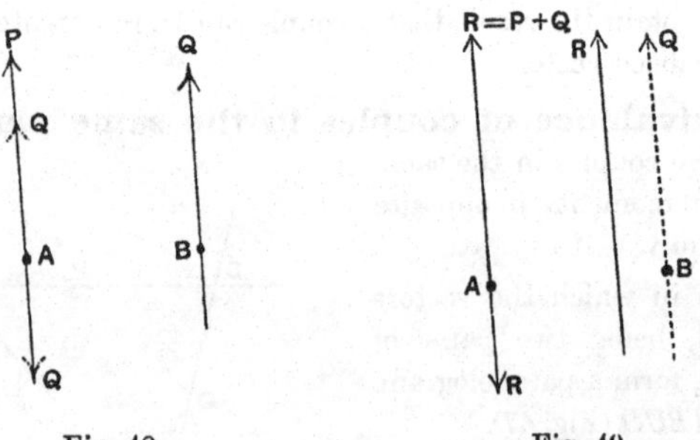

Fig. 48. Fig. 49.

When P and Q are in *unlike senses*, let Q be the greater. Introduce two vectors each of magnitude Q into the line of action of P. Then the vectors P and Q are equivalent to a vector of magnitude $Q-P$ localized in the line of P, and having the opposite sense to P, and a couple of moment Qd. See Fig. 50. Replace the couple of moment Qd by two vectors each of magnitude $Q-P$ localized in parallel lines, one of which is the line of P, and let the sense of the vector in this line be the same as that of P. The line of the other vector is at a distance from the line of P which is equal to $Qd/(Q-P)$, it lies on the side of the line of Q which is remote from the line of P, and the sense of the vector $Q-P$ in it is that of Q. See Fig. 51. The two vectors P and Q are equivalent to a single vector $Q-P$ in this line.

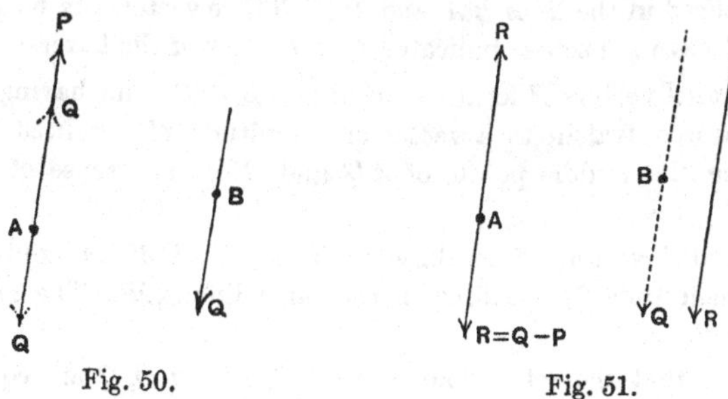

Fig. 50. Fig. 51.

Hence two vectors localized in parallel lines, when they are not equal and opposite, are equivalent to a resultant vector localized in a parallel line, and the moment of the resultant about any axis is equal to the sum of the moments of the components about the same axis.

(*d*) **Equivalence of couples in parallel planes.** We shall prove that two couples in parallel planes having equal moments and opposite senses are equivalent to zero.

Let the vectors of one couple be of magnitude P, and be localized in the lines AB, CD; and let the vectors of the other couple be of magnitude Q, and be localized in the lines $A'D'$, $C'B'$.

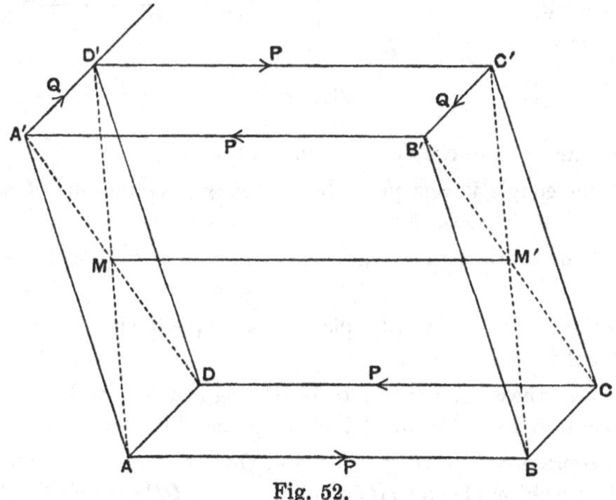

Fig. 52.

Through $A'D'$ and $B'C'$ let there pass a pair of parallel planes meeting the lines of the couple P in the points A, D, B, C.

Through AB and CD let there pass a pair of parallel planes meeting the lines of the couple Q in the points A', B', C', D'.

These two pairs of planes with the planes of the two couples form a parallelepiped.

Replace the couple Q in its plane by an equivalent couple consisting of vectors localized in the lines $B'A'$ and $D'C'$. These vectors are both of magnitude P, and have the senses indicated by the order of the letters.

Now parallel vectors P localized in lines AB, $D'C'$, and having the senses indicated, are equivalent to a vector of magnitude $2P$ localized in the line MM' joining the middle points of AD' and BC'. The sense of this vector is MM'.

Also parallel vectors P localized in lines CD, $B'A'$ are equivalent to a vector of magnitude $2P$ localized in the same line MM'. The sense of this vector is $M'M$.

It follows that the set of four vectors P, P and Q, Q are equivalent to zero.

This theorem shows that a couple may be replaced by any couple of the same moment in any parallel plane.

(*e*) **Composition of couples.**

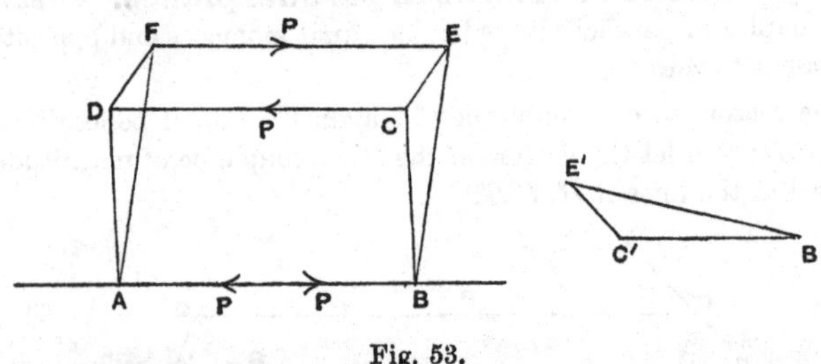

Fig. 53.

Let the planes of two couples meet in the line AB.

Replace the couple in one plane by any couple having one of its vectors localized in AB in the sense AB.

Let the two vectors be of magnitude P, and let the other be localized in the line CD.

Replace the couple in the other plane by a couple having one of its vectors localized in BA in the sense BA.

We can take these vectors also to be of magnitude P, and then the other will be localized in a certain line FE in the plane of the second couple.

Let AB represent P in magnitude, and through the points AB let there pass planes at right angles to AB cutting the lines CD and EF in the points named C, D, E, F.

Then the two couples are seen to be equivalent to a single couple, whose vectors are of magnitude P, and are localized in the lines CD, FE.

The figures $ABCD$, $ABEF$, $CDFE$ are rectangles, and their areas are proportional to the moments of the couples. These areas are in the ratios of the lengths of BC, BE, CE.

Hence if we turn the triangle BCE through a right angle in its plane its sides will be parallel and proportional to the axes of the couples. Let $B'C'E'$ be the new triangle. See Fig. 53. It is clear that, if $E'B'$ represents the axis of the second couple in sense, the sense of the first is $B'C'$, and the sense of the resultant is $E'C'$.

Thus the axis of a couple which has the magnitude, direction, and sense of a line $E'C'$ is the axis of the resultant of two component couples, the axes of the components having the magnitudes, directions, and senses of two lines $E'B'$ and $B'C'$. This is the vector law.

It follows from the preceding theorems that a couple can be regarded as an unlocalized vector represented by its axis.

(*f*) **Systems of localized vectors in a plane.** Let a vector of any magnitude P be localized in a line AB, and let O be any point not in the line AB. Through O draw a line parallel to AB, and let there be two vectors each of magnitude P and of opposite senses localized in this line. Then the system of vectors is equivalent to a vector localized in the line through O parallel to AB, of magnitude P, and having the sense of the original vector in AB, together with a couple of moment Pp, where p is the distance of AB from O. This couple has a definite sense, and its axis is perpendicular to the plane AOB.

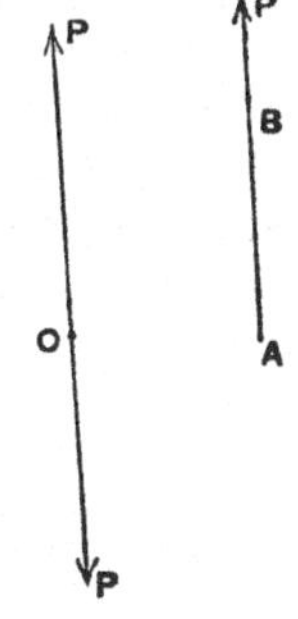

Fig. 54.

Any given system of vectors in a plane can in this way be replaced by a resultant vector localized in a line passing through a chosen point O in the plane together with a couple. The resultant vector is the resultant of vectors localized in lines through O, equal and parallel to the given vectors, and having the same senses as those vectors. The axis of the couple is perpendicular to the plane and its moment is $\Sigma(\pm Pp)$, where P is the magnitude of any one of the original vectors, p the perpendicular on its line from O, and the sign of each term is determinate.

Let R be the resultant of the vectors at O, and G the moment of the couple. If R is not zero, replace G by two localized vectors, each of magnitude R, one localized in the line of R through O and in the sense opposite to R, and the other in a parallel line at a distance G/R from O. The whole system is then equivalent to this last vector. See Fig. 55.

If R is zero the whole system is equivalent to the couple G.

If R and G are both zero the system is equivalent to zero.

Thus any system of vectors localized in lines lying in a plane is equivalent to a single vector localized in a line lying in the plane, or to a couple whose axis is perpendicular to the plane, or to zero.

The single vector or the couple, in the cases where the system is equivalent to a single vector or a couple, are determinate and unique.

The conditions of equivalence of two systems of vectors localized in lines lying in a plane are these: (1) When one system is equivalent to a single vector, the other is equivalent to a single vector, of the same magnitude and sense, localized in the same line. (2) When one system is equivalent to a couple, the other is equivalent to a couple, of the same magnitude and sense. (3) When one system is equivalent to zero, the other is equivalent to zero.

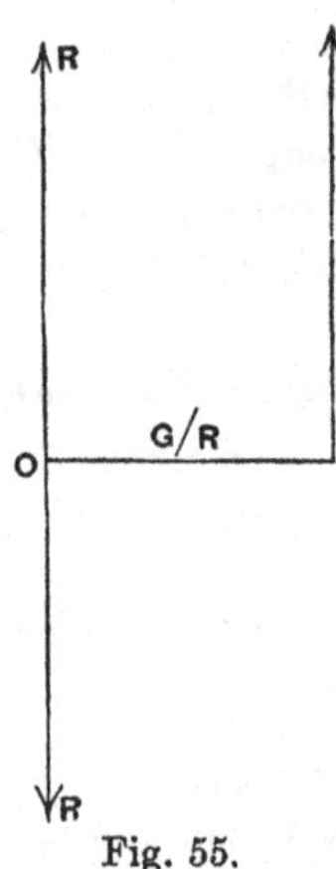

Fig. 55.

(*g*) **Reduction of a system of vectors localized in lines.** Take any origin O, and any rectangular axes of x, y, z. Let X, Y, Z be the resolved parts parallel to the axes of one of the vectors, and x, y, z the coordinates of a point on the line in which it is localized. Introduce a pair of equal and opposite vectors localized in a line through O parallel to the line of this vector, and resolve them into components localized in the axes. The magnitudes of these components are X, Y, Z. The original vector is thus replaced by vectors X, Y, Z localized in the axes, and by three couples about the axes, whose moments are

$$yZ-zY, \quad zX-xZ, \quad xY-yX$$

respectively. Cf. Art. 84.

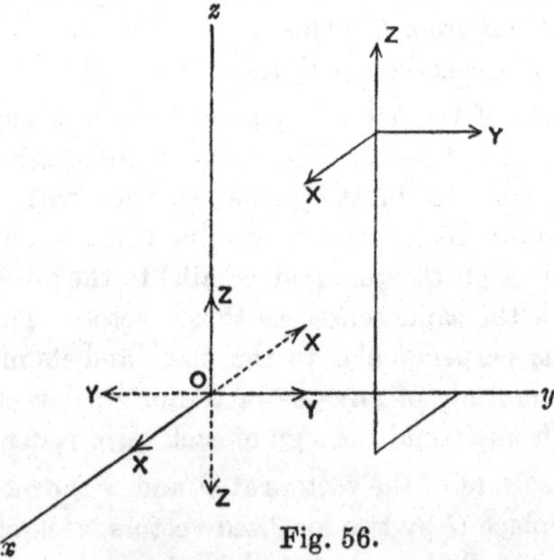

Fig. 56.

Hence any system of vectors localized in lines can be replaced by a resultant vector localized in a line through the origin, whose resolved parts parallel to the axes of coordinates are ΣX, ΣY, ΣZ, together with a couple equivalent to component couples about the axes, whose moments are $\Sigma(yZ-zY)$, $\Sigma(zX-xZ)$, $\Sigma(xY-yX)$, where X, Y, Z are the resolved parts of any one of the original vectors parallel to the axes, and x, y, z are the coordinates of any point on the line in which that vector is localized. The resultant vector, of which the components are ΣX, ..., is independent of the position of the origin; but the vector couple, of which the components are $\Sigma(yZ-zY)$, ..., takes different values for different origins.

CHAPTER VII

MISCELLANEOUS METHODS AND APPLICATIONS

191. WE propose in this Chapter to bring together a number of methods and theories relating to general classes of problems which can be solved by the principles laid down in previous Chapters. One of the great difficulties of our subject is the integration of the differential equations of motion of a system of bodies, but there are a number of cases in which all the desired information can be obtained without any integration. Such cases include sudden changes of motion, and initial motions, or the motions which ensue upon release from constraint. There are other cases in which the method of integration is known. Such cases include small oscillations, and problems in which the principles of energy and momentum supply all the first integrals of the equations of motion.

SUDDEN CHANGES OF MOTION

192. Nature of the action between impinging bodies. When two bodies collide, at first their surfaces come into contact at a point of each, but a little observation shows that, before separation, they must be in contact over a finite area; for example, if one body is smeared over with soot, the other, after separation, will show a sooty patch. It is clear therefore that during the impact the bodies undergo deformation. There are numberless cases in which the deformation is permanent, there are others in which the recovery of form is practically complete. Now it is clear that, if the bodies are rigid, no deformation can take place, and accordingly we shall be unable to give an account of the circumstances if we treat the bodies as rigid. On the other hand, the problem of calculating the deformation from the elastic properties of the bodies is generally beyond our power. Further, we shall find that one inevitable result of every impulsive action between parts of a system is a loss of kinetic energy in the system, and this apparent loss of energy can frequently be calculated. Nor have we far to seek for the form of energy that is developed in compensation for the apparent loss. It is a fact of observation that, when one body strikes against another, the temperature of both is raised, and it has been abundantly proved that the production of thermal effects of this kind is of the nature of a transformation of energy. We must therefore expect that in impulsive changes of motion some mechanical energy will be transformed into heat. In order to formulate in a simple and general manner the mechanical effects produced in two bodies by collision it is necessary to have recourse to special experiments and subsidiary hypotheses.

193. Newton's experimental Investigation. Newton made an elaborate series of experiments* on the impact of spheres which come into contact when their centres are moving in the line joining them. He found that the relative velocity of the two spheres after impact was oppositely directed to that before impact, and that the magnitude of the velocity of separation bears to the velocity of approach a ratio which is less than unity. He found that this ratio depends upon the materials of which the spheres are made.

To express this result, let U and U' be the velocities of the two spheres in the line of centres, and in the same sense, before impact, u and u' their velocities in the same line and in the same sense after impact, then

$$u - u' = -e(U - U'),$$

where e is a positive number less than unity.

194. Coefficient of restitution. The number e is called the "coefficient of restitution." For very hard elastic solids, such as glass and ivory, e is little different from unity; for very soft materials, such as wool or putty, it approaches zero. The connexion between e and the elasticity of the impinging bodies has led to its being sometimes called the "coefficient of elasticity," but we avoid this phrase because it is sometimes used (in a different meaning) in the Theory of Elasticity. For a like reason we avoid the phrase "coefficient of resilience" which has also been sometimes used. Materials for which e is zero or unity may be regarded as ideal limits to which some bodies approach. We shall speak of such materials as being "without restitution" and "of perfect restitution" respectively, ordinary materials we shall speak of as

* *Loc. cit. ante*, p. 137.

having "imperfect restitution." It is, of course, to be understood that any such phrase refers to an action between two bodies of the same or different materials. The coefficient e depends on both the materials, just as the coefficient of friction between two bodies depends on the materials and degree of polish of both.

195. Direct impact of elastic spheres. Let the masses of the spheres be m, m'; let the velocities of their centres just before impact be U, U', and just after impact, u, u', these velocities being parallel to the line of centres. We suppose all the velocities to be estimated in the same sense, which is that *from* the centre of the sphere m *to* the centre of the sphere m'.

For the determination of u, u' we have the equation given by Newton's experimental result, viz.

$$u - u' = -e(U - U'),$$

and the equation of constancy of momentum of the system, viz.

$$mu + m'u' = mU + m'U'.$$

Hence we find

$$u = \frac{(m - m'e)U + m'(1+e)U'}{m+m'},$$

$$u' = \frac{(m' - me)U' + m(1+e)U}{m+m'}.$$

Let R be the *impulsive pressure* between the spheres. R is regarded as the impulse of a force acting on the sphere m in the direction opposite to that of U. Then we have

$$R = -m(u - U) = m'(u' - U') = (1+e)\frac{mm'}{m+m'}(U - U').$$

The *kinetic energy lost* in the impact is

$$(\tfrac{1}{2}mU^2 + \tfrac{1}{2}m'U'^2) - (\tfrac{1}{2}mu^2 + \tfrac{1}{2}m'u'^2),$$

or

$$\tfrac{1}{2}m(U - u)(U + u) + \tfrac{1}{2}m'(U' - u')(U' + u'),$$

or

$$\tfrac{1}{2}R[(U + u) - (U' + u')].$$

This expression accords with the result of Art. 174.

In virtue of the equation

$$u - u' = -e(U - U'),$$

the expression for the kinetic energy lost becomes

$$\tfrac{1}{2}R(1-e)(U-U');$$

and, when we substitute for R, we find that this is equal to

$$\tfrac{1}{2}\frac{mm'}{m+m'}(1-e^2)(U-U')^2.$$

196. Generalized Newton's rule. For the purpose of applications to problems of collision in which the circumstances are less simple than in the case of direct impact of spheres we state the following generalization of Newton's experimental result:—

The relative velocities, after and before impact, of the points of two impinging bodies that come into contact, resolved along the common normal to their surfaces at these points, are in the ratio $-e:1$, *where* e *is the coefficient of restitution.*

197. Oblique impact of smooth elastic spheres. Let two smooth uniform spheres, of masses m, m', impinge.

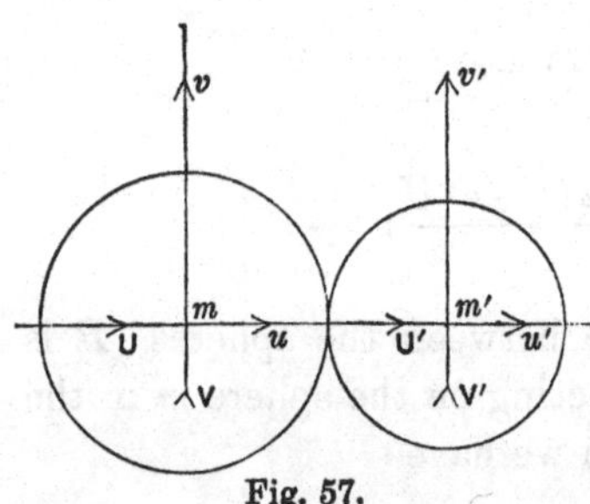

Fig. 57.

Let U, V be the resolved velocities of m in the line of centres and at right angles thereto before impact, U', V' corresponding velocities of m', and let u, v and u', v' be corresponding velocities for m and m' after impact.

The spheres being smooth, there is no friction between them, and the pressure between them is directed along the line of centres. Hence the resolved part of the momentum of either sphere at right angles to the line of centres is unaltered by the impact. We have therefore the equations

$$v=V, \quad v'=V'.$$

The generalized Newton's rule gives the equation

$$u-u'=-e(U-U'),$$

and the equation of constancy of momentum parallel to the line of centres is

$$mu+m'u'=mU+m'U'.$$

Solving these equations we find

$$u = \frac{(m - m'e)\,U + m'(1+e)\,U'}{m+m'},$$

$$u' = \frac{(m' - me)\,U' + m(1+e)\,U}{m+m'}.$$

Hence the velocity of each sphere after impact is determined.

The impulsive pressure between the spheres is found in the same way as in Art. 195 to be

$$(1+e)\frac{mm'}{m+m'}(U-U'),$$

and the kinetic energy lost is found in the same way as in that Article to be

$$\tfrac{1}{2}\frac{mm'}{m+m'}(1-e^2)(U-U')^2.$$

198. Deduction of Newton's rule from a particular assumption. In the motion before impact, let $\bar{u}$, $\bar{v}$ denote the components of velocity of the centre of mass of the two spheres parallel to the line of centres and at right angles to this line, W, η the components of the velocity of m relative to m' parallel to the same directions. Then $\bar{u}$, $\bar{v}$, η are unaltered by the impact. Let W be changed into w by the impact. The quantities W and w are the "relative velocity of approach" and the "relative velocity of separation." The kinetic energy before impact is equal to

$$\tfrac{1}{2}(m+m')(\bar{u}^2+\bar{v}^2)+\tfrac{1}{2}\frac{mm'}{m+m'}(W^2+\eta^2).$$

Cf. Art. 159, Ex. 1. The kinetic energy after impact can be expressed in a similar form. Hence the kinetic energy lost in the impact is

$$\tfrac{1}{2}\frac{mm'}{m+m'}(W^2-w^2).$$

If we assume that the kinetic energy lost is proportional to the square of the relative velocity of approach, we have the result that w has a constant ratio to W, and this is Newton's rule.

199. Elastic systems. The method followed in applying the above rule is to treat the impact as instantaneous, and the impinging bodies as rigid both before and after it. This method is adequate for the discussion of many questions. It cannot however give an exact account of the effects of impact in elastic systems. In such systems no internal forces are developed except after *some* deformation has taken place, so that at the beginning of a motion which is suddenly produced some part of the system yields at once, and starts to move with a finite velocity; after a finite time a finite deformation is produced, and is opposed by finite elastic forces, which continue to act as long as there is any deformation. This statement may conveniently be summed up in the proposition:—*An elastic system cannot support an impulse.* It is now clear that the method founded on Newton's result is of the nature of a compromise, the time of the action in which the elasticity of the bodies is concerned being treated as negligible. An example of the statement that an elastic system cannot support an impulse will be found in the action of elastic strings attached to rigid bodies whose motion is altered suddenly. There is no impulsive tension in such a string, and the motion of the body immediately after the impulse is exactly the same as if the string were not attached to it (cf. Art. 213). On the other hand, an inextensible string is regarded as capable of supporting an impulsive tension.

200. General theory of sudden changes of motion. So far we have been confining our attention to the impulsive action between impinging bodies, but there are many other changes of motion which take place so rapidly that it is convenient to regard them as suddenly produced. The general method of treating such changes of motion depends simply on repeated applications of the principle that for every particle in a connected system, and for each rigid body in such a system, the changes of momentum are a system of vectors equivalent to the impulses that produce them. We shall illustrate the application of this principle by means of some problems.

201. Illustrative problems.

I. *Two equal smooth balls, whose centres are A and B, lie nearly in contact on a smooth table, and a third ball of equal size and mass impinges directly on A, so that the line joining its centre C to A makes with the line AB an angle CAB,* $=\pi-\theta$. *Prove that, if* $\sin\theta>(1-e)/(1+e)$, *the ball A will start off in a direction making with AB an angle* $\tan^{-1}\{2(1-e)^{-1}\tan\theta\}$, *e being the coefficient of restitution for either pair of balls.*

Let V be the velocity of C before striking A; since the impact is direct, V is localized in CA. Let w be its velocity after striking A; the direction of w is that of V. Let u' be the velocity of A immediately after C strikes it, u its velocity just after A strikes B, v the velocity of B after A strikes it, then the direction of u' makes an angle θ with AB. Suppose the direction of u to make an angle ϕ with AB. The direction of v is AB.

We have the equations of momentum

$$V=u'+w,\quad u'\cos\theta=u\cos\phi+v,\quad u'\sin\theta=u\sin\phi,$$

and the equations given by Newton's rule

$$u'-w=eV,\quad u\cos\phi-v=-eu'\cos\theta;$$

whence $\quad 2w=V(1-e),\quad 2u'=V(1+e),\quad 2u\cos\phi=(1-e)\,u'\cos\theta,$

and
$$\tan\phi=\frac{2\tan\theta}{1-e}.$$

Thus A moves off as stated, provided that there is no second impact between A and C. The condition for this is $u \cos(\phi - \theta) > w$,

or $$\tfrac{1}{2}(1-e)\, u' \cos^2\theta + u' \sin^2\theta > \frac{1-e}{1+e} u',$$

which leads to $$\sin\theta > (1-e)/(1+e).$$

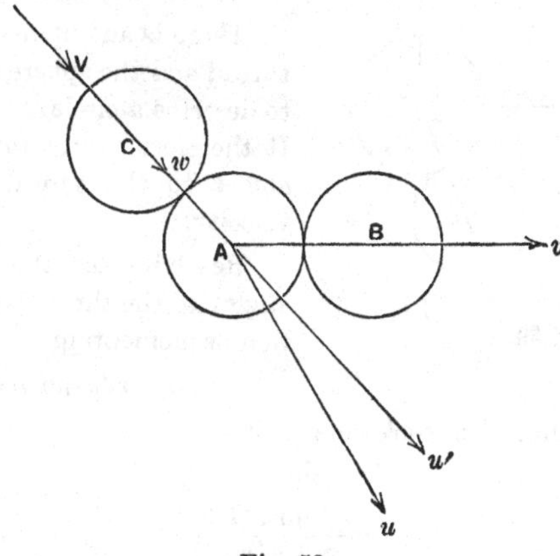

Fig. 58.

II. *A particle is projected with velocity. V from the foot of a smooth fixed plane of inclination θ in a direction making an angle a with the horizon ($a > \theta$). Find the condition that it may strike the plane n times, striking it at right angles at the nth impact, e being the coefficient of restitution between the plane and the particle.*

Since the velocity parallel to the plane is unaltered by impact, the motion of the particle parallel to the plane is determined by the same equation as if there were no impacts, thus at the end of any interval t from the beginning of the motion the velocity parallel to the plane is $V\cos(a-\theta) - gt\sin\theta$.

Let $t_1, t_2, \ldots t_n$ be the times of flight before the first impact, between the first and second, and so on. Then t_1 is given by

$$Vt_1 \sin(a-\theta) - \tfrac{1}{2} g t_1^2 \cos\theta = 0,$$

and thus $t_1 = 2V\sin(a-\theta)/g\cos\theta$. The velocity perpendicular to the plane at time t_1 is $V\sin(a-\theta) - gt_1\cos\theta$ or $-V\sin(a-\theta)$. Immediately after the impact the velocity at right angles to the plane becomes $eV\sin(a-\theta)$ away from the plane. We thus find that $t_2 = et_1$, $t_3 = et_2$, ...

Hence $t_1 + t_2 + \ldots + t_n, = \dfrac{1-e^n}{1-e} \dfrac{2V\sin(a-\theta)}{g\cos\theta}$, is the interval from the beginning of the motion till the nth impact. By supposition, at the end of this interval the velocity parallel to the plane vanishes, or this interval is $V\cos(a-\theta)/g\sin\theta$. The required condition is therefore

$$\cot\theta = 2\tan(a-\theta)(1-e^n)/(1-e).$$

III. *A smooth sphere of mass m is tied to a fixed point by an inextensible thread, and another sphere of mass m′ impinges directly on it with velocity v in a direction making an acute angle α with the thread. Find the velocity with which m begins to move.*

The impulse between the spheres acts in the line of centres so that the direction of motion of m' is unaltered. Let its velocity after impact be v'.

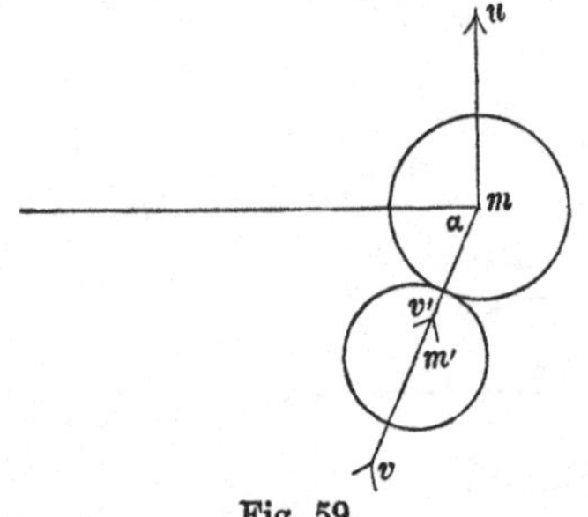

Fig. 59.

There is an impulsive tension in the thread and the sphere m is constrained to describe a circle about the fixed end. It therefore starts to move at right angles to the thread. Let u be its velocity.

Resolving for the system at right angles to the thread we have the equation of momentum

$$mu + m'v' \sin \alpha = m'v \sin \alpha.$$

By the generalized Newton's rule we have

$$v' - u \sin \alpha = -ev.$$

Whence
$$u = \frac{m' \sin \alpha \, (1+e)}{m + m' \sin^2 \alpha} v.$$

IV. *Two particles A, B of equal mass are connected by a rigid rod of negligible mass, and a third equal particle C is tied to a point P of the rod at distances a, b from the two ends. C is projected with velocity u perpendicular to AB. Find the velocity of C immediately after the string becomes tight.*

Let v be the velocity of C immediately after the string becomes tight.

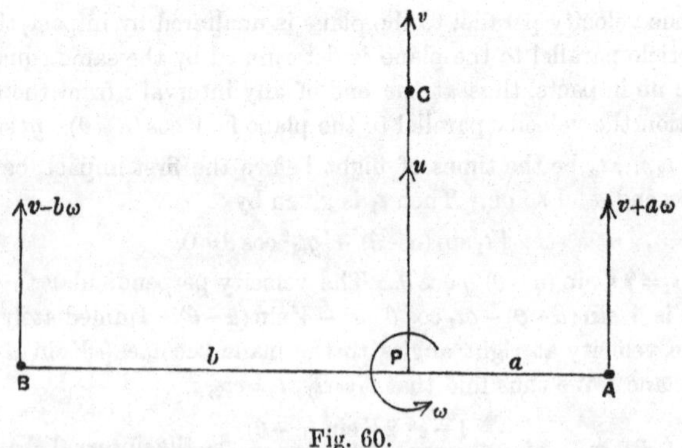

Fig. 60.

Since the impulse on C is along the string its direction of motion is unaltered. The velocity with which P starts to move is v along the string.

Let ω be the angular velocity with which the rod begins to turn. The velocity of A is compounded of the velocity of P and the velocity of A

relative to P. Thus A starts with velocity $v+a\omega$. So B starts with velocity $v-b\omega$.

The equation of momentum parallel to the string is

$$mv+m(v+a\omega)+m(v-b\omega)=mu,$$

m being the mass of either particle.

The equation of moment of momentum about P is

$$ma(v+a\omega)-mb(v-b\omega)=0,$$

giving
$$\omega=(b-a)v/(a^2+b^2).$$

Eliminating ω we find

$$3v-\frac{(a-b)^2}{a^2+b^2}v=u,$$

or
$$v=\frac{1}{2}\frac{a^2+b^2}{a^2+b^2+ab}u.$$

202. Examples.

[In these examples e is the coefficient of restitution between two bodies.]

1. The sides of a rectangular billiard table are of lengths a and b. If a ball is projected from a point on one of the sides of length b to strike all four sides in succession and continually retrace its path, show that the angle of projection θ with the side is given by $ae\cot\theta=c+ec'$, where c and c' are the parts into which the side is divided at the point of projection.

2. Prove that, in order to produce the greatest deviation in the direction of a smooth billiard ball of diameter a by impact on another equal ball at rest, the former must be projected in a direction making an angle

$$\sin^{-1}\left(\frac{a}{c}\sqrt{\frac{1-e}{3-e}}\right)$$

with the line (of length c) joining the two centres.

3. A particle is projected from a point at the foot of one of two smooth parallel vertical walls so that after three reflexions it may return to the point of projection; and the last impact is direct. Prove that $e^3+e^2+e=1$, and that the vertical heights of the three points of impact are in the ratios

$$e^2 : 1-e^2 : 1.$$

4. A particle is projected from the foot of an inclined plane and returns to the point of projection after several rebounds, one of which is at right angles to the plane; prove that, if it takes r more leaps in coming down than in going up, the inclination θ of the plane and the angle of projection a are connected by the equation

$$\cot\theta\cot(a-\theta)=2\{\sqrt{(1-e^r)}-(1-e^r)\}/\{e^r(1-e)\}.$$

5. A particle is projected from the foot of a plane of inclination γ in a direction making an angle β with the normal to the plane, in a plane through this normal making an angle a with the line of greatest slope on the inclined

plane. Prove that, for the particle to be on the horizontal through the point of projection when it meets the plane for the nth time, the angles α, β, γ must satisfy the equation

$$(1-e^n)\tan\gamma=(1-e)\cos\alpha\tan\beta.$$

6. Three equal spheres are projected simultaneously from the corners of an equilateral triangle with equal velocities towards the centre of the triangle, and meet near the centre. Prove that they return to the corners with velocities diminished in the ratio $e : 1$.

7. A smooth uniform hemisphere of mass M is sliding with velocity V on a plane with which its base is in contact; a sphere of smaller mass m is dropped vertically, and strikes the hemisphere on the side towards which it is moving, so that the line joining their centres makes an angle $\pi/4$ with the vertical. Show that, if the coefficient of restitution between the plane and the hemisphere is zero, and that between the sphere and the hemisphere is e, the height through which the sphere must have fallen if the hemisphere is stopped dead is

$$\frac{V^2}{2g}\frac{(2M-em)^2}{(1+e)^2m^2}.$$

8. A particle of mass M is moving on a smooth horizontal table with uniform speed in a circle, being attached to the centre by an inextensible thread, and strikes another particle of mass m at rest. Show that, if the two particles adhere, the tension of the thread is diminished in the ratio

$$M/(M+m).$$

If there is restitution between the particles and the second one is describing the same circle as the first, prove that the tensions T and t in the two threads after impact are connected with their values T_0 and t_0 before impact by the equation

$$T+t=T_0+t_0-(1-e^2)\{\surd(mT_0)-\surd(Mt_0)\}^2/(M+m).$$

9. A bucket and a counterpoise, of equal mass M, connected by a chain of negligible mass passing over a smooth pulley, just balance each other, and a ball, of mass m, is dropped into the centre of the bucket from a height h above it; find the time that elapses before the ball ceases to rebound, and show that the whole distance descended by the bucket during this interval is

$$4meh/\{(2M+m)(1-e)^2\}.$$

10. Three equal particles are attached to the ends and middle point of a rod of negligible mass, and one of the end ones is struck by a blow so that it starts to move at right angles to the rod. Prove that the magnitudes of the velocities of the particles at starting are in the ratios 5 : 2 : 1.

11. An impulsive attraction acts between the centres of two spheres which are approaching each other so as to generate kinetic energy E. If v is their relative velocity before the impulse, and θ, θ' the angles which the

directions of the relative velocity, before and after, make with the line of centres, then

$$\sin\theta = \sin\theta' \sqrt{\left(1+\frac{4E}{Mv^2}\right)},$$

where M is the harmonic mean of the masses.

12. Two small bodies of equal mass are attached to the ends of a rod of negligible mass; the rod is supported at its centre and is turning uniformly, so that each of the bodies is describing a horizontal circle, when one of the bodies is struck by a vertical blow equal in magnitude to twice its momentum. Prove that the direction of motion of each of the bodies is instantaneously deflected through half a right angle.

Initial Motions

203. Nature of the problems. We suppose that a system is held in some definite position in a field of force, and that at a particular instant some one of the constraints ceases to be applied; then the system begins to move, each particle of it with a certain acceleration. Our first object in such a case is to determine the accelerations with which the parts of the system begin to move. When the accelerations have been found there is generally no difficulty in determining the initial values of the reactions of supports, or internal actions between different bodies of the system; and the determination of the unknown reactions is our second object.

The senses of the accelerations with which a conservative system moves away from a position of instantaneous rest can sometimes be determined by help of the observation that the motion must be one by which the potential energy is diminished. This is evident since the kinetic energy must be increased above the value (zero) which it has in the position of rest.

The problem of determining the curvature of the path of a particle whose velocity is not zero offers no difficulty when the velocity and acceleration are known, since the resolved acceleration along the normal to the path is the product of the square of the resultant velocity and the curvature. This remark enables us easily to determine the initial curvature of the path of a particle when its motion is changed suddenly.

204. Method for initial accelerations. It is always possible to determine expressions for the accelerations of all the points of a connected system in terms of a small number of independent accelerations, and there is always the same number of equations of motion free from unknown reactions, so that all the accelerations can be found. The expression of the initial accelerations in the proposed manner is facilitated by observing (1) that the velocity of every particle initially vanishes, (2) that every composition and resolution may be effected by taking the position of the system to be that from which it starts. The method will be better understood after the study of an example. We purposely choose one of a somewhat complicated character in order to illustrate the various details of the method.

205. Illustrative Problem. *Four equal rings A, B, C, D are at equal distances on a smooth fixed horizontal rod, and three other equal and similar rings P, Q, R are attached by pairs of equal inextensible threads to the pairs of rings (A, B), (B, C), (C, D). The system is held so that all the threads initially make the same angle α with the horizontal, and is let go. It is required to find the acceleration of each ring.*

From the symmetry of the system the accelerations of A, D are equal and

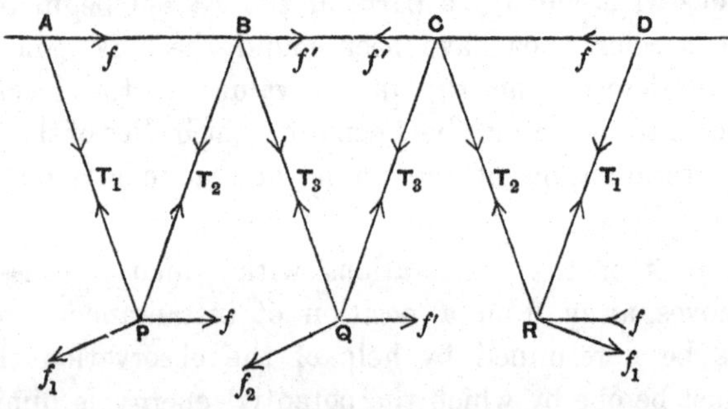

Fig. 61.

opposite, so are those of B, C, and those of P, R. Also the acceleration of Q is vertical.

Let f, f' be the accelerations of A, B along the smooth horizontal rod.

Now relatively to A, P describes a circle, and thus the acceleration of P relative to A is made up of a tangential acceleration f_1 at right angles to AP, and a normal acceleration proportional to the square of the angular velocity of AP. Since the initial angular velocity vanishes, we have, as the relative acceleration, f_1 at right angles to AP. Again, since the threads AP, BP are equal, the particle P is always vertically under the middle point of AB and thus its horizontal acceleration is $\frac{1}{2}(f+f')$.

Hence $$\tfrac{1}{2}(f+f')=f-f_1\sin\alpha,$$

giving $$f_1\sin\alpha=\tfrac{1}{2}(f-f').$$

Again, the horizontal acceleration of Q vanishes, and we have therefore the acceleration f_2 of Q relative to B given by the equation

$$f_2 \sin a = f'.$$

Thus the accelerations of the particles are expressed in terms of f and f'; in particular the vertical accelerations of P and Q are $\frac{1}{2}(f-f')\cot a$ and $f'\cot a$ downwards.

Now let m be the mass of each particle and T_1, T_2, T_3 the tensions in the threads as shown in the figure. Then resolving horizontally for A, P, and B we have

$$mf = T_1 \cos a,\quad \tfrac{1}{2}m(f+f') = (T_2 - T_1)\cos a,\quad mf' = (T_3 - T_2)\cos a \ldots(1);$$

and resolving vertically for P and Q we have

$$\tfrac{1}{2}m(f-f')\cot a = -(T_1 + T_2)\sin a + mg,\quad mf'\cot a = -2T_3 \sin a + mg \ldots(2).$$

From the set of equations (1) we have

$$T_1 \cos a = mf,\quad T_2 \cos a = m(\tfrac{3}{2}f + \tfrac{1}{2}f'),\quad T_3 \cos a = m\tfrac{3}{2}(f+f');$$

and from (2), on substituting for T_1, T_2, T_3, we have

$$(f-f')\cot a + (5f+f')\tan a = 2g,\quad f'\cot a + 3(f+f')\tan a = g;$$

whence
$$\frac{f}{4-\cos 2a} = \frac{f'}{\cos 2a} = \frac{g\sin 2a}{12 - 11\cos 2a + \cos^2 2a}.$$

206. Initial curvature. As an example of initial curvatures when the motion does not start from rest we take the following problem:

Two particles of masses m, m' connected by an inextensible thread of length l are placed on a smooth table with the thread straight, and are projected at right angles to the thread in opposite senses. It is required to find the initial curvatures of their paths.

Let u, v be the initial velocities of the particles, and ω the initial angular velocity of the thread, then

$$u + v = l\omega.$$

Let G be the centre of mass of the two particles. Then G moves uniformly on the table with velocity

$$(mu - m'v)/(m+m').$$

The acceleration of G vanishes, and the acceleration of m relative to G is that of a particle describing a circle of radius $m'l/(m+m')$ with angular velocity ω; thus the acceleration of m along the thread is $m'l\omega^2/(m+m')$, and this is the acceleration of m along the normal to its path. Hence, if ρ is the initial radius of curvature of the path of m,

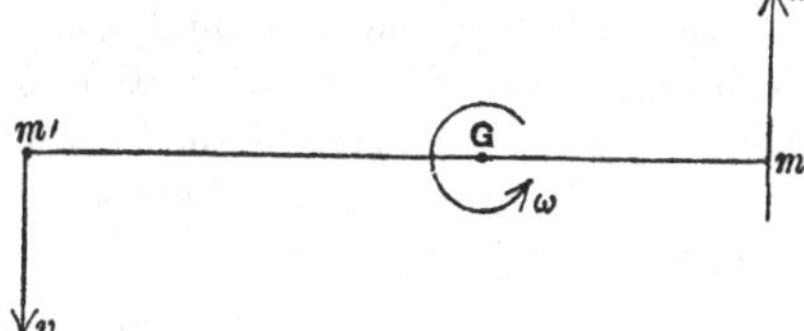

Fig. 62.

$$\frac{u^2}{\rho} = \frac{m'l}{m+m'}\left(\frac{u+v}{l}\right)^2,$$

giving
$$1/\rho = m'(u+v)^2/\{(m+m')lu^2\}.$$

In like manner the initial curvature of the path of m' is

$$m(u+v)^2/\{(m+m')lv^2\}.$$

207. Examples.

1. Two bodies A and B of equal weight are suspended from the chains of an Atwood's machine; A is rigid, while B consists of a vessel full of water in which is a cork attached to the bottom by a string. Supposing the string to be destroyed in any manner, determine the sense in which A begins to move.

2. A particle is supported by equal threads inclined at the same angle a to the horizontal. One thread being cut, prove that the tension in the remaining thread is suddenly changed in the ratio $2\sin^2 a : 1$.

3. Particles of equal mass are attached to the points of trisection C, D of a thread $ACDB$ of length $3l$, and the system is suspended by its ends from points A, B distant $l(1+2\sin a)$ apart in a horizontal line, so that CD is horizontal and equal to l. Prove that, if the portion DB of the thread is cut, the tension of AC is instantly changed in the ratio $2\cos^2 a : 1+\cos^2 a$, and that the initial direction of motion of D is inclined to the vertical at an angle ϕ such that

$$\cot\phi = \tan a + 2\cot a.$$

4. Three small equal rings rest on a smooth vertical circular wire at the corners of an equilateral triangle with one side vertical, the uppermost being connected with the other two by inextensible threads. Prove that, if the vertical thread is cut through, the tension in the other thread is instantly diminished in the ratio 3 : 4.

5. A set of $2n$ equal particles are attached at equal intervals to a thread, and the ends of the thread are attached to equal small smooth rings which can slide on a horizontal rod. The rings are initially held in such a position that the lowest part of the thread is horizontal and the highest parts make equal angles γ with the horizontal, and the rings are let go. Prove that in the initial motion (i) the acceleration of each particle is vertical, (ii) the tension in the lowest part of the thread is to what it was in equilibrium in the ratio $m' : mn\cot^2\gamma + m'$, where m is the mass of a particle and m' the mass of a ring.

6. Three particles A, B, C of equal masses are attached at the ends and middle point of a thread so that $AB = BC = a$, and the particles are moving at right angles to the thread, which is straight, with the same velocities, when B impinges directly on an obstacle. Prove that, if there is perfect restitution, the radii of curvature of the paths which A and C begin to describe are equal to $\frac{1}{4}a$.

7. Two particles, of masses M and nM, are attached respectively to a point of a thread distant a from one end and to that end, and the other end is fixed to a point on a smooth table on which the particles rest, the thread being in two straight pieces containing an obtuse angle $\pi - a$. Prove that, if

the particle nM is projected on the table at right angles to the thread, the initial radius of curvature of its path is $a(1+n\sin^2 a)$.

8. Two particles P, Q, of equal mass, are connected by a thread of length l which passes through a small hole in a smooth table. P being at a distance c from the hole and Q hanging vertically, P is projected on the table at right angles to the thread with velocity v; prove that the initial radius of curvature of P's path is $2cv^2/(v^2+cg)$. Prove also that, if Q is projected horizontally with velocity v, while P is not moved, the initial radius of curvature of Q's path is

$$2v^2(l-c)/\{g(l-c)\sim v^2\}.$$

Applications of the Energy Equation

208. Equilibrium. The possible positions of equilibrium of a system are distinguished from other positions by the condition that, if the system is at rest in an equilibrium position, so that all the velocities vanish there, the accelerations also vanish there.

Now let the equations of motion be taken in the forms

$$m\ddot{x} = X + X'$$

of Art. 160; and let the system pass through a position of equilibrium with any velocities denoted typically by $\dot{x}'$, $\dot{y}'$, $\dot{z}'$. The equation which expresses the result that the rate of change of kinetic energy (per unit of time) is equal to the rate at which work is done (Art. 173) is

$$\Sigma[m(\ddot{x}\dot{x}' + \ddot{y}\dot{y}' + \ddot{z}\dot{z}')] = \Sigma[(X+X')\dot{x}' + (Y+Y')\dot{y}' + (Z+Z')\dot{z}'].$$

Since, by hypothesis, the position is one of equilibrium, the left-hand member of this equation vanishes. Hence the right-hand member also vanishes, or we have the result:—

The rate at which work is done when a system passes through a position of equilibrium with any velocity vanishes.

This result is usually stated in a form involving infinitesimals, and is called the "Principle of Virtual Work" or of "Virtual Velocities."

In forming the expression for the rate at which work is done, or the expression for the virtual work, the velocities must be such as are compatible with the connexions of the system. Further, if there are any resistances which depend upon velocities, and vanish with the velocities, the rate at which these resistances would do work is to be omitted, for manifestly such resistances do not affect the positions of equilibrium.

When there is a work function W, the rate at which work is done is $\frac{dW}{dt}$. If W is a function of any quantities which define the position of the system, say $\theta, \phi, \ldots$, then

$$\frac{dW}{dt} = \frac{\partial W}{\partial \theta}\dot{\theta} + \frac{\partial W}{\partial \phi}\dot{\phi} + \ldots.$$

If the position is one of equilibrium, this vanishes for all values of $\dot{\theta}, \dot{\phi}, \ldots$. Hence we have the equations

$$\frac{\partial W}{\partial \theta} = 0, \ \frac{\partial W}{\partial \phi} = 0, \ \ldots;$$

and the values of $\theta, \phi, \ldots$ which satisfy these equations determine the positions of equilibrium.

If we sought the positions in which W is a maximum or minimum, we should have to begin by solving the equations

$$\frac{\partial W}{\partial \theta} = 0, \ \ldots,$$

and then proceed to determine which among the various sets of solutions make W a true maximum or a true minimum. In the positions in which

$$\frac{\partial W}{\partial \theta} = 0, \ \ldots$$

we should say that W is *stationary*, whether it is a true maximum or minimum or not. Since the potential energy of the system in any position is $-W$, we have the result:

The equilibrium positions of a conservative system are those positions in which the potential energy is stationary.

209. Machines. In all the so-called "simple machines" or "mechanical powers" the positions of all the parts can be expressed in terms of a single variable, and consequently the potential energy is determined in terms of a single variable. The condition that the potential energy is stationary in the position of equilibrium becomes a relation between the masses of two moving parts: the "power" and the "weight." This result is worked out in books on Statics.

In any conservative system in which the positions of all the parts can be expressed in terms of a single variable, the equation of energy determines the whole motion. We had an example in Atwood's machine [Ex. 1 of Art. 74].

210. Examples.

1. Two bodies are supported in equilibrium on a wheel and axle, and a body whose mass is equal to that of the greater body is suddenly attached to that body. Prove that the acceleration with which it moves is $bg/(2b+a)$, a and b being the radii of the wheel and the axle respectively, and the inertia of the machine being neglected.

2. In any machine without friction and inertia a body of weight P supports a body of weight W, both hanging by vertical cords. These bodies are replaced by bodies of weights P' and W', which, in the subsequent motion, move vertically. Prove that the centre of mass of P' and W' will descend with acceleration

$$g\,(WP' - W'P)^2/(W^2P' + W'P^2)\,(W' + P').$$

211. Small oscillations. We have to consider the small motion of a system which is slightly displaced from a position of equilibrium. We confine our attention to cases where any position of the system is determined by assigning the value of a single geometrical quantity θ, as in the case of the simple circular pendulum (Article 119). We can always choose θ to vanish in the position of equilibrium; for, if it has been chosen in any other way so that its value in the position of equilibrium is θ_0, then $\theta - \theta_0$ can be used instead of θ.

Now the velocity of each particle of the system can be expressed in terms of θ and $\dot{\theta}$, and the kinetic energy T is thus of the form $\frac{1}{2}A\dot{\theta}^2$, where A may depend upon θ, but does not vanish with θ.

Also the potential energy V vanishes with θ, if the standard position is the position of equilibrium. Thus V is a function of θ which may be expanded in powers of θ and the series contains no term independent of θ. Again, the principle of virtual work shows that $\frac{dV}{d\theta}$ vanishes with θ, or that the term of the first order is missing from the series for V. Thus V can be expressed as a series beginning with the term in θ^2, and more generally we may say that, when θ is sufficiently small, $V = \frac{1}{2}C\theta^2$, where C is a function of θ which is finite when $\theta = 0$.

The equation of energy accordingly is

$$\tfrac{1}{2}A\dot{\theta}^2 + \tfrac{1}{2}C\theta^2 = \text{const.},$$

and on differentiating we have

$$A\ddot{\theta} + \tfrac{1}{2}\left(\frac{dA}{d\theta}\right)\dot{\theta}^2 + C\theta + \tfrac{1}{2}\left(\frac{dC}{d\theta}\right)\theta^2 = 0.$$

Omitting small quantities of an order higher than the first we have

$$A\ddot{\theta} + C\theta = 0,$$

where A and C have their values for $\theta = 0$. Thus, if these two quantities have the same sign, the motion in θ is simple harmonic with period $2\pi\sqrt{(A/C)}$.

Now A must be positive since otherwise the expression $\frac{1}{2}A\dot{\theta}^2$ could not represent an amount of kinetic energy. Hence there are oscillations in a real period if C is positive.

The value of C for $\theta = 0$ is the value of $\dfrac{d^2V}{d\theta^2}$ for $\theta = 0$, and thus the conditions for a real period of oscillation are the same as the conditions that V may have a minimum value in the position of equilibrium.

If the period is real, the motion can be small enough for the approximation to be valid; otherwise it soon becomes so large that we cannot simplify the equation of motion by neglecting θ^2. In the former case the equilibrium is *stable* and in the latter *unstable*.

We learn that *in a position of stable equilibrium the potential energy is a minimum**.

The process which has been adopted shows that we might have reduced the expression for T by substituting zero for θ in A, and the expression for V might have been taken to be simply the term of the series which contains θ^2. These simplifications might have been made before differentiating the energy equation. If we express the kinetic energy correctly to the second order of small quantities in the form $\frac{1}{2}A\dot{\theta}^2$, and the potential energy also correctly to the second order of small quantities in the form $\frac{1}{2}C\theta^2$, the period of the small oscillations is $2\pi\sqrt{(A/C)}$. In the case of a simple pendulum of mass m and length l, A is ml^2 and C is mgl, so that

$$A/C = l/g.$$

In any other case we may compare the motion with that of a simple pendulum, and then the quantity gA/C is the length of a simple pendulum which oscillates in the same time as the system. It is called the length of the *equivalent simple pendulum* for the small oscillations of the system.

* This result, here proved for a special class of cases, is true for all conservative systems.

212. Examples.

1. Two rings of masses m, m' connected by a rigid rod of negligible mass are free to slide on a smooth vertical circular wire of radius a, the rod subtending an angle α at the centre. Prove that the length of the equivalent simple pendulum for the small oscillations of the system is

$$(m+m')\,a/\sqrt{(m^2+m'^2+2mm'\cos\alpha)}.$$

2. One end of an inextensible thread is attached to a fixed point A, and the thread passes over a small pulley B fixed at the same height as A and at a distance $2a$ from it and supports a body of mass P. A ring of mass M can slide on the thread and the system is in equilibrium with M between A and B. Prove that the time of a small oscillation is

$$4\pi\sqrt{\{aMP(M+P)/g(4P^2-M^2)^{\frac{3}{2}}\}}.$$

3. A particle is suspended from two fixed points at the same level by equal elastic threads of natural length a, and hangs in equilibrium at a depth h with each thread of length l. Prove that, if it is slightly displaced parallel to the line joining the fixed ends of the threads, the length of the equivalent simple pendulum for the small oscillations is

$$hl^2(l-a)/(l^3-h^2a).$$

4. Prove that, if the fixed points in Ex. 3 are at a distance $2c$ apart, and the particle is displaced vertically, the length of the equivalent simple pendulum is

$$hl^2(l-a)/(l^3-c^2a).$$

5. A pulley of negligible mass is hung from a fixed point by an elastic cord of modulus λ and natural length a, and an inextensible cord passing over the pulley carries at its ends bodies of masses M and m. Prove that the time of a small oscillation in which the pulley moves vertically is

$$4\pi\sqrt{\{Mma/(M+m)\lambda\}}.$$

213. Principles of Energy and Momentum. We have remarked that there are numerous cases in which the principles of energy and momentum supply all the first integrals of the equations of motion of a system, and thus suffice to determine the velocities of the parts of the system in any position.

To illustrate these principles further we take the following problem:

Two particles A and B, placed on a smooth horizontal table, are connected by an elastic string of negligible mass. When the string is straight, and of its natural length, one of the particles is struck by a blow in the line of the string and away from the other particle; determine the subsequent motion.

Let m be the mass of the particle struck, m' that of the other, V the velocity with which m begins to move. There is no tension in the string until it is extended, and thus at first m' has no velocity.

The centre of mass moves on the table with uniform velocity u, $=mV/(m+m')$, in the line of the string. Let x be the increase in the length of the string at time t, then the velocities of the particles are

$$u+\frac{m'\dot{x}}{m+m'},\quad u-\frac{m\dot{x}}{m+m'}.$$

Hence the kinetic energy is $\frac{1}{2}(m+m')\,u^2+\frac{1}{2}\frac{mm'}{m+m'}\dot{x}^2$.

The potential energy is $\frac{1}{2}\frac{\lambda}{a}x^2$ so long as x is positive, a being the natural length of the string, and λ the modulus of elasticity.

Thus the energy equation is

$$\frac{1}{2}\frac{m^2}{m+m'}V^2+\frac{1}{2}\frac{mm'}{m+m'}\dot{x}^2+\frac{1}{2}\frac{\lambda}{a}x^2=\frac{1}{2}mV^2,$$

showing that the motion in x is simple harmonic motion of period

$$2\pi\sqrt{\{mm'a/(m+m')\lambda\}},$$

so long as x remains positive. Whenever the string is unstretched we have $\dot{x}=\pm V$. When $\dot{x}$ vanishes the string has its greatest length

$$a+V\sqrt{\{mm'a/(m+m')\lambda\}}.$$

We can thus describe the whole motion:—m moves off with velocity V which gradually diminishes, and m' moves in the same direction from rest with gradually increasing velocity; the string begins to extend, and continues to do so until it attains its greatest length; this happens at the end of a quarter of the period of the simple harmonic motion, and at this instant the particles have equal velocities u. The velocity of m continues to diminish until it is reduced to $V(m-m')/(m+m')$, and the velocity of m' continues to increase until it has the value $2mV/(m+m')$, these values being attained at the same instant; in the meantime the string contracts to its natural length a, which it attains at the instant in question, and this happens at the end of half a period from the beginning of the motion. The particles then move with the velocities they have attained until m' overtakes m, when a collision takes place. The subsequent motion depends on the coefficient of restitution. If this is unity, the relative motion is reversed. In any case the description of the subsequent motion involves nothing new.

214. Examples.

1. A shot of mass m is fired from a gun of mass M placed on a smooth horizontal plane and elevated at an angle a. Prove that, if the muzzle velocity of the shot is V, the range is

$$\frac{2V^2}{g}\frac{(1+m/M)\tan a}{1+(1+m/M)^2\tan^2 a}.$$

2. A smooth wedge of mass M whose base angles are a and β is placed on a smooth table, and two particles of masses m and m' move on the faces, being connected by an inextensible thread which passes over a smooth pulley at the summit. Prove that the wedge moves with acceleration

$$g\frac{(m\sin a\sim m'\sin\beta)(m\cos a+m'\cos\beta)}{(m+m')(M+m+m')-(m\cos a+m'\cos\beta)^2}.$$

3. Two bodies of masses m_1, m_2 are connected by a spring of such strength that when m_1 is held fixed m_2 makes n complete vibrations per second. Prove that, if m_2 is held, m_1 will make $n\sqrt{(m_2/m_1)}$, and that, if both are free, they will make $n\sqrt{\{(m_1+m_2)/m_1\}}$ vibrations per second, the vibrations in all cases being in the line of the spring.

4. Three equal particles are attached at equal intervals to an inextensible thread, and, when the thread is straight, the two end ones are projected with equal velocities in the same sense at right angles to the thread. Prove that, if there are no external forces, the velocity of each of the end particles (at right angles to the part of the thread which is attached to it) at the instant when they impinge is $\frac{1}{3}\sqrt{3}$ of their initial velocity.

5. A particle is attached by an elastic thread of natural length a to a point of a smooth plank which is free to slide on a horizontal table, and the thread is stretched to a length $a+c$, in a horizontal line passing over the centre of mass of the plank, and the system is let go from rest. Prove that, if the plank and particle have equal masses, and the modulus of elasticity of the thread is equal to the weight of the particle, the velocity of the particle relative to the plank when the thread has its natural length is that due to falling through a height c^2/a.

6. A spherical shell of radius a and mass m contains a particle of the same mass, which is attached to the highest point by an elastic thread of natural length a, stretched to length $a+c$, and is also attached to the lowest point by an inextensible thread; and the shell rests on a horizontal plane. Suddenly the lower thread breaks, the particle jumps up to the highest point of the shell and adheres there, and it is observed that the shell jumps up through a height h. Prove that the modulus of elasticity of the upper thread is

$$2mga\,(a+c+4h)/c^2.$$

What external forces produce momentum in the system as a whole?

7. Three equal particles are connected by an inextensible thread of length $a+b$, so that the middle one is at distances a and b from the other two. The middle one is held fixed and the other two describe circles about it with the same uniform angular velocity so that the two portions of the thread are always in a straight line. Prove that, if the middle particle is set free, the tensions in the two parts of the thread are altered in the ratios $2a+b : 3a$ and $2b+a : 3b$, there being no external forces.

8. Two equal particles are connected by an inextensible thread of length l; one of them A is on a smooth table and the other is just over the edge, the thread being straight and at right angles to the edge. Find the velocities of the particles immediately after they have become free of the table, and prove (i) that in the subsequent motion the tension of the thread is always half the weight of either particle, and (ii) that the initial radius of curvature of the path of A immediately after it leaves the table is $\frac{5}{12}\sqrt{5}\,.\,l$.

MISCELLANEOUS EXAMPLES

1. Two equal balls lie in contact on a table. A third equal ball impinges on them, its centre moving along a line nearly coinciding with a horizontal common tangent. Assuming that the periods of the impacts do not overlap, prove that the ratio of the velocities which either ball will receive according as it is struck first or second is $4 : 3-e$, where e is the coefficient of restitution.

2. Two unequal particles are attached to a thread which passes over a smooth pulley. Initially the smaller is in contact with a fixed horizontal plane, and the other at a height h above the plane. Prove that, if the coefficient of restitution for each impact is e, and if e is a root of any equation of the form $e^n - 2e + 1 = 0$ with n integral, the system will come to rest after a time $2h(1+e)/v(1-e)$, where v is the velocity of the particle of greater mass immediately before its first impact on the plane.

3. Two balls of masses M, m and of equal radii, connected by an inextensible thread, lie on a smooth table with the thread straight, and a ball of the same radius, and of mass m', moving parallel to the thread with velocity v, strikes the ball m so that the line of centres (m', m) makes an acute angle a with the line of centres (M, m). Prove that, if e is the coefficient of restitution between m and m', M starts with velocity

$$vmm'(1+e)\cos^2 a/\{Mm'\sin^2 a + m(M+m+m')\}.$$

4. Two balls are attached by inextensible threads to fixed points, and one of them, of mass m, describing a circle with velocity u, impinges on the other, of mass m', at rest, so that the line of centres makes an angle a with the thread attached to m, and the threads cross each other at right angles. Prove that m' will start to describe a circle with velocity

$$mu \sin a \cos a (1+e)/(m\cos^2 a + m'\sin^2 a),$$

where e is the coefficient of restitution between the balls.

5. A shell of mass M is moving with velocity V. An internal explosion generates an amount E of energy, and thereby breaks the shell into two fragments whose masses are in the ratio $m_1 : m_2$. The fragments continue to move in the original line of motion of the shell. Prove that their velocities are

$$V + \surd(2m_2E/m_1M), \quad V - \surd(2m_1E/m_2M).$$

6. Three particles A, B, C of equal mass are placed on a smooth plane inclined at an angle a to the horizontal, and B, C are connected with A by threads of length $h\sec a$ which make equal angles a with the line of greatest slope through A on opposite sides of it, the line BC being below the level of A. If A is struck by a blow along the line of greatest slope, so as to start to move down this line with velocity V, find when the threads become tight, and prove that the velocity of A immediately afterwards is

$$V/(3-2\sin^2 a) + 2gh\sin a/V.$$

7. Three particles of equal mass are attached at equal intervals to a rigid rod of negligible mass, and, the system being at rest, one of the extreme particles is struck by a blow at right angles to the rod. Prove that the kinetic energy imparted to the system, when the other extreme particle is fixed, and the rod turns about it, is less than it would be if the system were free in the ratio 24 : 25.

8. Two equal rigid rods AB, BC of negligible masses carry four equal particles, attached at A, C and at the middle points of the rods. The rods being freely hinged at B, and laid out straight, the end A is struck with an impulse at right angles to the rods. Prove that the magnitudes of the velocities of the particles are in the ratios 9 : 2 : 2 : 1.

9. Four particles of equal masses are tied at equal intervals to a thread, and the system is placed on a smooth table so as to form part of a regular polygon whose angles are each $\pi - a$. Prove that, if an impulse is applied to one of the end particles in the direction of the thread attached to it, the kinetic energy generated is greater than it would be if the particles were constrained to move in a circular groove, and the impulse were applied tangentially, in the ratio $\cos^2 a + 4 \sin^2 a : \cos^2 a + 2 \sin^2 a$.

---

10. Four small smooth rings of equal mass are attached at equal intervals to a thread and rest on a circular wire in a vertical plane. The radius of the wire is one-third of the length of the thread, and the rings are at the four upper corners of a regular hexagon inscribed in the circle, the two lower rings being at the ends of the horizontal diameter. Prove that, if the thread is cut between one of the extreme particles and one of the middle ones, the tension in the horizontal part is suddenly diminished in the ratio 5 : 9.

11. Particles of masses m and m' are fastened to the ends of a thread, which rests in a vertical plane on the surface of a smooth horizontal circular cylinder of mass M. The cylinder can slide on a horizontal plane. The system is initially held at rest so that the radii of the circular section, which pass through the particles, make angles a and β with the vertical. Prove that, when the system is released, the tension of the thread immediately becomes

$$mm'g \frac{M(\sin a + \sin \beta) + (m \sin a + m' \sin \beta)\{1 - \cos(a+\beta)\}}{(m+m')(M + m \sin^2 a + m' \sin^2 \beta) + mm'(\cos a - \cos \beta)^2}.$$

12. A particle P, of mass M, rests in equilibrium on a smooth horizontal table, being attached to three particles of masses m, m', m'' by cords which pass over smooth pulleys at points A, B, C at the edge of the table. Prove that, if the cord supporting m'' is cut, M will begin to move in a direction making with CP an angle

$$\tan^{-1} \frac{\mu(m \sim m')\{(m+m')^2 - m''^2\}}{4Mmm'm''^2 + (m+m')\mu^2},$$

where $$\mu^2 = 2m'^2m''^2 + 2m''^2m^2 + 2m^2m'^2 - m^4 - m'^4 - m''^4.$$

13. A circular wire of mass M is held at rest in a vertical plane, so as to touch at its lowest point a smooth table; and a particle of mass m rests against it, being supported by an inextensible thread, which passes over the wire, and is secured to a fixed point in the plane of the wire at the same level as the highest point of the wire. Prove that, if the wire is set free, the pressure of the particle upon it is immediately diminished by an amount $m^2 g \sin^2 a/(M+4m\sin^2 \frac{1}{2}a)$, where a is the angular distance of the particle from the highest point of the wire.

14. Four particles A, B, C, D of equal mass, connected by equal threads, are placed on a smooth plane of inclination $a\,(<\frac{1}{4}\pi)$ to the horizontal, so that AC is a line of greatest slope, and AB, AD make angles a with AC on opposite sides of it. If the uppermost particle A is held, and the particles B and D are released, prove that the tension in each of the lower threads is instantly diminished in the ratio

$$(1-2\sin^2 a)/(1+2\sin^2 a).$$

15. A particle of mass m on a smooth table is joined to a particle of mass m' hanging just over the edge by a thread of length a at right angles to the edge. Prove that, if the system starts from rest, the radius of curvature of the path of m immediately after it leaves the table is

$$\frac{2m'a}{(m+m')^2}\,\frac{\{(m+m')^2+m'^2\}^{\frac{3}{2}}}{(m+m')^2+2m'^2}.$$

16. Two particles A, B are connected by a fine string; A rests on a rough horizontal table (coefficient of friction$=\mu$) and B hangs vertically at a distance l below the edge of the table. If A is on the point of motion, and B is projected horizontally with velocity u, show that A will begin to move with an acceleration $\mu u^2/\{(\mu+1)\,l\}$, and that the initial radius of curvature of B's path will be $(\mu+1)\,l$.

17. A particle of mass m is attached to one end of a thread which passes through a bead of mass M and the other end is secured to a point on a smooth horizontal table on which the whole rests. Initially the two portions of the thread are straight and contain an obtuse angle a, the portion between m and M being of length a, and m is projected at right angles to this portion. Prove that the initial radius of curvature of the path of m is

$$a\,(1+4mM^{-1}\cos^2 \tfrac{1}{2}a).$$

---

18. A window is supported by two cords passing over pulleys in the framework of the window (which it loosely fits), and is connected with counterpoises each equal to half the weight of the window. One cord breaks and the window descends with acceleration f. Show that the coefficient of friction between the window and the framework is

$$a\,(g-3f)/b\,(g+f),$$

where a is the height and b the breadth of the window.

19. A bucket of mass M is raised from the bottom of a shaft of depth h by means of a cord which is wound on a wheel of mass m. The wheel is driven by a constant force, which is applied tangentially to its rim for a certain time and then ceases. Prove that, if the bucket just comes to rest at the top of the shaft t seconds after the beginning of the motion, the greatest rate of working is

$$2hM^2g^2t/\{Mgt^2 - 2h(M+m)\},$$

the mass of the wheel being regarded as condensed uniformly on its rim.

20. An engine is pulling a train, and works at a constant power doing H units of work per second. If M is the mass of the whole train and F the resistance (supposed constant), prove that the time of generating velocity v from rest is

$$\left(\frac{MH}{F^2}\log\frac{H}{H-Fv} - \frac{Mv}{F}\right) \text{ seconds.}$$

21. Two pulleys each of mass $8m$ hang at the ends of a chain of negligible mass which passes over a fixed pulley; a similar chain passes over each of the two suspended pulleys and carries at its ends bodies of mass $2m$. A mass m is now removed from one of the bodies and attached to one of those which hang over the other pulley; prove that the acceleration of each pulley is $\frac{1}{11}g$. Prove also that the two descending bodies move with the same velocity, and that the velocity of one of the ascending bodies is five times that of the other.

22. A chain of negligible mass passes over a fixed pulley B, and supports a body of mass m at one end and a pulley C of mass p at the other. A similar chain is fastened to a point A below B, passes over C, and supports a body of mass m'. Prove that the acceleration of the pulley is

$$g(2m' - m + p)/(4m' + m + p).$$

---

23. Two equal particles of mass $P\sin\alpha$ are attached, at a distance $2a\sin\alpha$ apart, to a thread, to the ends of which particles of mass P are attached. The thread is hung over two pegs distant $2a$ apart in a horizontal line. Prove that the period of the small oscillations about the position of equilibrium is the same as that for a simple pendulum of length $a\tan\alpha$.

24. A particle of mass M is placed at the centre of a smooth circular horizontal table of radius a; cords are attached to the particle and pass over n smooth pulleys placed symmetrically round the circumference, and each cord supports a mass M. Show that the time of a small oscillation of the system is $2\pi\{a(n+2)/gn\}^{\frac{1}{2}}$.

25. A triangle ABC is formed of equal smooth rods each of length $2a$, and small equal rings rest on the rods at the middle points of AB, AC, being attached to A by equal elastic threads of natural length l, and connected together by an inextensible thread passing through a fixed smooth ring at

the middle point of BC. Prove that, if there are no external forces, and if one of the rings is slightly displaced, the period of the small oscillations is

$$2\pi\sqrt{\{2alm/E(5a-3l)\}},$$

where m is the mass of each ring and E is the modulus of elasticity.

26. A circular hoop of negligible mass and of radius b carries a particle rigidly attached to it at a point distant c from its centre, and its inner surface is constrained to roll on the outer surface of a fixed circle of radius a, $(b > a)$, under the action of a repulsive force, directed from the centre of the fixed circle and equal to μ times the distance. Prove that the period of small oscillations of the hoop will be

$$2\pi\frac{b+c}{a}\sqrt{\frac{b-a}{c\mu}}.$$

---

27. An equilateral wedge of mass M is placed on a smooth table, with one of its lower edges in contact with a smooth vertical wall, and a smooth ball of mass M' is placed in contact with the wall and with one face of the wedge, so that motion ensues without rotation of the wedge. Prove that the ball will descend with acceleration

$$3M'g/(M+3M').$$

28. Two particles A, B of masses $2m$ and m are attached to an inextensible thread OAB, so that $OA=AB$, and lie on a smooth table with the thread straight and the end O fixed. The particle B is projected on the table at right angles to AB. Prove that, in the subsequent motion, when OAB is again a straight line, the velocity of B is half that of A.

29. A gun is suspended freely at an inclination α to the horizontal by two equal parallel vertical cords in a vertical plane containing the axis of the gun, and a shot whose mass is $1/n$ of that of the gun is fired from it. Prove that the range on a horizontal plane through the muzzle is $4n(1+n)h\tan\alpha$, where h is the height through which the gun rises in the recoil.

30. A railway carriage of mass M, moving with velocity v, impinges on a carriage of mass M' at rest. The force necessary to compress a buffer through the full extent l is equal to the weight of a mass m. Assuming that the compression is proportional to the force, prove that the buffers will not be completely compressed if

$$v^2 < 2mgl(1/M+1/M').$$

Prove also that, if v exceeds this limit, and if the backing against which the buffers are driven is inelastic, the ratio of the final velocities of the carriages is

$$Mv-\sqrt{\{2mM'gl(1+M'/M)\}} : Mv+\sqrt{\{2mMgl(1+M/M')\}}.$$

31. Two particles of masses m and m', joined by an elastic thread of natural length l and modulus λ, are placed on a smooth table with m at the edge and m' at a distance l in a line perpendicular to the edge. The particle

m is then just pushed over the edge. Prove that, if the length of the thread at any time is $l+s$, then

$$\dot{s}^2 = 2gs - \lambda s^2 (m+m')/mm'l.$$

Also, if at time t, m has fallen through z and m' is at a distance x from the edge, prove that

$$m'(l-x) + mz = \tfrac{1}{2}mgt^2.$$

32. Two particles each of mass m are connected by a rod of negligible mass and of length l, and lie on a rough horizontal plane (coefficient of friction μ). One of the particles is projected vertically upwards with velocity V, prove that the other particle will begin to move when the rod makes with the plane an angle α, where α is the least angle which satisfies the equation

$$(V^2 - 3gl \sin \alpha)(\cos \alpha + \mu \sin \alpha) = \mu gl,$$

provided that V^2/gl is less than $3 \sin \alpha + \operatorname{cosec} \alpha$. Find also the radius of curvature of the path immediately afterwards.

33. Two particles, each of mass m, are connected by an inextensible thread of length l, passing over a smooth pulley at the top of a smooth plane of inclination α, on which one of the particles rests at a distance a from the top $(a < l)$. Prove that, in the motion which ensues after the system is free of the plane, the tension of the thread is constant and equal to

$$\tfrac{1}{2}mgal^{-1}\cos^2 \alpha (1 - \sin \alpha),$$

and that the radius of curvature of the path of the upper particle immediately after it leaves the plane is

$$a\,\frac{1-\sin \alpha}{\cos \alpha}\,\frac{[\cos^2 \alpha + \frac{1}{4}(1-\sin \alpha)^2]^{\frac{3}{2}}}{1+\frac{1}{2}al^{-1}\cos^2 \alpha (1-\sin \alpha)}.$$

34. A spherical shell contains a particle of mass equal to $1/k$ times that of the shell, supported by springs of equal length and strength, which are attached at opposite ends of a diameter; and the system, all parts of which are moving in the line of the springs with the same velocity, strikes directly a fixed plane. Show that, if the coefficient of restitution between the shell and the plane is unity, the shell will or will not strike the plane again according as $k <$ or $> 1 + 2\cos \alpha$, where α is the least positive root of the equation $\tan \alpha = \alpha + \pi$.

35. In a smooth table are two small holes A, B at a distance $2a$ apart; a particle of mass M rests on the table at the middle point of AB, being connected with a particle of mass m hanging beneath the table by two inextensible threads, each of length $a(1+\sec \alpha)$, passing through the holes. A blow J is applied to M at right angles to AB. Prove that, if

$$J^2 > 2Mmag \tan \alpha,$$

M will oscillate to and fro through a distance $2a \tan \alpha$, but if

$$J^2 = 2Mmag(\tan \alpha - \tan \beta),$$

where $\tan \beta$ is positive, the distance through which M oscillates will be

$$2a\sqrt{\{(\sec \alpha - \sec \beta)(\sec \alpha - \sec \beta + 2)\}}.$$

CHAPTER VIII†

MOTION OF A RIGID BODY IN TWO DIMENSIONS

215. IN this Chapter we propose to discuss the motion of a rigid body in cases where every particle of the body moves parallel to a fixed plane, for example the plane (x, y) of a frame of reference. In such a case the x and y of a particle of the body vary with the time, but the z of each particle remains constant throughout the motion. The motion is said to be "in two dimensions," or "in one plane." Now we saw in Art. 180 that to determine the position of a rigid body it is requisite and sufficient to determine the positions of a particle of the body, of a line of particles passing through that particle, and of a plane of particles passing through that line. In the case now under discussion we may take the line and plane in question to be parallel to the plane (x, y). Then the position of the plane is invariable; and the position of the line is determined by the angle which it makes with a fixed line in the plane, for instance the axis of x; further, the position of the chosen particle is determined by its coordinates x and y. Thus the determination of the position of the rigid body (moving in two dimensions) requires the determination of three numbers, representing the coordinates of the position of one of the particles, and the angle which a line of the body drawn through that particle, and moving in the plane of its motion, makes with a fixed line.

We can now see what is meant by the angular velocity of a rigid body moving in one plane. Let one line of particles, fixed in the body, and parallel to the plane, make an angle θ at time t with a line fixed in the plane. Then this angle is increasing at a rate $\dot{\theta}$. Let any other line of particles be drawn also parallel to the plane, and let α be the angle which it makes with the first line. Then α is invariable, for if it were to change the body would be deformed. Now the second line of particles makes an angle $\theta + \alpha$

† Articles in this Chapter which are marked with an asterisk (*) may be omitted in a first reading.

with the fixed line, and this angle also increases at a rate $\dot{\theta}$. We thus see that every line of particles parallel to the plane turns with the same angular velocity, and this is the angular velocity of the rigid body.

216. Moment of Inertia. Consider a rigid body turning about an axis with angular velocity ω. Let m be the mass of a particle of the body at a distance r from the axis. Then this particle is describing a circle of radius r with velocity $r\omega$. Hence its moment of momentum about the axis is $mr^2\omega$, and its kinetic energy is $\frac{1}{2}mr^2\omega^2$.

It follows that the moment of momentum of the rigid body about the axis is

$$\omega\Sigma mr^2,$$

and the kinetic energy is

$$\tfrac{1}{2}\omega^2\Sigma mr^2,$$

the summations referring to all the particles.

These expressions become

$$\omega\iiint\rho\,(x^2+y^2)\,dx\,dy\,dz,$$

and

$$\tfrac{1}{2}\omega^2\iiint\rho\,(x^2+y^2)\,dx\,dy\,dz,$$

for a body of density ρ at a point (x, y, z), the axis of rotation being the axis of z.

The integrals are volume integrals taken through the volume of the body; that is to say we must divide the volume of the body into a very large number of volumes, very small in all their dimensions, multiply the value of $\rho\,(x^2+y^2)$ at a point in one of these volumes by this volume, sum the products for all the volumes, and pass to a limit by diminishing the volumes indefinitely. The process will be exemplified in Art. 218.

The multiplier of ω and $\frac{1}{2}\omega^2$ in these expressions is called the *moment of inertia* of the body about the axis. We shall see presently that it enters into the expressions for the kinetic energy and moment of momentum of a rotating body, whether the axis of rotation is fixed or not.

The moment of inertia of a body about an axis depends only on the shape of the body, its situation with reference to the axis, and the distribution of density within it.

217. Theorems concerning Moments of Inertia. I. The moment of inertia of a system about any axis is equal to the moment of inertia about a parallel axis through the centre of mass together with the moment of inertia about the original axis of the whole mass placed at the centre of mass.

Let x, y, z be the coordinates of any particle of the system, m its mass, $\bar{x}, \bar{y}, \bar{z}$ the coordinates of the centre of mass, x', y', z' those of the particle m relative to the centre of mass.

Then
$$\left.\begin{array}{c} x=\bar{x}+x', \quad y=\bar{y}+y', \quad z=\bar{z}+z', \\ \Sigma mx'=0, \quad \Sigma my'=0, \quad \Sigma mz'=0. \end{array}\right\}$$

Now
$$\Sigma mx^2=\Sigma m(\bar{x}+x')^2=\bar{x}^2\Sigma m+\Sigma mx'^2+2\bar{x}\Sigma mx'$$
$$=\bar{x}^2\Sigma m+\Sigma mx'^2.$$

So
$$\Sigma my^2=\bar{y}^2\Sigma m+\Sigma my'^2.$$

Hence
$$\Sigma m(x^2+y^2)=\Sigma m(x'^2+y'^2)+(\bar{x}^2+\bar{y}^2)\Sigma m,$$
which is the theorem stated.

II. The moment of inertia of a plane lamina, of any form, about any axis perpendicular to its plane, is the sum of those about any two rectangular axes in the plane which meet in any point on the first axis.

For, if the axes are taken to be those of z, x, y, the moments of inertia about the axes of x and y are respectively Σmy^2 and Σmx^2, and the moment of inertia about the axis of z is $\Sigma m(x^2+y^2)$.

III. To compare the moments of inertia of a lamina about different axes in its plane.

For parallel axes we can use Theorem I and it will therefore be sufficient to consider axes in different directions through the origin. Let θ be the angle which any line makes with the axis x. The distance of any point (x, y) from this line is $-x\sin\theta+y\cos\theta$, and thus the moment of inertia about the line is $\Sigma m(y\cos\theta-x\sin\theta)^2=\sin^2\theta\Sigma(mx^2)+\cos^2\theta\Sigma(my^2)-2\sin\theta\cos\theta\Sigma mxy$.

The expression for the moment of inertia about a perpendicular line would be
$$\cos^2\theta\Sigma(mx^2)+\sin^2\theta\Sigma(my^2)+2\sin\theta\cos\theta\Sigma(mxy).$$

The quantity $\Sigma(mxy)$ is known as the *product of inertia* with respect to the axes of x and y (in two dimensions). For new axes obtained by turning through an angle θ it has the value
$$(\cos^2\theta-\sin^2\theta)\Sigma(mxy)+\sin\theta\cos\theta\{\Sigma(my^2)-\Sigma(mx^2)\}.$$

We can always choose the axes of (x, y) so that this quantity $\Sigma(mxy)$ vanishes. When this is done the axes of x and y are called *Principal axes* of the lamina. The directions of the principal axes vary with the point chosen as origin.

Now let the axes of x and y be principal axes of the lamina at the origin. Let $A, =\Sigma(my^2)$, be the moment of inertia about the axis x, and $B, =\Sigma(mx^2)$, be the moment of inertia about the axis y. Then the moment of inertia about a line through the origin making an angle θ with the axis x is

$$A\cos^2\theta + B\sin^2\theta.$$

If an ellipse whose equation is $Ax^2+By^2=\text{const.}$ is drawn on the lamina, then the moment of inertia about any diameter of it is inversely proportional to the square of the length of that diameter. This ellipse is known as the *ellipse of inertia.*

IV. If two plane systems in the same plane have the same mass, the same centre of mass, the same principal axes at the centre of mass, and the same moments of inertia about these principal axes, they have the same moment of inertia about any axis in or perpendicular to the plane.

For, in the first place, the two systems have by Theorem III the same moment of inertia about any axis lying in the plane and passing through the common centre of mass, by Theorem I they have the same moment of inertia about any other axis in the plane, and by Theorem II they have the same moment of inertia about any axis perpendicular to the plane.

Such systems are described as *momental equivalents.*

It is clear that two plane systems are momental equivalents if they have the same mass, and the same centre of mass, and if their moments of inertia about any three assigned axes in the plane are equal.

218. Calculations of moments of inertia.

I. *Uniform ring. Radius of gyration of a body.* For a circular ring of mass m and radius a, and of very small section, the moment of inertia about the axis is ma^2, since every element of the mass can be taken to be at the same distance a from the axis.

In the case of a body of any shape, and of mass m, we can always express the moment of inertia about any axis in the form mk^2, where k represents the length of a line; and thus we see that k is the radius of a ring such that, if the mass of the body were condensed uniformly upon the ring, the moment of inertia of the ring about its axis would be the same as the moment of inertia of the body about the axis in question. The quantity k for any body and any axis is known as the *radius of gyration* of that body about that axis.

II. *Uniform rod.* Let m be the mass of the rod, and $2a$ its length, r the distance of any section from the middle point. The mass of the part between the sections r and $r+\delta r$ is $\frac{m}{2a}\delta r$. Therefore, if the thickness of the rod is disregarded, the moment of inertia about an axis through the middle point at right angles to the rod is

$$\int_{-a}^{a} \frac{m}{2a} r^2 dr = \tfrac{1}{3}ma^2.$$

The radius of gyration of the rod is $a/\sqrt{3}$.

III. *Circular disk.* The mass per unit of area of a uniform thin circular disk of radius a and mass m is $m/\pi a^2$. The area of the narrow ring contained between two concentric circles of radii r and $r+\delta r$ is $2\pi(r+\frac{1}{2}\delta r)\,\delta r$. All the particles in such a ring are at distances from the centre which lie between r and $r+\delta r$. Hence the moment of inertia of the disk about an axis drawn through its centre at right angles to its plane is

$$\int_0^a \frac{m}{\pi a^2} r^2 \,.\, 2\pi r dr,$$

which is $\frac{1}{2}ma^2$. The radius of gyration of the disk about this axis is $a/\sqrt{2}$.

IV. *Uniform sphere.* Let a be the radius of the sphere, ρ the (constant) density of the material, and let the origin of coordinates be the centre of the sphere. According to the general formula of Art. 216 we must integrate (x^2+y^2) through the volume of the sphere. Now it follows from the symmetry of the sphere that

$$\iiint x^2 dx\, dy\, dz = \iiint y^2 dx\, dy\, dz = \iiint z^2 dx\, dy\, dz,$$

where the integrations are taken through the volume of the sphere. Hence each of these integrals is equal to

$$\tfrac{1}{3}\iiint (x^2+y^2+z^2)\, dx\, dy\, dz \quad \text{or} \quad \tfrac{1}{3}\iiint r^2 dx\, dy\, dz,$$

where the integration is taken through the volume of the sphere, and r denotes the distance of the point (x, y, z) from the centre.

To evaluate this integral we have first to divide the sphere into a very large number of very small volumes, next to multiply the value of r^2 for a point within one of the small volumes by this volume, then to sum the products so formed, and finally to pass to a limit by diminishing all the small volumes indefinitely.

Now the volume contained between two concentric spheres of radii r and $r+\delta r$ is $4\pi\{r^2+r\,\delta r+\frac{1}{3}(\delta r)^2\}\,\delta r$, and the distances from the centre of all the points in this volume lie between r and $r+\delta r$. Hence the required integral

$$\iiint r^2 dx\, dy\, dz = \int_0^a r^2 \,.\, 4\pi r^2 dr = \frac{4\pi a^5}{5}.$$

The moment of inertia of the sphere about any diameter is therefore

$$\tfrac{2}{3}\rho\frac{4\pi a^5}{5}, \text{ or } \tfrac{2}{5}ma^2,$$

where $m, = \frac{4}{3}\pi\rho a^3$, is the mass of the sphere.

219. Examples.

1. Prove that a momental equivalent of a thin rod of mass m consists of three particles: one of mass $\frac{2}{3}m$ at the middle point, and one of mass $\frac{1}{6}m$ at each of the ends.

2. Prove that the moments of inertia of a uniform rectangular lamina of mass m and sides $2a$, $2b$ about axes through its centre parallel to its edges are $\frac{1}{3}mb^2$ and $\frac{1}{3}ma^2$.

3. Prove that the radius of gyration of a circular disk about a diameter is half the radius. Hence evaluate the integral $\iint x^2 dx dy$ taken over the area of a circle of radius a, the origin being at the centre of the circle. (Cf. II of Art. 217 and IV of Art. 218.)

4. To evaluate the integral $\iint x^2 dx dy$ taken over the area within an ellipse which is given by the equation $x^2/a^2+y^2/b^2=1$, change the variables by putting $x=a\xi$, $y=b\eta$. We have to find the value of $a^3b\iint \xi^2 d\xi d\eta$ where the integration extends over a range of values given by the inequality $\xi^2+\eta^2 \not> 1$. This is the same thing as an integration over the area of a circle of unit radius. Hence prove that the moments of inertia of a uniform thin elliptic lamina of semi-axes a, b and mass m about its principal axes are $\frac{1}{4}mb^2$ and $\frac{1}{4}ma^2$.

5. An ellipsoid is given by an equation of the form $x^2/a^2+y^2/b^2+z^2/c^2=1$. To find the value of $\iiint x^2 dx dy dz$ taken through the volume of the ellipsoid, change the variables by putting $x=a\xi$, $y=b\eta$, $z=c\zeta$. We get

$$a^3bc\iiint \xi^2 d\xi d\eta d\zeta,$$

where the integration extends over a range of values given by the inequality $\xi^2+\eta^2+\zeta^2 \not> 1$. This is the same thing as an integration through the volume of a sphere of unit radius. According to IV of Art. 218 the result is $\frac{4}{15}\pi$. Hence prove that the moments of inertia of the ellipsoid (supposed to be of uniform density ρ) about the axes of x, y, z are

$$\frac{m}{5}(b^2+c^2),\quad \frac{m}{5}(c^2+a^2),\quad \frac{m}{5}(a^2+b^2),$$

where m, $=\frac{4}{3}\pi\rho abc$, is the mass of the ellipsoid.

6. Prove that a momental equivalent of a uniform triangular lamina consists of three particles, each one-third of its mass, placed at the middle points of its sides.

7. Prove that the moment of inertia of a uniform cube of mass m and side $2a$ about an axis through its centre parallel to an edge or at right angles to an edge is $\frac{2}{3}ma^2$.

[It can be shown that the same formula holds for any axis drawn through the centre of the cube.]

220. Velocity and Momentum of rigid body.

Let G be the centre of mass of a rigid body moving in two dimensions, and let u and v be resolved parts of the velocity of G parallel to the axes x and y. Let P be any other particle of the body, r its distance from G, and x', y' its coordinates relative to G

at time t. Then the line GP is turning with the angular velocity ω of the rigid body, and the velocity of P relative to G is $r\omega$ at right angles to GP; the resolved parts of this relative velocity parallel to the axes are $-\omega y'$ and $\omega x'$, since the line GP makes with the axis x an angle whose cosine is x'/r and whose sine is y'/r.

Hence the resolved velocities of P parallel to the axes are

$$u - \omega y' \text{ and } v + \omega x'.$$

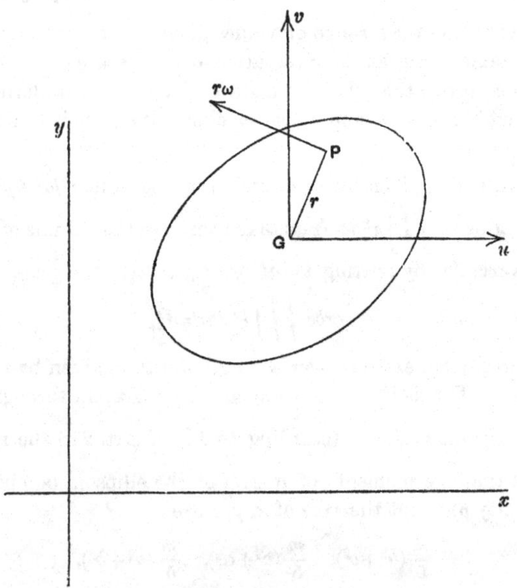

Fig. 63.

Let m be the mass of the particle at P. Then the resultant momentum of the body parallel to the axis x is

$$\Sigma m (u - \omega y'),$$

which is equal to Mu, where $M, = \Sigma m$, is the mass of the body. Similarly the momentum of the body parallel to the axis y is Mv. Thus the resultant momentum of the body is the same as the momentum of a particle of mass equal to the mass of the body placed at the centre of mass and moving with it. (Art. 153.)

The moment of momentum of the body about an axis through the centre of mass perpendicular to the plane of motion is

$$\Sigma m \{x' (v + \omega x') - y' (u - \omega y')\},$$

which is equal to $\omega\Sigma m(x'^2 + y'^2)$ or to $Mk^2\omega$, where k is the radius of gyration about the axis.

The moment of momentum about any parallel axis is the moment about that axis of the momentum of the whole mass placed at the centre of mass and moving with it together with the moment $Mk^2\omega$ (Art. 156). Thus the momentum of the rigid body is specified by the resultant and couple of a system of vectors localized in lines. The resultant is localized in a line through G, and has resolved parts Mu, Mv in the two chosen directions; and the moment of the couple is $Mk^2\omega$.

Again, the kinetic energy of the body is

$$\tfrac{1}{2}\Sigma m\{(u - \omega y')^2 + (v + \omega x')^2\}$$
$$= \tfrac{1}{2}M(u^2 + v^2 + k^2\omega^2),$$

which is the kinetic energy of the whole mass, moving with the centre of mass, together with the kinetic energy of the rotation about the centre of mass (Art. 158).

The formulæ for the velocity of a point show that at each instant the point whose coordinates relative to G are $-v/\omega$ and u/ω has zero velocity, so that the motion of the body at the instant is a motion of rotation about an axis through this point perpendicular to the plane of motion. The point is called the *instantaneous centre of no velocity*, or frequently "the instantaneous centre." The fact that the motion of a rigid plane figure in its plane is equivalent to rotation about a point is of importance in many geometrical investigations.

221. Kinetic Reaction of rigid body. With the notation of the last Article, the point P moves relatively to G in a circle of radius r with angular velocity equal to ω at time t; its acceleration relative to G may therefore be resolved into $r\dot{\omega}$ at right angles to GP, and $r\omega^2$ along PG. Hence the resolved parts of the acceleration of P parallel to the axes are

$$\dot{u} - \dot{\omega}y' - \omega^2 x' \text{ and } \dot{v} + \dot{\omega}x' - \omega^2 y'.$$

The kinetic reactions may be reduced to a resultant kinetic reaction localized in a line through the centre of mass and a couple. The resultant in question has resolved parts parallel to the axes which are

$$\Sigma m(\dot{u} - \dot{\omega}y' - \omega^2 x') \text{ and } \Sigma m(\dot{v} + \dot{\omega}x' - \omega^2 y'),$$

and these are $M\dot{u}$ and $M\dot{v}$.

The couple is the moment of the kinetic reactions about a line through the centre of mass perpendicular to the plane of motion; this moment is

$$\Sigma m\{x'(\dot{v} + \dot{\omega}x' - \omega^2 y') - y'(\dot{u} - \dot{\omega}y' - \omega^2 x')\},$$

and this is $Mk^2\dot{\omega}$.

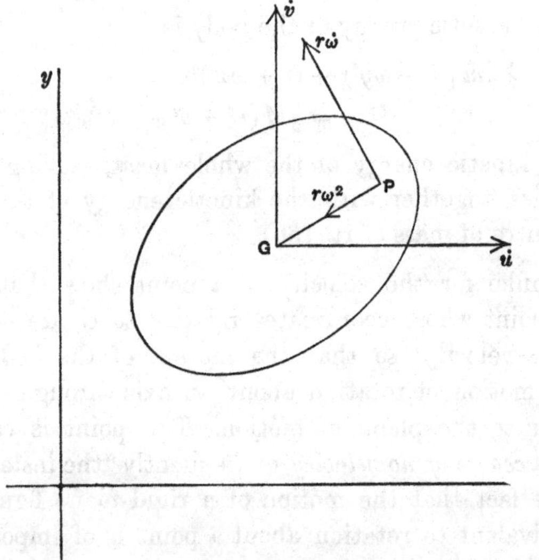

Fig. 64.

The moment of the kinetic reactions about any axis perpendicular to the plane of motion is the moment about that axis of the kinetic reaction of a particle of mass equal to the mass of the body, moving with the centre of mass, together with the moment of the couple $Mk^2\dot{\omega}$. (Art. 157.)

The formulæ for the acceleration of any point of the body show that at each instant there is a point which has zero acceleration. This point is called the *instantaneous centre of no acceleration.* It is of much less importance than the instantaneous centre of no velocity.

222. Examples.

1. Prove that, at any instant, the normal to the path of every particle passes through the instantaneous centre (of no velocity).

[It follows that this centre can be constructed if we know the directions of motion of two particles.]

2. Calculation of the moment of the kinetic reactions about the instantaneous centre (of no velocity).

The coordinates of the instantaneous centre I being $-v/\omega$ and u/ω referred to axes through the centre of mass G parallel to the axes of reference, the moment in question is

$$\frac{v}{\omega} m\dot{v} + \frac{u}{\omega} m\dot{u} + mk^2\dot{\omega}.$$

The velocity of G is $r\omega$ at right angles to the line joining it to I, where $r = IG$, or we have $u^2 + v^2 = r^2\omega^2$.

Hence the above is $$\frac{1}{\omega}\frac{d}{dt}(\tfrac{1}{2}mr^2\omega^2) + mk^2\dot{\omega},$$

or $$\frac{1}{\omega}\frac{d}{dt}\{\tfrac{1}{2}m(k^2+r^2)\omega^2\}.$$

If we take an angle θ such that $\dot{\theta} = \omega$, and write K for the moment of inertia about the instantaneous centre I, then $K = m(k^2 + r^2)$ by I of Art. 217, and the result obtained may be written $\frac{d}{d\theta}(\frac{1}{2}K\omega^2)$.

When the point I is fixed in the body this can be replaced by $K\dot{\omega}$. Other cases in which this formula can be used are noted in Arts. 235 and 236 *infra*.

3. Prove that those particles which at any instant are at inflexions on their paths lie on a circle.

[This circle is called the "circle of inflexions."]

4. Prove that the curvature of the path of any particle which is not on the circle of inflexions is $\omega^3 p^2/V^3$, where p^2 is the power with respect to the circle of the position of the particle, ω is the angular velocity of the body, and V is the resultant velocity of the particle.

5. Prove that, in general, that particle which is at the instantaneous centre (of no velocity) is at a cusp on its path.

223. Equations of motion of rigid body. The equations of motion express the conditions that the kinetic reactions and the external forces may be equivalent systems of vectors.

Let M be the mass of the body, f_1, f_2 the resolved accelerations of the centre of mass in any two directions at right angles to each other in the plane of motion, ω the angular velocity of the body.

Let the forces acting on the body be reduced to a resultant force at its centre of mass and a couple. Let P, Q be the resolved parts of the force in the directions in which the acceleration of the centre of mass was resolved, and let N be the couple.

Then the system of vectors expressed by Mf_1, Mf_2, $Mk^2\dot{\omega}$ has the same resolved part in any direction, and the same moment about any axis, as the system P, Q, N.

In particular we have

$$Mf_1 = P,\ Mf_2 = Q,\ Mk^2\dot{\omega} = N,$$

and the equations of motion of the body can always be written in this form.

In the formation of equations of motion diversity can arise from the choice of directions in which to resolve, and of axes about which to take moments. As in the case of Dynamics of a Particle, the equations arrived at are differential equations, and no rules can be given for solving them in general. If however the circumstances are such that there is an equation of energy, or an equation of conservation of momentum, such equations are first integrals of the equations of motion.

224. Continuance of motion in two dimensions. The question arises whether a body, which at some instant is moving in two dimensions parallel to a certain plane, continues to move parallel to that plane or will presently be found to be moving in a different manner. A general answer to this question cannot be given here, but it is clear that there is a class of cases in which the motion in two dimensions persists. This class includes all the cases in which the body is symmetrical with respect to a plane and the forces applied to it are directed along lines lying in that plane, or, more generally, when the forces can be reduced to a single resultant in the plane of symmetry and a couple about an axis perpendicular to that plane.

225. Rigid Pendulum†. A heavy body free to rotate about a fixed horizontal axis is known as a "compound pendulum" to

† Ch. Huygens was the first to solve the problem of the motion of the pendulum, and the principles which he invoked were among the considerations which ultimately led to the establishment of the Theory of Energy. His work, *De horologio oscillatorio*, was first published in 1673.

distinguish it from the "simple pendulum" whose motion was discussed in Arts. 95 and 119.

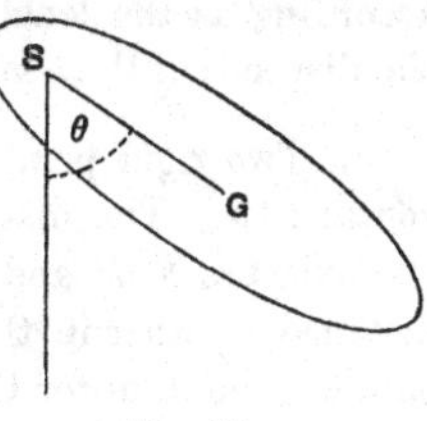

Fig. 65.

Let G be the centre of mass of the body, GS the perpendicular from G to the axis, θ the angle which GS makes with the vertical at time t. Then the whole motion takes place in the vertical plane which passes through G and is at right angles to the axis; and the position of the pendulum at any time depends only on the angle θ.

Let $GS = h$. Let M be the mass of the body, k its radius of gyration about an axis through G perpendicular to the plane of motion.

The velocity of the centre of mass is $h\dot{\theta}$, and the kinetic energy is

$$\tfrac{1}{2}M(h^2 + k^2)\,\dot{\theta}^2.$$

The potential energy of the body in the field of the earth's gravity is

$$Mgh(1 - \cos\theta),$$

the standard position being the equilibrium position.

Hence the energy equation can be written

$$\tfrac{1}{2}M(h^2 + k^2)\,\dot{\theta}^2 = Mgh\cos\theta + \text{const.}$$

Comparing this equation with that obtained in Art. 119, we see that the motion is the same as that of a simple pendulum of length $(k^2 + h^2)/h$.

A point in the line SG at this distance from S is known as the "centre of oscillation," S is called the "centre of suspension." The distance between these centres is the "length of the equivalent simple pendulum."

226. Examples.

1. A rigid pendulum, for which S and O are respectively a centre of suspension and the corresponding centre of oscillation, is hung up so that it can oscillate in the same vertical plane as before, but with O as centre of suspension instead of S; prove that S will be the centre of oscillation.

2. A uniform rod moves with its ends on a smooth circular wire fixed in a vertical plane. Prove that, if it subtends an angle of 120° at the centre, the length of the equivalent simple pendulum is equal to the radius of the circle.

3. A compound pendulum consists of a rod, which can turn about a fixed horizontal axis, and a spherical bob, which can slide on the rod. Prove that the period of oscillation will be prolonged by sliding the bob up or down, according as the length of the equivalent simple pendulum is $>$ or $<$ twice the distance of the centre of gravity of the bob from the axis of rotation.

4. Two rigid pendulums of masses m and m' turn about the same horizontal axis. The distances of the centres of mass and of oscillation from the axis are h, h' and l, l' respectively. Prove that, if the pendulums are fastened together in the position of equilibrium, the length of the equivalent simple pendulum for the compound body will be $(mhl+m'h'l')/(mh+m'h')$.

227. Illustrative Problems. We exemplify the application of the principles that have been laid down by partially working out some problems. The most important matters to be illustrated are actions between two rigid bodies whether smooth or rough, and the expression of the effects of the inertia of a rigid body by means of the moment of inertia. Other matters of subsidiary interest are the kinematical expression of velocities and accelerations in terms of a small number of independent geometrical quantities, the expression of kinematical conditions, and the calculation of resultant stresses.

I. *Inertia of machines.* We shall consider Atwood's machine. To avoid having to take account of the motion of the pulley in our preliminary notice of Atwood's machine (Art. 73) we assumed the pulley to be perfectly smooth or that the rope slides over it without frictional resistance and without setting it in motion. It will now be most convenient, in order to get some idea of the way in which the motion of the pulley affects the result, to suppose the pulley to be so rough that the particles of the rope and the pulley in contact move with the same velocity along the tangents to the pulley.

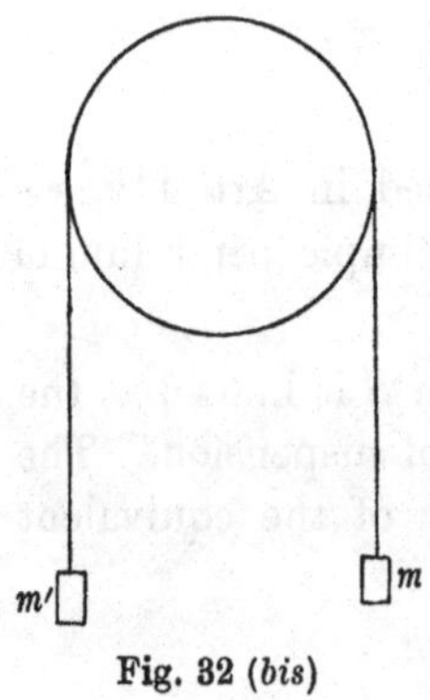

Fig. 32 (*bis*)

Now let M be the mass of the pulley, a its radius, k its radius of gyration about its axis, θ the angle through which it has turned up to time t.

Let m and m' be the masses of the bodies attached to the rope, and x the distance through which m has fallen at time t. Then $x=a\theta$.

The mass of the rope being neglected, the kinetic energy is

$$\tfrac{1}{2}Mk^2\dot{\theta}^2+\tfrac{1}{2}(m+m')\,\dot{x}^2,$$

and the work done is $\qquad (m-m')\,gx,$

so that the energy equation is

$$\tfrac{1}{2}M\frac{k^2}{a^2}\,\dot{x}^2+\tfrac{1}{2}(m+m')\,\dot{x}^2=(m-m')\,gx+\text{const.}$$

Thus the acceleration with which m descends is

$$\frac{m-m'}{m+m'+Mk^2/a^2}g.$$

It appears that the effect of the inertia of the pulley is equivalent to an increase of each of the masses in the simple problem (where the pulley is regarded as smooth and its mass is neglected) by $\frac{1}{2}Mk^2/a^2$.

II. *Wheel set in motion by couple.* Let a wheel, the plane of which is vertical, be in contact with rough horizontal ground; and let the wheel be set in motion by a couple about its axis.

Let a be the radius of the wheel, k the radius of gyration about the axis, m the mass, G the applied couple, F the friction and R the pressure at the point of contact with the ground.

Let ω be the angular velocity with which the wheel turns, v the velocity with which its centre moves.

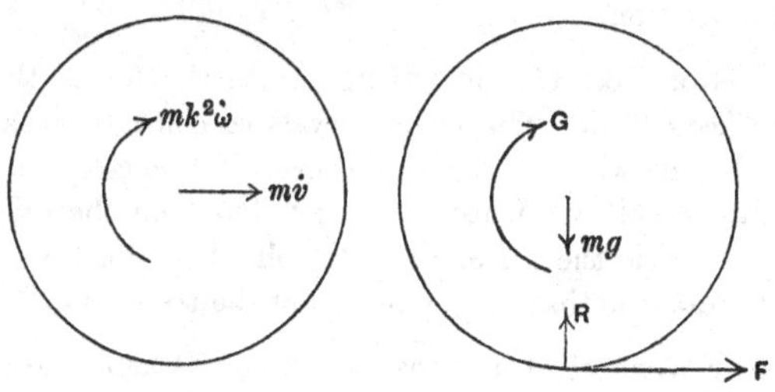

Fig. 66.

The left-hand figure is the diagram of the kinetic reactions, and the right-hand figure is the diagram of the applied forces.

The equations of motion, obtained by resolving horizontally and vertically and taking moments about the centre, are

$$m\dot{v}=F, \quad 0=R-mg, \quad mk^2\dot{\omega}=G-Fa.$$

We have drawn the figure, and written down the equations, on the supposition that v does not exceed $a\omega$. When $v<a\omega$, the point of contact slips on the plane in the sense opposite to that of v, and then the friction acts in the sense shown.

If $v=a\omega$, so that the wheel rolls, we may eliminate F from two of our equations, and obtain the equation

$$m(k^2+a^2)\dot{\omega}=G.$$

The sense of $\dot{\omega}$ is the same as that of G; and therefore, if the motion starts from rest, the sense of ω is the same as that of G. In the same case $F=Ga/(k^2+a^2)$, which is positive, so that the friction acts in the sense in which the centre of the wheel moves (the sense shown in Fig. 66).

In order that this motion may take place it is necessary that F/R or $Ga/\{(k^2+a^2)mg\}$ should not exceed the coefficient of friction.

We conclude that, if the ground is sufficiently rough, the wheel will begin to roll along the road, and that the friction at the point of contact is the horizontal force which produces the horizontal momentum.

III. *Wheel set in motion by force.* Again let the wheel of No. II be set in motion by a horizontal force P applied at its centre in its plane. With the same notation as before, we have the equations of motion

$$m\dot{v}=P+F, \quad 0=R-mg, \quad mk^2\dot{\omega}=-Fa.$$

If the wheel rolls, so that $v=a\omega$, we have, on eliminating F,

$$m(k^2+a^2)\dot{\omega}=Pa.$$

Hence $\dot{\omega}$ is positive, and F is negative, and equal to $-Pk^2/(k^2+a^2)$. The friction in this case acts in the sense opposite to that in which P acts, or the centre of the wheel moves (*i.e.* in the sense opposite to that shown in Fig. 66). The motion will be one of rolling if $Pk^2/\{mg(k^2+a^2)\}$ is less than the coefficient of friction.

The problems of Nos. II and III illustrate the forces that affect the motion of a *railway train.* The machinery is so contrived that a couple is exerted on the driving wheel of the locomotive. If this couple is too great, or the friction is too small, the wheel slips or "skids" on the rail; but, if the friction is great enough, the wheel starts to roll. The direction of the friction at the point of contact is that of the motion of the train as in No. II.

The motion of a wheel of any coach or truck attached to the train is of the character considered in No. III. The tension in the coupling is a horizontal force setting the vehicle in motion, and the frictions at the points of contact of the wheels with the rails act as resistances.

It appears that the "pull of the engine" (Art. 71) is really the friction of the rails on the driving wheel. This is the "force" which sets the train in motion, and keeps it in motion against the resistances. The condition for the production of the motion is the existence of a source of internal energy, which can be transformed into work done by the couple acting on the driving wheel. The way in which a source of internal energy may result in the production of motion, through the agency of external forces, has already been illustrated in simple cases in Ex. 1 of Art. 207 and Ex. 6 of Art. 214. All the characteristic motions of machines and of living creatures are examples of the same principles, but the working out of the details is in general a matter of difficulty. The external forces, such as the friction in this problem, are necessary to the successful action of the animal or machine. (Cf. R. S. Ball, *Experimental Mechanics*, 2nd Edition, London, 1888, pp. 83, 84.)

IV. *Rolling and sliding.* We take the problem presented by a uniform cylinder of mass M and radius a which is set rolling and sliding on a rough horizontal plane, the angular velocity being initially such that the points on the lowest generator have the greatest velocity.

Let V be the velocity of the axis, and ω the angular velocity at time t, the senses being those shown in Fig. 67.

The system of kinetic reactions reduces to $M\dot{V}$ horizontally through the centre of mass, in the sense of V, and a couple $Mk^2\dot{\omega}$ in the sense of ω, where k is the radius of gyration about the axis of the cylinder.

Taking moments about the point of contact we have

$$Ma\dot{V} - Mk^2\dot{\omega} = 0.$$

Fig. 67.

Now let F be the friction between the cylinder and the plane. The particles on the lowest generator have velocity $V + a\omega$ in the sense of V, and therefore F has the opposite sense.

Resolving horizontally we have

$$M\dot{V} = -F,$$

where F is positive. Hence $\dot{V}$ is negative and $\dot{\omega}$ is also negative.

The velocity V diminishes and the angular velocity ω also diminishes according to the equation

$$k^2\omega - aV = k^2\omega_0 - aV_0,$$

where V_0 and ω_0 are the values of V and ω in the beginning of the motion. We shall proceed with the case where $V_0 < \omega_0 k^2/a$. Then there must come an instant at which V vanishes, and at this instant ω has the value $\omega_0 - aV_0/k^2$. At this instant the lowest point has velocity $a\omega_0 - V_0 a^2/k^2$ in the same sense as before, the friction is still finite and in the same sense as before, and a velocity of the centre in the opposite sense begins to be generated.

At any later stage of the motion let U be the velocity in the sense opposite to V_0. See Fig. 68. Then so long as $a\omega > U$ the friction F acts in the same sense, and we have

$$\left.\begin{aligned} M\dot{U} &= F, \\ Ma\dot{U} + Mk^2\dot{\omega} &= 0, \end{aligned}\right\}$$

whence U increases and ω diminishes according to the equation

$$aU + k^2\omega = k^2\omega_0 - aV_0.$$

When U becomes equal to $a\omega$ the value of either is

$$a(k^2\omega_0 - aV_0)/(a^2 + k^2),$$

and at this instant the cylinder is rolling on the plane. Thereafter the cylinder rolls on the plane uniformly.

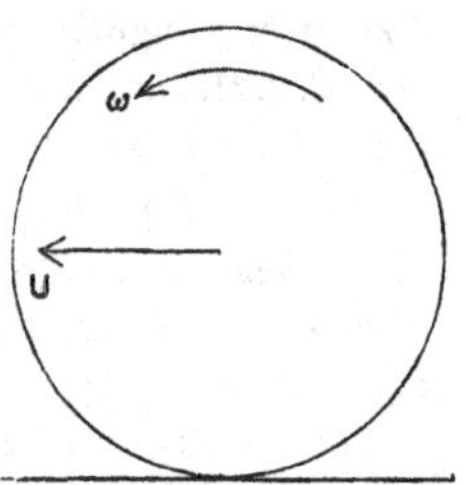

Fig. 68.

It is to be noticed that, in this problem, so long as the cylinder slips, the friction is constantly equal to μMg, where μ is the coefficient of friction between the cylinder and the plane.

228. Examples.

1. In the problem just considered prove that the time from the beginning of the motion until the motion becomes uniform is $\frac{k^2}{a^2+k^2}\frac{V_0+a\omega_0}{\mu g}$.

2. A homogeneous cylinder of mass M and radius a is free to turn about its axis which is horizontal, and a particle of mass m is placed upon it close to the highest generator. Prove that, when the particle begins to slip, the angle θ which the radius through it makes with the vertical is given by the equation

$$\mu\{(M+6m)\cos\theta-4m\}=M\sin\theta,$$

where μ is the coefficient of friction between the particle and the cylinder.

3. A uniform thin circular hoop of radius a spinning in a vertical plane about its centre with angular velocity ω is gently placed on a rough plane of inclination a equal to the angle of friction between the hoop and the plane so that the sense of rotation is that for which the slipping at the point of contact is down a line of greatest slope. Prove that the hoop will remain stationary for a time $a\omega/g\sin a$ before descending with acceleration $\frac{1}{2}g\sin a$.

4. A locomotive engine of mass M has two pairs of wheels of radius a such that the moment of inertia of either pair with its axle about its axis of rotation is A. The engine exerts a couple G on the forward axle. Prove that, if both pairs of wheels bite at once when the engine starts, the friction between one of the forward wheels and the line capable of being called into play must not be less than $\frac{1}{2}G(A+Ma^2)/a(2A+Ma^2)$. Prove also that, if the only action between an axle and its bearings is a frictional couple varying as the angular velocity of the axle, the final friction called into play between either forward wheel and the line is $G/4a$.

5. A uniform sphere rolls down a rough plane of inclination a to the horizontal. Prove that the acceleration of its centre is $\frac{5}{7}g\sin a$, and that the ratio of the friction to the pressure is $\frac{2}{7}\tan a$.

***229. Kinematic condition of rolling.** Consider the following problem :—

A cylinder of radius b rolls on a cylinder of radius a, which rolls on a horizontal plane. It is required to determine the motion.

Let m and m' be the masses, A and B the centres, V the horizontal velocity of m, Ω the angular velocity of m, θ the angle which AB makes with the vertical, ω the angular velocity of m', k and k' the radii of gyration of m and m' about their axes.

The condition that m rolls on the plane is $V=a\Omega$ ....(1).

The velocity of B relative to A is $(a+b)\dot{\theta}$ at right angles to AB, and the velocity of B is therefore compounded of this velocity and V horizontally. (Fig. 69.)

The velocity of P (considered as a point of m') relative to B is $b\omega$ at right angles to AB, in the sense of $(a+b)\dot{\theta}$.

The velocity of P (considered as a point of m) relative to A is $a\Omega$ at right angles to AB, but in the opposite sense.

The condition of rolling is that the particles of m and m' that are at P have the same velocity along the common tangent to the two circles.

We therefore have $$(a+b)\dot{\theta}+b\omega=-a\Omega \quad\text{.............(2).}$$

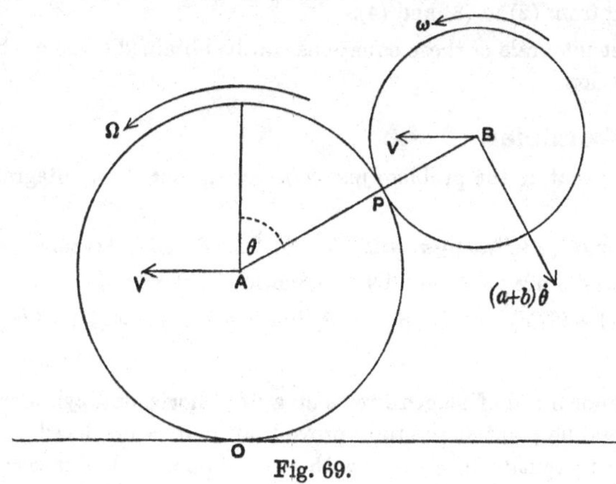

Fig. 69.

In the diagram of accelerations (Fig. 70) we have introduced the value of V from equation (1).

Since B describes a circle relative to A with angular velocity $\dot{\theta}$, the acceleration of B relative to A is compounded of $(a+b)\ddot{\theta}$ at right angles to AB, and $(a+b)\dot{\theta}^2$ in BA. This gives us the diagram.

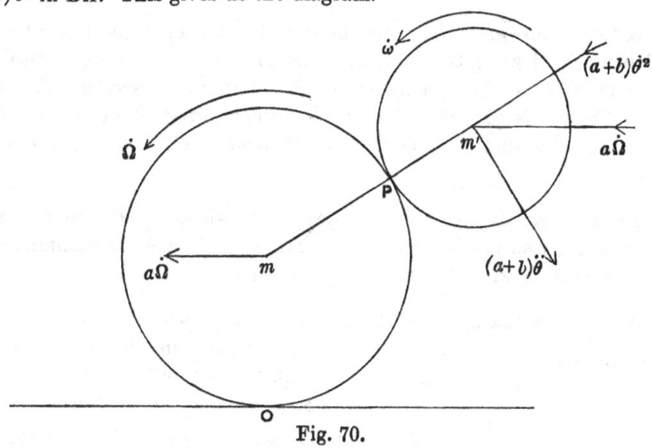

Fig. 70.

Now, to form the equations of motion, take moments about P for m', and about O for the system. We have

$$-m'b(a+b)\ddot{\theta}+m'a\dot{\Omega}b\cos\theta+m'k'^2\dot{\omega}=-m'gb\sin\theta \quad\text{...........(3),}$$

and

$$\left.\begin{aligned}&mk^2\dot{\Omega}+ma^2\dot{\Omega}+m'a\dot{\Omega}\{a+(a+b)\cos\theta\}+m'k'^2\dot{\omega}\\&-m'(a+b)\ddot{\theta}(a+b+a\cos\theta)+m'(a+b)\dot{\theta}^2a\sin\theta=-m'g(a+b)\sin\theta\end{aligned}\right\}\text{...(4).}$$

One of the quantities ω and Ω can be eliminated by means of equation (2), and there then remain two unknown quantities in terms of which the motion can be completely expressed by solving the equations that are obtained by substituting from (2) in (3) and (4).

Two first integrals of these equations can be obtained; one of them is the energy equation.

*230. Examples.

1. Prove that, in the problem just considered, there is an integral equation of the form

$$ma\Omega\,(1+k^2/a^2)+m'\,\{a\Omega-(a+b)\,\dot\theta\cos\theta-\omega k'^2/b\}=\text{const.},$$

and that $\dot\theta$ and θ are connected by an equation of the form

$$\tfrac{1}{2}(a+b)\,\dot\theta^2\,[(1+k'^2/b^2)-m'\,(\cos\theta-k'^2/b^2)^2/\{m\,(1+k^2/a^2)+m'\,(1+k'^2/b^2)\}]+g\cos\theta=\text{const.}$$

2. A uniform rod of length l rests on a fixed horizontal cylinder of radius a with its middle point at the top; prove that, if it is displaced in a vertical plane, so as to remain in contact with the cylinder, and if it rocks without slipping, the angle θ which it makes with the horizontal at time t is given by the equation

$$\tfrac{1}{2}\left(\tfrac{1}{12}l^2+a^2\theta^2\right)\dot\theta^2+ga\,(\cos\theta+\theta\sin\theta)=\text{const.},$$

and the length of the equivalent simple pendulum for small oscillations is

$$\tfrac{1}{12}l^2/a.$$

3. A thread unwinds from a reel of radius a, the uppermost point of the thread being held fixed, the unwound part of the thread being vertical, and the axis of the reel being horizontal. Prove that the acceleration of the centre of the reel is $ga^2/(a^2+k^2)$, where k is the radius of gyration of the reel about its axis, and that the tension of the thread is $k^2/(k^2+a^2)$ of the weight of the reel.

4. A thread passes over a smooth peg and unwinds itself from two cylindrical reels freely suspended from it and having their axes horizontal. Prove that each reel descends with uniform acceleration.

5. A ball is at rest in a cylindrical garden roller, when the roller is seized and made to roll uniformly on a level walk; to find the motion of the ball, assuming that it does not slip on the roller.

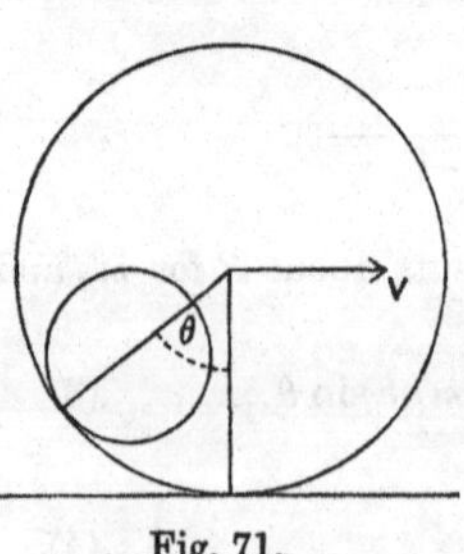

Fig. 71.

Let a be the radius of the ball, b of the roller, θ the angle which the line of centres makes with the vertical, V the velocity of the roller.

Prove (i) that the angular velocity of the roller is V/b,

(ii) that the angular velocity ω of the ball is

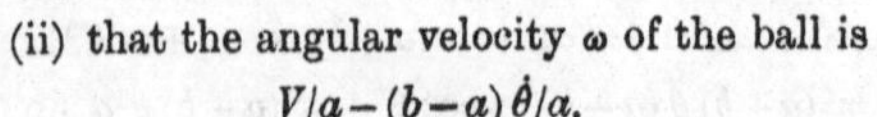

$$V/a-(b-a)\,\dot\theta/a.$$

Let k be the radius of gyration of the ball, supposed uniform, about an axis through its centre, m the mass of the ball. Initially all the impulsive forces acting on the ball pass through the point of contact, and therefore the moment of momentum of the ball about any axis through this point is zero initially. Hence obtain the equation

$$mk^2\omega_0 - ma\{(b-a)\dot{\theta}_0 - V\} = 0$$

for the initial values ω_0 of ω and $\dot{\theta}_0$ of $\dot{\theta}$. Prove also that ω_0 vanishes, and find the value of $\dot{\theta}_0$.

Obtain the equations of motion

$$mk^2\dot{\omega} - ma(b-a)\ddot{\theta} = mga \sin\theta,$$

$$m(b-a)\dot{\theta}^2 = R - mg\cos\theta,$$

where R is the pressure of the roller on the ball. Prove that the motion in θ is the same as that of a simple pendulum of length $\frac{7}{5}(b-a)$. Prove also that the value of R in any position is

$$mg\left(\tfrac{17}{7}\cos\theta - \tfrac{10}{7}\right) + mV^2/(b-a).$$

Deduce the condition that the ball may roll quite round the interior of the roller.

6. A cube containing a spherical cavity slides without friction down a plane of inclination α, and a homogeneous sphere rolls in the cavity. Prove that the angle θ, between the normal to the plane and the common normal to the sphere and the cavity, is connected with the angular velocity ω of the sphere by the equation $(a-b)\dot{\theta} = b\omega$, where a is the radius of the cavity, and b is the radius of the sphere.

Further, taking M and m for the masses of the cube and sphere, and x for the distance described by the cube in time t, obtain the equations of motion by resolving for the system down the plane and at right angles to it and taking moments for the sphere about its point of contact with the cavity.

Finally obtain the equation

$$\tfrac{1}{2}\{\tfrac{7}{5}(M+m) - m\cos^2\theta\}\dot{\theta}^2 - (M+m)\cos\alpha\cos\theta\, g/(a-b) = \text{const.}$$

7. Prove that, when the plane of Ex. 6 is rough, and ϵ is the angle of friction between it and the cube, the value of θ at time t is given by the equation

$$\frac{1}{2}\frac{d}{d\theta}[\{\tfrac{7}{5}(M+m)\cos\epsilon - m\cos\theta\cos(\theta-\epsilon)\}\dot{\theta}^2] - \tfrac{1}{2}m\dot{\theta}^2\sin\epsilon$$
$$+ (M+m)\cos\alpha\sin(\theta-\epsilon)\, g/(a-b) = 0.$$

8. Motion of a circular disk rolling on a given curve under gravity.

Let c be the radius of the disc, ϕ the angle which the normal at the point of contact makes with the vertical, ρ the radius of curvature of the curve at this point. The centre of the disk describes a curve

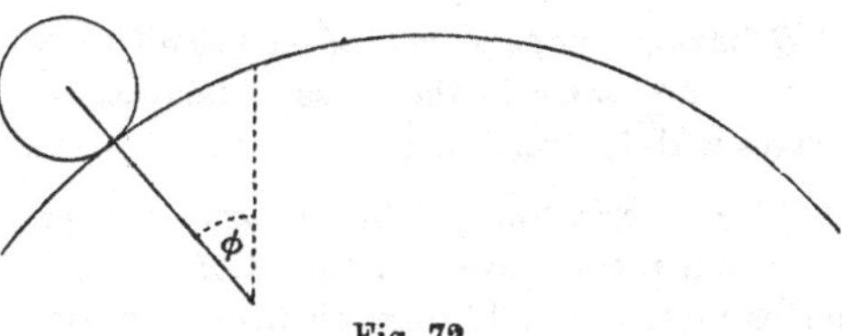

Fig. 72.

parallel to the given curve and at a distance c from it, and the instantaneous centre of rotation of the disk is at the point of contact, so that if ω is the angular velocity of the disk, we have

$$\text{Velocity of centre} = c\omega = (\rho + c)\,\dot{\phi}.$$

Hence obtain the equation of energy

$$\tfrac{1}{2}(\rho+c)^2(1+k^2/c^2)\,\dot{\phi}^2 = g\int(\rho+c)\sin\phi\,d\phi,$$

where k is the radius of gyration of the disk about its centre of mass, supposed to coincide with its centre of figure. Investigate the corresponding equation when the curve is concave to the disk.

Prove that the disk can roll inside a cycloid the radius of whose generating circle is a and whose vertex is lowest so that the angular velocity $\dot{\phi}$ is uniform and equal to

$$\tfrac{1}{2}\sqrt{\{g/a\,(1+k^2/c^2)\}}.$$

Prove that, when the disk is uniform and rolls outside a cycloid, the radius of whose generating circle is $\tfrac{1}{4}c$ and whose vertex is highest, the motion is determined by the equation

$$3c\dot{\phi}^2\cos^4\tfrac{1}{2}\phi = g\,(3+\cos\phi)\sin^2\tfrac{1}{2}\phi,$$

and that the disk leaves the cycloid when $\cos\phi = \tfrac{3}{5}$.

9. A uniform rod slides in a vertical plane between a smooth vertical wall and a smooth horizontal plane. To determine the motion.

Let AB be the rod, $2a$ its length, m its mass, and let the end A move vertically in contact with the wall and the end B horizontally in contact with the plane. The instantaneous centre I is the intersection of the horizontal through A and the vertical through B, and the figure $OBIA$ is a rectangle, so that the centre of mass G, which is the middle point of AB, is always at a distance a from O.

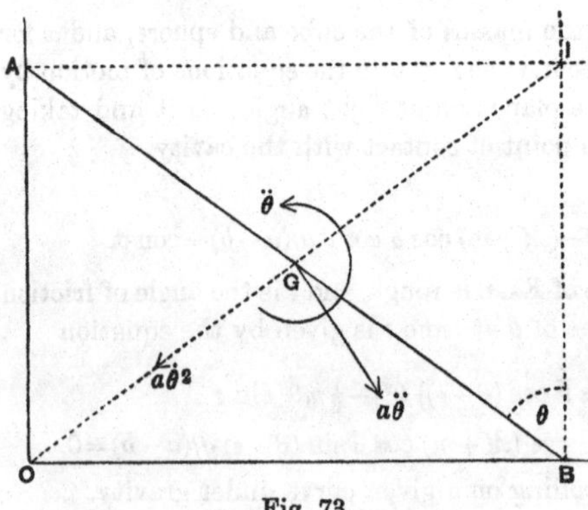

Fig. 73.

The system of kinetic reactions is therefore equivalent to a resultant kinetic reaction at G having components $ma\ddot{\theta}$ and $ma\dot{\theta}^2$ perpendicular to OG and along GO, and a couple $mk^2\ddot{\theta}$ in the sense of increase of the angle θ which the rod BA makes with the vertical BI.

The forces acting on the rod are its weight at G, the horizontal pressure at A, and the vertical pressure at B. The lines of action of the two latter forces meet in I. If then we take moments about I the pressures do not enter into the equation.

Hence prove that the motion in θ is the same as that of a simple pendulum of length $\frac{4}{3}a$.

By resolving horizontally and vertically find the pressures at A and B, and show that the rod leaves the wall when $\cos\theta = \frac{2}{3}\cos\alpha$, α being the initial value of θ.

10. When the plane and the wall of Ex. 9 are both rough, with the same angle of friction ϵ, prove that the value of θ at time t is given by the equation

$$a\left(\tfrac{1}{3} + \cos 2\epsilon\right)\ddot{\theta} - a\dot{\theta}^2 \sin 2\epsilon = g\sin(\theta - 2\epsilon).$$

11. A wheel, whose centre of gravity is at its centre, rolls down a rough plane of inclination α, dragging a particle of mass m, which slides on the plane, and is connected with the centre of the wheel by a thread; the whole motion takes place in a vertical plane, and the thread makes an angle β with the line of greatest slope down which the particle slides. Prove that the system descends with uniform acceleration

$$\frac{M\sin\alpha\cos(\beta-\epsilon) + m\cos\beta\sin(\alpha-\epsilon)}{M(k^2+a^2)\cos(\beta-\epsilon) + ma^2\cos\beta\cos\epsilon}\,ga^2,$$

where a is the radius of the wheel, M its mass, k its radius of gyration about its axis, m the mass of the particle and ϵ the angle of friction between it and the plane.

12. Two smooth spheres are in contact, and the lower slides on a horizontal plane.

Let M, m be the masses, a and b the radii, θ the angle which the line of centres makes with the vertical at time t. If the whole system starts from rest, the centre of mass G descends vertically, for there is no resultant horizontal force on the system. Further, since all the forces acting on either sphere pass through its centre, neither acquires any angular velocity. Let x be the distance of the centre of the lower sphere (M) from the vertical through the centre of mass at time t, then the distance of G from the centre of M is $m(a+b)/(M+m)$, and thus the horizontal velocity of G is

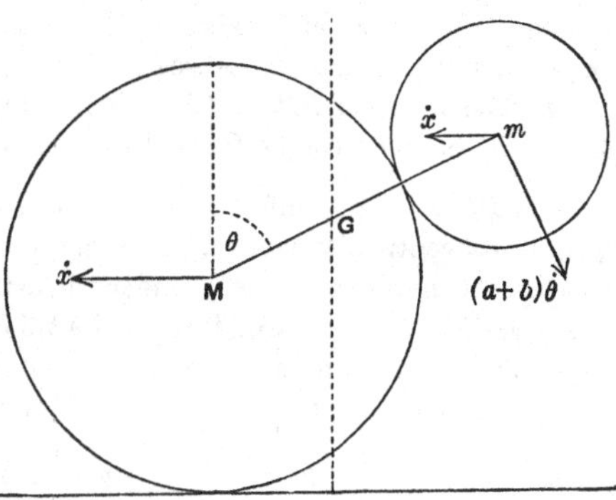

Fig. 74.

$$\dot{x} - \frac{m}{M+m}(a+b)\dot{\theta}\cos\theta.$$

By equating this to zero we express $\dot{x}$ in terms of θ and $\dot{\theta}$.

Hence prove that the equation of energy can be put in the form

$$\frac{1}{2}\left(1 - \frac{m}{M+m}\cos^2\theta\right)\dot{\theta}^2 + \frac{g}{a+b}\cos\theta = \text{const.}$$

Find the pressure between the spheres in any position, and prove that, if $\theta = \alpha$ initially, the spheres separate when

$$\cos\theta\left(3 - \frac{m}{M+m}\cos^2\theta\right) = 2\cos\alpha.$$

***231. Stress in a rod.** As an example of the resultant force between two parts of a body we consider the case of a rigid uniform rod swinging as a pendulum about one end.

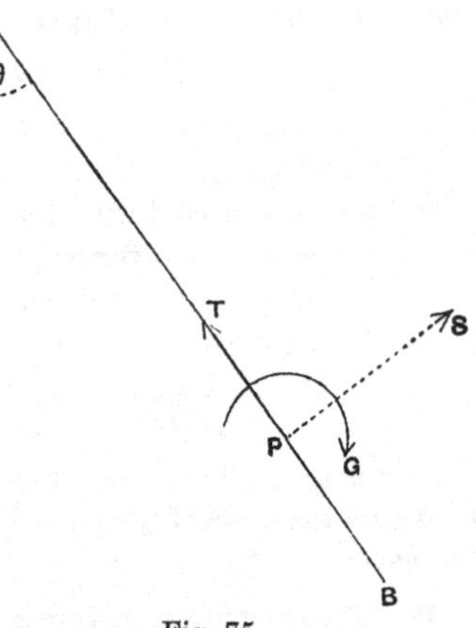

Fig. 75.

If m is the mass of the rod, $2a$ its length, θ the angle which it makes with the vertical at time t, we have, since the radius of gyration about the centre of mass is $a/\sqrt{3}$,

$$\tfrac{4}{3}a^2\ddot{\theta} = -ag\sin\theta,$$

and $$\tfrac{2}{3}a\dot{\theta}^2 = g(\cos\theta - \cos\alpha),$$

where α is the amplitude of the oscillations.

Now consider the action between the two parts of the rod exerted across a section distant $2x$ from the free end. Let P be the centroid of this section. We may suppose the action of AP on BP reduced to a force at P and a couple, and we may resolve the force into a tension T in the rod, and a shearing force S at right angles to it. We call the couple G, and suppose the senses of T, S, and G to be those shown in the figure. The action of BP on AP is then reducible to a force at P having components T, S, and a couple G, in the opposite senses to those shown.

Now BP is a rigid uniform rod of mass mx/a, turning with angular velocity $\dot{\theta}$, while its centre describes a circle of radius $2a - x$ with the same angular velocity. It moves in this way under the action of the forces T, S, the weight mgx/a vertically downwards through its middle point, and the couple G. By resolving along AB and at right angles to it, and by taking moments about P, we obtain the equations of motion of BP in the form

$$\left.\begin{aligned} m\frac{x}{a}(2a-x)\dot{\theta}^2 &= T - mg\frac{x}{a}\cos\theta, \\ m\frac{x}{a}(2a-x)\ddot{\theta} &= S - mg\frac{x}{a}\sin\theta, \\ m\frac{x}{a}\left\{x(2a-x)\ddot{\theta} + \frac{x^2}{3}\ddot{\theta}\right\} &= -G - mg\frac{x}{a}x\sin\theta, \end{aligned}\right\}$$

and by these equations T, S, and G are determined, $\ddot{\theta}$ and $\dot{\theta}^2$ being known. In particular the couple G resisting bending is

$$\tfrac{1}{2}mg\sin\theta\,\frac{x^2}{a^2}(a-x), \text{ or } \tfrac{1}{4}mg\sin\theta\,\frac{AP\,.\,BP^2}{AB^3}.$$

232. Impulsive motion. We apply the theory of sudden changes of motion of any system (Ch. VI, Art. 168) and the theory of the momentum of a rigid body (Art. 220).

We have three equations of impulsive motion expressing that the change of momentum of the body is equivalent to the impulses exerted upon it.

The momentum of the body was shown to be equivalent to a resultant momentum localized in a line through the centre of mass, and equal to the momentum of the whole mass of the body moving with the centre of mass, together with a couple, of amount equal to the product of the angular velocity of the body and the moment of inertia about an axis through the centre of mass perpendicular to the plane of motion.

Let m be the mass of the body, U, V the resolved velocities of the centre of mass in two directions (at right angles to each other) in the plane of motion, and Ω the angular velocity before impact; let u, v be the resolved velocities of the centre of mass in the same two directions after impact, and ω the angular velocity; also let k be the radius of gyration of the body about an axis through the centre of mass perpendicular to the plane of motion.

The change of momentum of the system can be expressed as a vector localized in a line through the centre of mass, whose resolved parts in the two specified directions are $m(u-U)$ and $m(v-V)$; together with a couple, in the plane of motion, of moment $mk^2(\omega-\Omega)$.

The impulses exerted on the body can be expressed as a single impulse at any origin and an impulsive couple.

The equations of impulsive motion express the equivalence of the two systems of vectors.

Thus if the impulses are reduced to an impulse at the centre of mass, whose resolved parts in the specified directions are $\underset{\cdot}{X}$ and $\underset{\cdot}{Y}$, together with a couple $\underset{\cdot}{N}$, we can take the equations of impulsive motion to be

$$m(u-U)=\underset{\cdot}{X},\quad m(v-V)=\underset{\cdot}{Y},\quad mk^2(\omega-\Omega)=\underset{\cdot}{N}.$$

More generally, the resolved part, in any direction, of the vector whose resolved parts, in the specified directions, are $m(u-U)$ and

$m(v-V)$ is equal to the resolved part, in the same direction, of the vector whose resolved parts, in the specified directions, are $\underset{\cdot}{X}$ and $\underset{\cdot}{Y}$; and the moment about any axis of the vector system determined by $m(u-U)$, $m(v-V)$, $mk^2(\omega-\Omega)$ is equal to the moment about the same axis of the vector system determined by $\underset{\cdot}{X}, \underset{\cdot}{Y}, \underset{\cdot}{N}$.

233. Kinetic energy produced by impulses. Let the body move in one plane. Let m be the mass of the body, U, V resolved velocities of its centre of mass parallel to the axes of reference, and Ω its angular velocity, just before the impulses act, u, v, ω corresponding quantities just after.

Let $\underset{\cdot}{X}$, $\underset{\cdot}{Y}$ be the resolved parts parallel to the axes of the impulse applied to the body at any point whose coordinates relative to the centre of mass are x, y.

The equations of impulsive motion are

$$\left.\begin{aligned} m(u-U) &= \Sigma \underset{\cdot}{X}, \\ m(v-V) &= \Sigma \underset{\cdot}{Y}, \\ mk^2(\omega-\Omega) &= \Sigma(x\underset{\cdot}{Y} - y\underset{\cdot}{X}). \end{aligned}\right\}$$

Multiply these equations in order by

$$\tfrac{1}{2}(u+U),\ \tfrac{1}{2}(v+V),\ \tfrac{1}{2}(\omega+\Omega),$$

and let T be the kinetic energy of the body after the impulses, T_0 that before. Then we have

$$T - T_0 = \Sigma \tfrac{1}{2}\{\underset{\cdot}{X}(u-\omega y+U-\Omega y)+\underset{\cdot}{Y}(v+\omega x+V+\Omega x)\}.$$

The right-hand member of this equation is the sum of the products of the external impulses and the arithmetic means of the velocities of their points of application resolved in their directions before and after.

Now the theorem of Art. 174 asserts that the change of kinetic energy is equal to the value of the like sum for all the impulses internal and external. It follows that the internal impulses between the parts of a rigid body, which undergoes a sudden change of motion, contribute nothing to this sum.

234. Examples.

1. A uniform rod at rest is struck at one end by an impulse at right angles to its length. Prove that, if the rod is free, it begins to turn about the point of it which is distant one-third of its length from the other end, and that the kinetic energy generated is greater than it would be if the other end were fixed in the ratio 4 : 3.

2. A free rigid body is rotating about an axis through its centre of mass, for which the radius of gyration is k, when a parallel axis at a distance c becomes fixed. Prove that the angular velocity of the body is suddenly diminished in the ratio $k^2 : c^2+k^2$.

3. An elliptic disk is rotating in its plane about one end P of a diameter PP', when P' is suddenly fixed. Find the impulse at P' and the angular velocity about P', and prove that, if the eccentricity exceeds $\sqrt{\frac{2}{3}}$, the diameter PP' may be so chosen that the disk is reduced to rest.

4. A uniform rod of length $2a$ and mass m is constrained to move with its ends on two smooth fixed straight wires which intersect at right angles, and is set in motion by an impulse of magnitude mV. Prove that the kinetic energy generated is $\frac{3}{8}mV^2p^2/a^2$, where p is the perpendicular from the intersection of the fixed wires on a line parallel to the line of the impulse and such that the centre of mass is midway between the two parallels.

235. Initial motions. No new method is required for the solution of problems concerning rigid bodies of the same kind as those which were considered in Arts. 203—206; but attention must be paid to the proper expression of the kinetic reaction of a rigid body. The kinetic reactions are equivalent as we saw in Art. 221 to a resultant kinetic reaction and a couple; and the resultant kinetic reaction is the same as that of a particle of mass equal to the mass of the body placed at the centre of mass and moving with the acceleration of the centre of mass.

Sometimes it is convenient to form an equation of motion by taking moments about the instantaneous centre. It is then to be remarked that, at an instant when the velocity of the centre of mass vanishes, the moment of kinetic reaction is $K\dot{\omega}$, where K is the moment of inertia about an axis drawn through the instantaneous centre at right angles to the plane of motion, and $\dot{\omega}$ is the angular acceleration. Cf. Ex. 3 of Art. 222.

236. Small oscillations. When the method of Art. 211 is applied, the most important matter to attend to is the expression of the potential energy correctly to the second order of the small quantity θ by which the displacement from the equilibrium position is specified.

As in the case of initial motions, so also in the case of small oscillations, it is sometimes convenient to form an equation of motion by taking moments about the instantaneous centre. If we take moments about the instantaneous centre in the position of equilibrium the equation is nugatory. This position is, of course, occupied by the body at one instant during the period of oscillation, and at any other instant during the period the instantaneous centre is in a slightly different position. The method which is now effective is to take moments about the instantaneous centre in a displaced position. The moment of the kinetic reaction about the instantaneous centre is expressed correctly to the first order in the displacement by the formula $K\dot{\omega}$, where the letters have the same meanings as in Art. 235. This approximation is sufficient for the purpose of forming the equation of oscillatory motion.

237. Illustrative problem.

A uniform rod can slide with its ends on two smooth straight wires which are equally inclined to the horizontal and fixed in a vertical plane. It is required to find the oscillations about the horizontal position.

Let OA, OB be the two wires, α the angle which each of them makes with the horizontal, AB the horizontal position of equilibrium of the rod. $A'B'$ a displaced position, θ the angle between AB and $A'B'$. Then $\dot{\theta}$ is the angular velocity, and $\ddot{\theta}$ the angular acceleration of the rod. The instantaneous centre in any position is the point of intersection of perpendiculars to OA, OB drawn

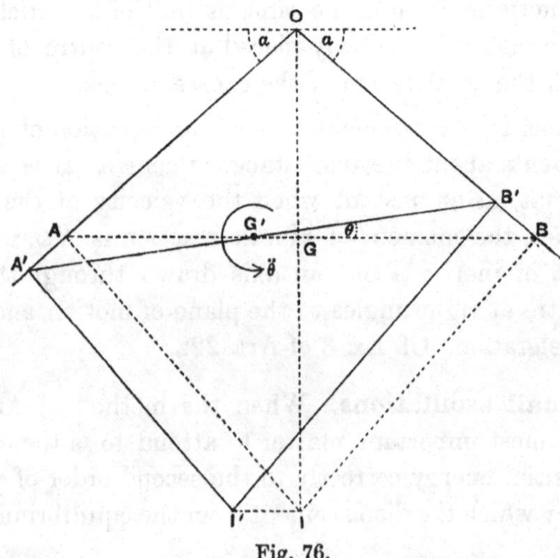

Fig. 76.

from the ends of the rod. We denote by I, I' the positions of the instantaneous centre corresponding to AB and $A'B'$, and by G, G' the corresponding positions of the centre of mass.

The moment of the kinetic reaction about I' is $m(k^2 + I'G'^2)\ddot{\theta}$, where m is the mass of the rod and k its radius of gyration about its centre of mass. With sufficient approximation we may put IG for $I'G'$.

The forces acting on the rod are its weight and the pressures at its ends, and the lines of action of the pressures pass through I'. Now OI' is a diameter of a circle of which $A'B'$ is a chord subtending an angle $\pi-2a$ at the circumference, and thus OI' is of constant length and II' is therefore ultimately at right angles to OI and horizontal. Also GG' being ultimately at right angles to IG is horizontal, and thus the moment of the weight about I' is $-mg(II'-GG')$. Hence we have the equation of moments

$$m(k^2+IG^2)\ddot{\theta}=-mg(II'-GG').$$

Now let $2a$ be the length of the rod. We find

$$II'=BB'\sec a=IB\theta\sec a=a\theta\operatorname{cosec} a\sec a,$$

$$GG'=IG\theta=a\theta\cot a,$$

and the equation becomes

$$ma^2(\tfrac{1}{3}+\cot^2 a)\ddot{\theta}=-mga\theta(\sec a\operatorname{cosec} a-\cot a).$$

The right-hand member is $-mga\theta\tan a$, and therefore the motion in θ is the same as that of a simple pendulum of length

$$a\cot a(\tfrac{1}{3}+\cot^2 a).$$

238. Examples.

1. A uniform rod of length $2a$ and mass m is supported in a horizontal position by two equal inextensible cords each of length l. The ends of the cords are attached, one to either end of the rod, and the other to a fixed point, so that the cords make equal angles a with the vertical. Prove that, if one cord is cut, the tension in the other immediately becomes

$$mg\cos a/(1+3\cos^2 a),$$

and that the initial angular accelerations of the remaining cord and the rod are in the ratio

$$a\sin a : 3l\cos^2 a.$$

2. A uniform triangular lamina is supported in a horizontal position by three equal vertical cords attached to its corners. Prove that, if one cord is cut, the tension in each of the others is instantly halved.

3. Into the top of a smooth fixed sphere of radius a is fitted a smooth vertical rod. A uniform rod of length $2b$ rests on the sphere with its upper end constrained to remain on the vertical rod, the centre of mass being at a distance c from the point of contact. Prove that, if the constraint is removed, the pressure on the sphere is instantly diminished in the ratio

$$b(b-c) : (b^2+3c^2).$$

4. A uniform rod of length $2a$ rests in a horizontal position in a smooth bowl in the form of a surface of revolution whose axis is vertical; the ends of the rod are at points where the radius of curvature of the meridian curve is ρ and the normal makes an angle a with the vertical. Prove that the length of

the equivalent simple pendulum for small oscillations in the vertical plane through the equilibrium position of the rod is

$$\tfrac{1}{3}a\rho\cos\alpha\,(1+2\cos^2\alpha)/(a-\rho\sin^3\alpha),$$

provided that this expression is positive.

5. A uniform rod of length $2a$ passes through a smooth ring, which is fixed at a height b above the lowest point of a smooth bowl in the form of a surface of revolution whose axis is vertical. The rod rests in a vertical position. Prove that, if c denotes the radius of curvature of the meridian curve at the lowest point, the length of the equivalent simple pendulum for small oscillations is

$$\tfrac{1}{3}c\,\{a^2+3\,(a-b)^2\}/(b^2-ac),$$

provided that this expression is positive.

6. A uniform rod of length $2a$ is supported in the way explained in Ex. 1, the distance between the fixed points of attachment of the cords being $2\,(a+l\sin\alpha)$. Prove that the length of the equivalent simple pendulum for small oscillations in the vertical plane through the cords is

$$\tfrac{1}{3}al\cos\alpha\,(1+2\cos^2\alpha)/(a+l\sin^3\alpha).$$

MISCELLANEOUS EXAMPLES

1. If any circle is drawn through the instantaneous centre of no acceleration, prove that the accelerations of all other points on this circle are directed to a common point.

2. A straight rod moves in any manner in its plane. Prove that, at any instant, the directions of motion of all its particles are tangents to a parabola.

3. A rope passes round a rough pulley, which moves in any manner in its plane, so that the rope remains tight. Prove that the directions of motion of all the points of the rope, which are in contact with the pulley at any instant, are tangents to a conic.

---

4. A uniform triangular lamina ABC is constrained to move in a vertical plane with its corners on a fixed circle. Prove that the motion is the same as that of a simple pendulum of length

$$R\,(1-2\cos A\cos B\cos C)/\sqrt{(1-8\cos A\cos B\cos C)},$$

where R is the radius of the circle circumscribing the triangle.

5. The pendulum of a clock consists of a rod with a moveable bob clamped to it, the position of the centre of mass of the bob on the central line of the rod being adjustable. Prove that, if x_1, x_2, x_3 are the distances of the centre of mass of the bob from the axis of suspension when the clock gains n_1, n_2, n_3 minutes a day respectively, the length of the equivalent simple pendulum when the clock keeps correct time is

$$\Sigma\,[x_l^2\theta_l^2\,(\theta_m^2-\theta_n^2)]/\Sigma\,[x_l\,(\theta_m^2-\theta_n^2)],$$

where l, m, n are the numbers 1, 2, 3 in cyclical order, $\theta_1 = 1 + n_1/1440, \ldots$, and each of the sums contains three terms obtained by putting 1, 2, 3 successively for l.

6. Two circular rings, each of radius a, are firmly joined together so that their planes contain an angle 2α and are placed on a rough horizontal plane. Prove that the length of the equivalent simple pendulum is

$$\tfrac{1}{2}a \cos\alpha \operatorname{cosec}^2\alpha\,(1 + 3\cos^2\alpha).$$

7. A thin uniform rod, one end of which can turn about a smooth hinge, is allowed to fall from a horizontal position. Prove that, when the horizontal component of the pressure on the hinge is a maximum, the vertical component is $\frac{11}{8}$ of the weight of the rod.

8. A uniform rectangular block, of mass M, stands on a railway truck with two faces perpendicular to the direction of motion, the lower edge of the front face being hinged to the floor of the truck. If the truck is suddenly stopped, find its previous velocity if the block just turns over. Prove that, in this case, the horizontal and vertical pressures on the hinge vanish when the angle which the plane through the hinge and the centre of mass of the block makes with the horizontal has the values $\sin^{-1}\frac{2}{3}$ and $\sin^{-1}\frac{1}{3}$ respectively, and that the total pressure is a minimum, and equal to $\frac{1}{4}Mg\sqrt{\frac{7}{11}}$, when the angle is $\sin^{-1}\frac{20}{33}$.

9. The door of a railway carriage, which has its hinges (supposed smooth) towards the engine, stands open at right angles to the length of the train when the train starts with an acceleration f. Prove that the door closes in time $\sqrt{\left(\frac{a^2+k^2}{2af}\right)}\int_0^{\frac{1}{2}\pi}\frac{d\theta}{\sqrt{(\sin\theta)}}$, with an angular velocity $\sqrt{\{2af/(a^2+k^2)\}}$, where $2a$ is the breadth of the door, and k the radius of gyration about a vertical axis through the centre of mass.

---

10. A uniform sphere is placed on the highest generator of a rough cylinder, which is fixed with its axis horizontal. Prove that, if slightly displaced, the sphere will roll on the cylinder until the plane through the centre of the sphere and the axis of the cylinder makes with the vertical an angle α satisfying the equation

$$17\mu\cos\alpha - 2\sin\alpha = 10\mu,$$

where μ is the coefficient of friction.

11. A uniform circular ring moves on a rough curve under no forces, the curvature of the curve being everywhere less than that of the ring. The ring is projected from a point A of the curve, and begins to roll at a point B. Prove that the angle between the normals at A and B is $\mu^{-1}\log 2$, where μ is the coefficient of friction.

12. A uniform sphere of mass M rests on a rough plank of mass M', which is on a rough horizontal plane; the plank is suddenly set in motion along its

length with velocity V. Prove that the sphere will first slide and then roll on the plank, and that the whole system will come to rest after a time

$$M'V/\mu g(M+M')$$

from the beginning of the motion, where μ is the coefficient of friction at each of the places of contact.

13. A reel of mass M and radius a rests on a rough floor, μ being the coefficient of friction. Fine thread is coiled on the reel so as to lie on a cylinder of radius $b\,(<a)$ and coaxal with the reel. The free end of the thread is carried in a vertical line over a smooth peg at a height h above the centre of the reel and supports a body of mass m. Prove that, if either

$$\mu < mb/(M-m)\,a, \text{ or if } M < m\,[1-b^2(1+a/h-a^2/bh)/(a^2+k^2)],$$

the thread will be unwound from the reel.

14. A garden roller, in which the mass of the handle may be neglected, is pulled with a force P in a direction making an angle α with the horizontal plane on which it rests. Show that it will not roll without slipping unless

$$P\,\{\sin\alpha\sin\phi+\cos\alpha\cos\phi\,.\,k^2/(a^2+k^2)\}\leqslant W\sin\phi,$$

where a, k, W are the radius, the radius of gyration about the axis, and the weight of the roller, and ϕ is the angle of friction between it and the ground.

15. Two rough cylinders of radii r_1, r_2 are put on a rough table, and on them is placed a rough plank. Prove that, under certain conditions, the system can start from rest and move so that each cylinder rolls on the table with the constant acceleration

$$Mg\sin 2\alpha/\{m_1(1+k_1^2/r_1^2)+m_2(1+k_2^2/r_2^2)+4M\cos^2\alpha\},$$

where $\sin\alpha=(r_1\sim r_2)/d$, and d is the initial distance between the axes of the cylinders.

---

16. On the top of a fixed smooth sphere rests a fine uniform ring with its centre in the vertical diameter, and its diameter subtends an angle 2α at the centre of the sphere. Prove that, if the ring is slightly displaced, it will first begin to leave the sphere when its plane has turned through an angle θ which is given by the equation

$$\sin(\alpha-\theta)\sin\alpha=2\cos^2\alpha\,(2-3\cos\theta).$$

[Assume that the pressure between the sphere and the ring acts only at the highest and lowest points of the ring.]

17. A uniform rod, lying at rest in a smooth sphere, is of such length that it subtends a right angle at the centre. The rod is set in motion so that its ends remain on the sphere and make complete revolutions in a vertical plane. Prove that, if V is the initial velocity of the centre, and a the radius of the sphere,

$$V^2 > ga\left(\tfrac{3}{4}\sqrt{2}+\tfrac{1}{8}\sqrt{202}\right).$$

18. Two equal uniform rods, each of mass m and length $2a$, are free to turn about their middle points, which are fixed at a distance $2a$ apart in a horizontal line. The rods being horizontal, a uniform sphere of mass M and radius c is gently placed upon them at the point where their ends meet. Prove that, if $9M\{a^2+c^2\}^2=2m\{a^2-c^2\}^2$, the sphere will, as it leaves the rods, have half the velocity which it would have had after falling freely through the same height.

19. An elastic thread of modulus λ is wound round the smooth rim of a homogeneous circular disk of mass m, one end being fastened to the rim, and the other to the top of a smooth fixed plane of inclination α to the horizontal, down which the disk moves in a vertical plane through a line of greatest slope, which is the line of contact of the straight portion of the thread with the plane. Initially the thread has its natural length l and is entirely wound on the rim of the disk which is at rest at the top. Prove that at any time t before the thread is entirely unwound the tension is

$$\tfrac{2}{3}mg\sin\alpha\sin^2\{\tfrac{1}{2}t\sqrt{(3\lambda/lm)}\}.$$

20. Two equal cylinders of mass m, bound together by a light elastic band of tension T, roll with their axes horizontal down a rough plane of inclination α. Show that their acceleration down the plane is

$$\tfrac{2}{3}g\sin\alpha\left(1-\frac{2\mu T}{mg\sin\alpha}\right),$$

μ being the coefficient of friction between the cylinders.

21. A rod AB, whose density varies in any manner, is swung as a pendulum about a horizontal axis through A. Prove that the couple resisting bending is greatest at a point P determined by the condition that the centre of mass of the part PB is the centre of oscillation of the pendulum.

---

22. A uniform rod of mass m has one extremity fastened by a pivot to the centre of a uniform circular disk of mass M, which rolls on a horizontal plane, the other extremity being in contact with a smooth vertical wall. The plane of the wall is at right angles to the plane containing the disk and the rod. Prove that the inclination θ of the rod to the vertical when it leaves the wall is given by the equation

$$9M\cos^3\theta+6m\cos\theta-4m\cos\alpha=0,$$

the system starting from rest in a position in which $\theta=\alpha$.

23. A smooth circular cylinder, of mass M and radius c, is at rest on a smooth horizontal plane; and a heavy straight rail, of mass m and length $2a$, is placed so as to rest with its length in contact with the cylinder, and to have one extremity on the ground. Prove that the inclination of the rail to the vertical in the ensuing motion (supposed to be in a vertical plane) is given by the equation

$$\tfrac{1}{2}\dot\theta^2\left[(\tfrac{1}{3}+\sin^2\theta)a^2+\frac{M}{M+m}\left(\frac{c}{1-\sin\theta}-a\cos\theta\right)^2\right]=ga(\cos\alpha-\cos\theta),$$

where α is the initial value of θ.

24. A circular cylinder, of radius a and radius of gyration k, rolls inside a fixed horizontal cylinder of radius b. Prove that the plane through the axes moves like a simple pendulum of length

$$(b-a)(1+k^2/a^2).$$

The second cylinder is free to turn about its axis; the first cylinder is of mass m, and the moment of inertia of the second about its axis is MK^2. Prove that the length of the equivalent simple pendulum is $(b-a)(1+n)/n$, where $n=a^2/k^2+mb^2/MK^2$; prove also that the pressure between the cylinders is proportional to the depth of the point of contact below a plane which is at a depth $2nb\cos\alpha/(1+3n)$ below the fixed axis, where 2α is the angle of oscillation.

25. A uniform circular hoop of radius a is so constrained that it can only move by rolling in a horizontal plane on a fixed horizontal line; and a particle whose mass is $1/\lambda$ of that of the hoop can slide on the hoop without friction. Prove that, if initially the hoop is at rest, and the particle is projected along it from the point furthest from the fixed line with velocity v, then the angle turned through by the hoop in time t will be

$$(vt/a-\sin\psi)/(2\lambda+1),$$

where ψ is the angle through which the diameter through the particle has turned in the same interval. Prove also that

$$vt\sqrt{(2\lambda)}=a\int_0^\psi \sqrt{(2\lambda+\sin^2\theta)}\,d\theta.$$

26. A uniform rod swings in a vertical plane, being suspended by two cords which are attached to its ends and to points A, B in a horizontal line. AB is equal to the length of the rod, and the cords are not crossed. Prove that, if the cords attached to A and B are of lengths a and $a+\lambda$ respectively, where λ is small, the angular velocity of the cord attached to A, when inclined to the vertical at an angle θ, is greater than it would be if λ were zero by

$$\lambda(g/2a^3)^{\frac{1}{2}}(\cos\theta-\cos\alpha)^{\frac{1}{2}}(\tan^2\theta-\tfrac{1}{2}\sec\theta\sec\alpha)$$

approximately, α being the value of θ in a position of rest, and not being nearly equal to a right angle.

27. A uniform rod, which is free to turn about a point fixed in it, touches, at a distance c from the fixed point, the rough edge of a disk of mass m, radius a, and radius of gyration k about its centre. The system being at rest on a smooth horizontal plane, an angular velocity Ω is suddenly communicated to the rod so that the disk also is set in motion. Prove that in the subsequent motion the distance r of the point of contact from the fixed point satisfies the equation

$$(MK^2+mr^2)(1+k^2/a^2)\dot{r}^2=(MK^2+mc^2)(k^2+a^2+r^2-c^2)\Omega^2,$$

where MK^2 is the moment of inertia of the rod about the fixed point, and the edge is rough enough to prevent slipping.

28. A uniform rod has its lower end on a smooth table and is released from rest in any position. Show that the velocity of its centre on arriving at the table is $\sqrt{(\frac{3}{2}gh)}$, where h is the height through which the centre has fallen. Prove also that, at the instant when the centre reaches the table, the pressure on the table is one quarter of the weight of the rod.

29. A wheel can turn freely about a horizontal axis; and a fly of mass m is at rest at the lowest point. If the fly suddenly starts off to walk along the rim of the wheel with constant velocity V relative to the rim, show that he cannot ever get to the highest point of the rim unless V is at least as great as

$$2\sqrt{\{ga\,(ma^2/MK^2)\,(1+ma^2/MK^2)\}},$$

where a is the radius of the wheel, and MK^2 its moment of inertia about its axis.

30. A hollow thin cylinder, of radius a and mass M, is maintained at rest in a horizontal position on a rough plane of inclination α; and an insect of mass m is at rest in the cylinder on the line of contact with the plane. The insect starts to crawl up the cylinder with velocity V, and the cylinder is released at the same instant. Prove that, if the relative velocity is maintained and the cylinder rolls uphill, then it will come to instantaneous rest when the angle which the radius through the insect makes with the vertical is given by the equation

$$V^2\{1-\cos(\theta-\alpha)\}+ag\,(\cos\alpha-\cos\theta)=(1+M/m)\,ag\,(\theta-\alpha)\sin\alpha.$$

31. A rigid square $ABCD$, formed of four uniform rods each of length $2a$, lies on a smooth horizontal table, and can turn freely about one angular point A, which is fixed. An insect, whose mass is equal to that of either rod, starts from the corner B to crawl along the rod BC with uniform velocity V relative to the rod. Prove that, in any time t before the insect reaches C, the angle through which the square turns is

$$\sqrt{\frac{3}{13}}\tan^{-1}\left(\frac{Vt}{a}\sqrt{\frac{3}{52}}\right).$$

---

32. The corners A, B of a uniform rectangular lamina $ABCD$ are free to slide on two smooth fixed rigid wires OA, OB at right angles to each other in a vertical plane and equally inclined to the vertical. The lamina being in a position of equilibrium with AB horizontal, find the velocity produced by an impulse applied along the lowest edge CD.

Prove that, if $AB=2a$, $BC=4a$, then AB will just rise to coincidence with a wire if the impulse is such as would impart to a mass equal to that of the lamina a velocity

$$\tfrac{2}{3}\sqrt{\{ag\,(4-2\sqrt{2})\}}.$$

33. A uniform rigid semicircular wire is rotating in its own plane about a hinge at one end, and is suddenly brought to rest by an impulse applied at the other end along the tangent at that end. Prove that the impulsive stress couple is greatest at a point whose angular distance from the hinge is ϕ, where $\phi\tan\frac{1}{2}\phi=1$.

34. A particle of mass m impinges directly on a smooth uniform spheroid of mass M and semi-axes a, b, the spheroid being at rest, and no energy being lost in the impact. Prove that, if

$$1 < M/m < 6 - 10ab/(a^2+b^2),$$

the point of impact may be so chosen that the particle is reduced to rest.

---

35. A uniform equilateral triangular board is suspended by three equal cords, which are attached to its corners and to the corners of a similar fixed triangle in a horizontal plane; the plane through any two cords makes an angle a with the horizontal. Prove that, if one of the cords is cut, the tensions in the remaining two are diminished in the ratio

$$3 \sin^2 a : 2+4 \sin^2 a.$$

36. A circular ring hangs in a vertical plane on two pegs which are in a horizontal line, and the line joining the pegs subtends an angle $2a$ at the centre. One peg is suddenly removed. Find the pressure on the remaining peg (1) when it is smooth, (2) when it is sufficiently rough to prevent slipping, and prove that these pressures are in the ratio $1 : (1+\frac{1}{4}\tan^2 a)^{\frac{1}{2}}$.

37. A sphere resting on a horizontal plane is divided into a very large number of segments by planes through the vertical diameter, and is kept in shape by a band round the horizontal great circle. Prove that, if the band is cut, the pressure on the plane is diminished by the fraction $45\pi^2/2048$ of itself.

---

38. The lower end of a uniform rod of length a slides on an inextensible thread of length $2a$ whose ends are fixed to two points distant $2\sqrt{(a^2-b^2)}$ apart in a horizontal line, and the upper end of the rod slides on a fixed smooth vertical rod which bisects the line joining the two fixed points. Prove that, if $2b > a$, the time of a small oscillation about the vertical position of equilibrium is

$$2\sqrt{2}\pi a/\sqrt{\{3g(2b-a)\}}.$$

39. In a heavy plane lamina, whose centre of gravity is G, are two narrow straight slits BA, AC, such that AG bisects the angle BAC. Through each slit passes a fixed peg, the pegs, P, Q, being in the same horizontal line. Prove that the time of a small oscillation of the lamina in its own plane, about a position of equilibrium in which the vertex A of the triangle APQ is upwards, is

$$2\pi\sqrt{\frac{2PQ\,(PQ^2+k^2\sin^2 A)}{g\sin A\,(4PQ^2-AG^2\sin^2 A)}},$$

where k is the radius of gyration of the lamina about a line through G perpendicular to its plane.

40. Two equal wheels, each of mass M, radius a, and radius of gyration k about its axis, are rigidly connected by an axle of length c and run on a horizontal plane. Two particles, each of mass m, are connected, one to each of the centres of the wheels, by cords which pass over smooth pegs in the line of

centres. Prove that, if the wheels are symmetrically placed between the pegs, and slightly displaced by rolling on the plane, the time of a small oscillation is

$$2\pi \sqrt{\{Mb(a^2+k^2)/mga^2\}},$$

where $2b+c$ is the distance between the pegs.

41. A solid circular cylinder, bounded by two planes making given angles with the axis, is laid on its curved surface on a rough horizontal plane. Find the position of stable equilibrium, and prove that, if l is the length of the equivalent simple pendulum for a small oscillation, and d the diameter of the cylinder, then the ratio of the longest and shortest generators is

$$l+4d : l-2d.$$

CHAPTER IX *

RIGID BODIES AND CONNECTED SYSTEMS

239. Impact of two solid bodies. To investigate the motion of solid bodies which collide, Poisson† introduced a certain hypothesis as to the motion which takes place while the bodies are in contact. In this short interval of time the bodies may not be regarded as rigid, but the deformation that occurs must be taken into account (Art. 192). Poisson supposed that this interval could be divided into two periods: during the first period the bodies are undergoing compression; during the second period the restitution of form takes place. Further Poisson supposed that the impulse of the pressure between the bodies during the period of restitution bears to the impulse of the pressure during the period of compression the ratio e, which is the coefficient of restitution.

This hypothesis leads to the following rule for solving the problem of impact:—First solve the problem on the supposition that there is no restitution, and find the impulsive pressure between the bodies. Multiply this pressure by $(1 + e)$. Now solve the problem again on the supposition that the impulsive pressure has the value so determined.

Let us apply this method to the problem of the direct impact of two spheres. With the notation of Art. 195 in Ch. VII, the equations of the problem, on the supposition that there is no restitution, are

$$u - u' = 0, \quad mu + m'u' = mU + m'U',$$

and the impulsive pressure R_0 between the bodies is

$$-m(u - U) \quad \text{or} \quad \frac{mm'}{m+m'}(U - U').$$

We multiply this by $(1+e)$. The equations of the problem, on the supposition that the impulsive pressure between the bodies is $(1+e)R_0$, are

$$-m(u - U) = \frac{mm'}{m+m'}(U - U')(1+e), \quad m'(u' - U') = \frac{mm'}{m+m'}(U - U')(1+e),$$

and the values of u and u' which are found from these equations are the same as those found in Art. 195.

* This Chapter may be omitted in a first reading.

† S. D. Poisson, *Traité de Mécanique*, 2nd ed., Paris 1833, t. 2, pp. 273 *et seq.*

In the case of the direct impact of smooth spheres the results that can be deduced from Poisson's hypothesis are the same as the results that can be deduced from Newton's experimental result We may show in like manner that, in the case of the oblique impact of smooth spheres (Art. 197), the results that can be deduced from Poisson's hypothesis are the same as those that can be deduced from the "generalized Newton's rule" stated in Art. 196. We shall show that this result holds for the impact of any two bodies, whether smooth or rough, provided that the friction is not great enough to prevent sliding.

240. Impact of smooth bodies. Let two rigid bodies moving in the same plane come into contact at a point P. Suppose the bodies to be smooth at P. Let R be the impulsive pressure between the bodies at P. The direction of R is the common normal at P to the two surfaces. Let the axis of x be taken in this direction, the axis of y being any fixed line in a perpendicular direction.

Let m and m' be the masses of the bodies, U, V, Ω the velocity system of m before impact, u, v, ω corresponding quantities after impact, and let accented letters denote similar quantities for m'. Also let x, y be the coordinates of the centre of mass of m and x', y' those of m' at the instant of impact, and let ξ, η be coordinates of P at the same instant. Also suppose that, as acting on m, the sense of R is the negative sense of the axis of x (Fig. 77).

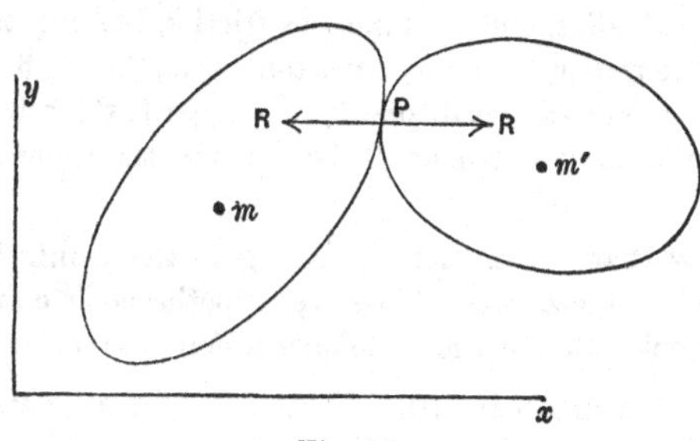

Fig. 77.

The velocity of P, considered as a point of m, has components

$$U-\Omega(\eta-y),\quad V+\Omega(\xi-x) \text{ before impact, and}$$

$$u-\omega(\eta-y),\quad v+\omega(\xi-x) \text{ after impact.}$$

The velocity of P, considered as a point of m', has components

$$U'-\Omega'(\eta-y'),\quad V'+\Omega'(\xi-x') \text{ before impact, and}$$

$$u'-\omega'(\eta-y'),\quad v'+\omega'(\xi-x') \text{ after impact.}$$

The equation provided by the generalized Newton's rule is accordingly

$$u-\omega(\eta-y)-u'+\omega'(\eta-y')=-e\{U-\Omega(\eta-y)-U'+\Omega'(\eta-y')\}.$$

The equations of impulsive motion of the two bodies, obtained by resolving parallel to the axis of x, are

$$m(u-U)=-R, \quad m'(u'-U')=R.$$

The equations obtained by resolving parallel to the axis of y are

$$m(v-V)=0, \quad m'(v'-V')=0.$$

The equations of moments about axes through the centres of mass perpendicular to the plane of motion are

$$mk^2(\omega-\Omega)=R(\eta-y), \quad m'k'^2(\omega'-\Omega')=-R(\eta-y'),$$

where k and k' are the radii of gyration of the bodies about the axes in question.

On substituting for u, u', ω, ω' in the equation containing e, we find

$$R\left[\frac{1}{m}\left\{1+\frac{(\eta-y)^2}{k^2}\right\}+\frac{1}{m'}\left\{1+\frac{(\eta-y')^2}{k'^2}\right\}\right]=(1+e)\left[U-U'-\Omega(\eta-y)+\Omega'(\eta-y')\right],$$

and this equation shows that the impulsive pressure with any value of e is $(1+e)$ times what it would be if e were zero.

The result of this Article can be expressed in the statement that the generalized Newton's rule and the rule derived from Poisson's hypothesis are equivalent for any two smooth bodies moving in one plane.

241. Impact of rough bodies. The impulsive action between two rough bodies which come into contact, when there is sliding at the point of contact, is assumed to be expressible by means of an impulsive pressure, as in the case of smooth bodies, and an impulsive friction tending to resist sliding, the friction and the pressure having a constant ratio, the coefficient of friction. We shall suppose the geometrical condition as regards the relative velocity to be the same as in the case of smooth bodies, viz. the generalized Newton's rule.

We shall show that, when there is sliding at the points that come into contact, the rule deduced from Poisson's hypothesis is equivalent to the generalized Newton's rule, for the impulsive action between rough bodies.

Writing F for the impulsive friction at the point of contact, and taking the same notation as in the last Article, we have the equations of impulsive motion

$$\left.\begin{array}{c} m(u-U)=-R, \quad m(v-V)=-F, \\ mk^2(\omega-\Omega)=-(\xi-x)F+(\eta-y)R \end{array}\right\} \text{.................(1)},$$

and

$$\left.\begin{array}{c} m'(u'-U')=R, \quad m'(v'-V')=F, \\ m'k'^2(\omega'-\Omega')=(\xi-x')F-(\eta-y')R \end{array}\right\} \text{.................(2)}.$$

Also we have the equation of sliding friction

$$F=\mu R \text{.......................................(3)},$$

and the equation provided by the generalized Newton's rule

$$u-\omega(\eta-y)-u'+\omega'(\eta-y')=-e\{U-\Omega(\eta-y)-U'+\Omega'(\eta-y')\} \text{...(4)}.$$

From these equations we obtain, by elimination of u, u', v, v', ω, ω', F, an equation for R, viz.

$$R\left[\left(\frac{1}{m}+\frac{1}{m'}\right)+\frac{\eta-y}{mk^2}\{(\eta-y)-\mu(\xi-x)\}+\frac{\eta-y'}{m'k'^2}\{(\eta-y')-\mu(\xi-x')\}\right]$$
$$=(1+e)\left[U-\Omega(\eta-y)-U'+\Omega'(\eta-y')\right].$$

This equation shows that R contains $(1+e)$ as a factor and is otherwise independent of e, and thus proves the equivalence of the two rules.

242. Case of no sliding. When the bodies are sufficiently rough to prevent sliding the problem is more complicated. The effects of the elasticity of the bodies cannot be so simple as in the previous cases*.

We may obtain a provisional solution by assuming that the generalized Newton's rule holds good. Then equations (1), (2), (4) of Art. 241 are still valid, but instead of equation (3) we have the condition that there is no sliding, viz.

$$v+\omega(\xi-x)=v'+\omega'(\xi-x') \quad\text{.........................(5)}.$$

From equations (1), (2), (4), (5) we can form two equations for R and F, viz.

$$R\left[\left(\frac{1}{m}+\frac{1}{m'}\right)+\frac{(\eta-y)^2}{mk^2}+\frac{(\eta-y')^2}{m'k'^2}\right]-F\left[\frac{(\xi-x)(\eta-y)}{mk^2}+\frac{(\xi-x')(\eta-y')}{m'k'^2}\right]$$
$$=(1+e)\left[U-\Omega(\eta-y)-U'+\Omega'(\eta-y')\right],$$

$$F\left[\left(\frac{1}{m}+\frac{1}{m'}\right)+\frac{(\xi-x)^2}{mk^2}+\frac{(\xi-x')^2}{m'k'^2}\right]-R\left[\frac{(\xi-x)(\eta-y)}{mk^2}+\frac{(\xi-x')(\eta-y')}{m'k'^2}\right]$$
$$=V+\Omega(\xi-x)-V'-\Omega'(\xi-x').$$

It is clear that the solution of these equations will give an expression for R consisting of two terms, one of them having $(1+e)$ as a factor and the other not containing that factor.

Since R is not in general proportional to $1+e$, the result which would be obtained from Poisson's hypothesis is not in general the same as that which would be obtained from the generalized Newton's rule.

The results would however be the same in any case in which either

$$V+\Omega(\xi-x)-V'-\Omega'(\xi-x')=0,$$

or

$$(\xi-x)(\eta-y)/mk^2+(\xi-x')(\eta-y')/m'k'^2=0.$$

The first of these equations expresses the condition that there is no relative velocity of sliding at the instant of impact, or that the impact is, in an obvious sense, "direct." The second is satisfied if $\eta=y=y'$, that is if the normal at the point of contact passes through the centres of mass of the two bodies, as it would if the bodies are spheres or circular disks. It is also satisfied if $\eta=y$ and $\xi=x'$, which would be the case if one body is a sphere or a circular disk and the other is a thin rod.

* Poisson himself did not suppose his hypothesis to be applicable to cases in which there is sufficient friction to prevent sliding. The question is not really of any practical interest because the motion must depend largely on accidental circumstances.

243. Examples.

1. A uniform sphere of radius a and mass m, moving without rotation, impinges directly on a smooth uniform cube of side $2a$ and mass m', the line of motion of the sphere being at a distance b from the centre of mass of the cube. Prove that, if there is no restitution, the kinetic energy lost in the impact is to that of the sphere before impact in the ratio

$$1 : 1+(m/m')(1+\tfrac{3}{2}b^2/a^2).$$

2. A uniform rod, falling without rotation, strikes a smooth horizontal plane. Prove that, for all values of the coefficient of restitution, the angular velocity of the rod immediately after impact is a maximum if the rod before impact makes with the horizontal an angle $\cos^{-1} 1/\sqrt{3}$.

3. A sphere whose centre of mass coincides with its centre of figure is moving in a vertical plane and rotating about an axis perpendicular to that plane when it strikes against a horizontal plane which is sufficiently rough to prevent sliding. Prove that the sphere will rebound at an angle greater or less than if there were no friction according as the lowest point of it at the instant of impact is moving forward or backward.

4. A disk of any form, of mass m, moving in its plane without rotation and with velocity V at right angles to a fixed plane, strikes the plane, so that the distances of the centre of mass from the point of impact and from the plane are r and p. Prove that, if the plane is sufficiently rough to prevent sliding, the impulsive pressure is

$$mV(1+e)(k^2+p^2)/(k^2+r^2),$$

where k is the radius of gyration of the disk about its centre of mass.

5. A ball spinning about a vertical axis moves on a smooth table, and impinges on a vertical cushion, the centre moving directly towards the cushion. Prove that, if θ is the angle of reflexion, the kinetic energy is diminished in the ratio

$$10+14\tan^2\theta : 10e^{-2}+49\tan^2\theta,$$

the cushion being sufficiently rough to prevent sliding.

6. A circular disk of mass M and radius c, moving in its own plane without rotation, impinges on a rod of mass m and length $2a$ which is free to turn about a pivot at its centre, and the point of impact is distant b from the pivot. Prove that, if the direction of motion of the centre of the disk makes angles α and β with the rod before and after collision, then

$$2(3Mb^2+ma^2)\tan\beta = 3(3Mb^2-ema^2)\tan\alpha,$$

the edges in contact being sufficiently rough to prevent sliding.

244. Impulsive motion of connected systems. In illustration of the application of the equations of impulsive motion to systems of rigid bodies with invariable connexions we take the following problems. In the first it will be observed that we do not need to introduce explicitly the reactions between the connected bodies. The second illustrates the choice of equations; for, although some of the unknown reactions must be introduced, it is unnecessary to form equations for each body separately.

I. *Three uniform rods of masses proportional to their lengths are freely jointed together and laid out straight, and one of the end rods is struck at the free end at right angles to its length. It is required to find how they begin to move.*

Let $2a$, $2b$, $2c$ be the lengths of the rods, the last being struck, and let x/a,

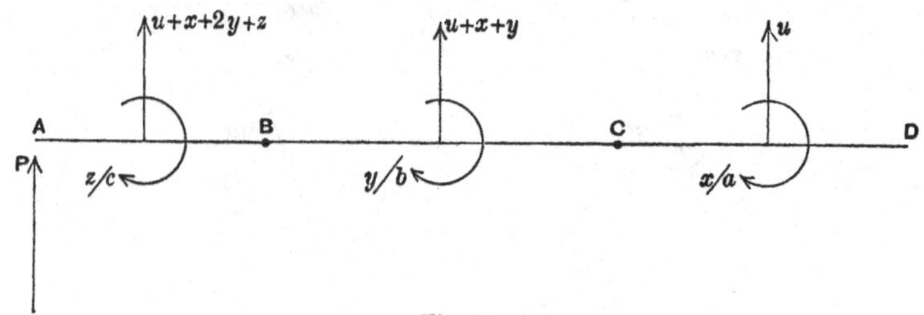

Fig. 78.

y/b, z/c be the angular velocities with which they begin to move, u the velocity of the centre of mass of the first. Then the system of velocities is as shown in the figure. Let P be the impulse applied at the end A, and κa, κb, κc the masses of the rods.

We take moments about C for the rod CD, about B for the rods BC, CD, and about A for the three rods, and we resolve for the whole system at right angles to the rods. We thus obtain the equations

$$\left.\begin{aligned}
&u-\tfrac{1}{3}x=0,\\
&b\left[b(u+x+y)-\tfrac{1}{3}by\right]+a\left[(2b+a)u-\tfrac{1}{3}ax\right]=0,\\
&c\left[c(u+x+2y+z)-\tfrac{1}{3}cz\right]+b\left[(2c+b)(u+x+y)-\tfrac{1}{3}by\right]\\
&\qquad\qquad+a\left[(2c+2b+a)u-\tfrac{1}{3}ax\right]=0,\\
&\kappa c(u+x+2y+z)+\kappa b(u+x+y)+\kappa au=P.
\end{aligned}\right\}$$

Subtracting the second and third we get, on dividing by c,

$$c(u+x+2y+\tfrac{2}{3}z)+2b(u+x+y)+2au=0,$$

and, on simplifying this and the second by using the first, we get

$$x(a+4b+2c)+y(3c+3b)+zc=0,$$

and

$$(2b+a)x+by=0.$$

Hence we have

$$\frac{u}{\frac{1}{3}b}=\frac{x}{b}=\frac{y}{-(2b+a)}=\frac{cz}{2ab+3ac+2b^2+4bc}$$

$$=\frac{P}{\kappa\left(\frac{4}{3}ab+ac+\frac{4}{3}bc+\frac{4}{3}b^2\right)}.$$

II. *A rhombus formed of four equal uniform rods freely jointed at the corners is set in motion by an impulse applied to one rod at right angles to it. To find how the rhombus begins to move.*

Let $2a$ be the length of each side of the rhombus $ABCD$, α the angle DAB, x the distance of the point struck from the middle point of the side AB containing it, P the impulse, m the mass of each rod.

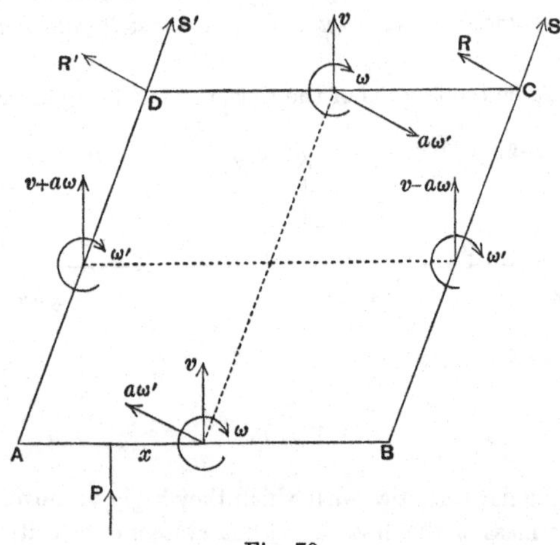

Fig. 79.

The centre of mass of the system is the point of intersection of the lines joining the middle points of opposite sides. Since the figure is always a parallelogram, opposite sides have the same angular velocities, and the lines joining the middle points of opposite sides are of constant length $2a$ and turn with the angular velocities of the sides to which they are parallel. Let these angular velocities be ω and ω', and let v be the velocity of the centre of mass. Then the velocities of the centres of mass of the rods and their angular velocities are as shown.

Now let the impulsive reaction of the hinge at C be resolved into S parallel to BC and R at right angles to BC, and the impulsive reaction of the hinge at D into S', R' in the same directions. These impulses act in opposite senses on the two rods which meet at a hinge. The figure shows the senses in which we take them to act on the rod CD.

We form two equations of motion by resolving for the system in the direction of the impulse and by taking moments about the centre of mass. We thus obtain

$$\left.\begin{aligned} 4mv &= P, \\ \tfrac{8}{3}ma^2(\omega+\omega') &= P(x+a\cos\alpha). \end{aligned}\right\}$$

Again, we can form three equations containing R and R' by resolving for CD at right angles to BC, and taking moments for BC and AD about B and

A respectively. We thus obtain

$$\left.\begin{aligned} m\,(v\cos\alpha - a\omega') &= R+R',\\ m\,[(v-a\omega)\,a\cos\alpha - \tfrac{1}{3}a^2\omega'] &= -2aR,\\ m\,[(v+a\omega)\,a\cos\alpha - \tfrac{1}{3}a^2\omega'] &= -2aR', \end{aligned}\right\}$$

from which, on elimination of R and R', we get

$$v\cos\alpha = \tfrac{2}{3}a\omega'.$$

Hence $\qquad v=\tfrac{1}{4}P/m,\quad \omega=\tfrac{3}{8}Px/ma^2,\quad \omega'=\tfrac{3}{8}P\cos\alpha/ma.$

245. Examples.

1. Two equal rods AB, AC freely jointed at A are at rest with the angle BAC a right angle, and AC is struck at C by an impulse in a direction parallel to AB. Prove that the velocities of the centres of mass of AB and AC in the direction of AB are in the ratio 2 : 7.

2. Two equal uniform rods freely hinged at a common end are laid out straight, and one end of one of them is struck by an impulse at right angles to their length. Prove that the kinetic energy generated is greater than it would be if the rods were firmly fastened together so as to form a single rigid body in the ratio 7 : 4.

3. Four equal uniform rods are freely hinged together so as to form a rhombus of side $2a$ with one diagonal vertical, and the system falling in a vertical plane with velocity V strikes against a fixed horizontal plane. Taking α to be the angle which each rod makes with the vertical and assuming no restitution, prove that (i) the impulsive action between the two upper rods is directed horizontally, (ii) the angular velocity of each rod after the impulse is $\tfrac{3}{2}(V/a)\sin\alpha/(1+3\sin^2\alpha)$, (iii) the impulsive action between the two upper rods is to the momentum of the system before impact in the ratio

$$\sin\alpha\,(3\cos^2\alpha \sim 1) : 8\cos\alpha\,(1+3\sin^2\alpha),$$

(iv) the impulsive action at either of the hinges in the horizontal diagonal makes with the horizontal an angle $\tan^{-1}\{(3\cos^2\alpha \sim 1)\cot\alpha\}$.

4. In Example 3, prove that, if the coefficient of restitution between the rhombus and the ground is e, the angular velocity of each rod after the impulse is

$$\tfrac{3}{2}(1+e)\,(V/a)\sin\alpha/(1+3\sin^2\alpha).$$

5. A square framework $ABCD$ is formed of uniform rods freely jointed at B, C, and D, the ends at A being in contact but free. Prove that, if AB is struck by a blow at A in the direction DA, the initial velocity of A is 79 times that of D.

6. A rectangle formed of four uniform rods, of lengths $2a$ and $2b$ and masses m and m', freely hinged together, is rotating in its plane about its centre with angular velocity n when a point in one of the sides of length $2a$ becomes suddenly fixed. Prove that the angular velocity of the sides of length $2b$ instantly becomes $\tfrac{1}{2}n\,(3m+m')/(3m+2m')$, and find the angular velocity of the sides of length $2a$.

246. Initial motions and initial curvatures. The kinetic reactions of the parts of a connected system of particles and rigid bodies can always be expressed in terms of a finite number of geometrical quantities which are unconnected by any geometrical equations. This can usually be effected by methods similar to those used in Art. 205.

It may however happen that such methods are difficult of application. When this is the case we may begin by writing down the geometrical equations which hold between the coordinates of the points *in any position*. If we differentiate these equations twice with respect to the time, and, in the results, substitute for every first differential coefficient of a geometrical quantity the value 0, and for every geometrical quantity the value that it has in the initial position, we shall obtain the relations between the initial accelerations of the various geometrical quantities involved. Thus if x, y are the coordinates of any particle whose acceleration is required, and $\theta, \phi, \ldots$ are a series of geometrical quantities which define the position of the system, there will be certain values $\theta_0, \phi_0, \ldots$ for these quantities in the initial position. Now the geometrical equations provide the means of expressing the x and y of the particle in any position in terms of the values of $\theta, \phi, \ldots$ for that position. Let $x = f(\theta, \phi, \ldots)$ be the form of one of the equations we can obtain. On differentiating we have

$$\dot{x} = \frac{\partial f}{\partial \theta}\dot{\theta} + \frac{\partial f}{\partial \phi}\dot{\phi} + \ldots,$$

$$\ddot{x} = \left(\frac{\partial^2 f}{\partial \theta^2}\dot{\theta} + \frac{\partial^2 f}{\partial \theta \partial \phi}\dot{\phi} + \ldots\right)\dot{\theta} + \left(\frac{\partial^2 f}{\partial \theta \partial \phi}\dot{\theta} + \frac{\partial^2 f}{\partial \phi^2}\dot{\phi} + \ldots\right)\dot{\phi} + \ldots$$
$$+ \frac{\partial f}{\partial \theta}\ddot{\theta} + \frac{\partial f}{\partial \phi}\ddot{\phi} + \ldots.$$

Reducing in the way that has been explained we obtain

$$\ddot{x}_0 = \left(\frac{\partial f}{\partial \theta}\right)_0 \ddot{\theta}_0 + \left(\frac{\partial f}{\partial \phi}\right)_0 \ddot{\phi}_0 + \ldots,$$

where $\ddot{x}_0, \ddot{\theta}_0, \ldots$ denote the initial values of $\ddot{x}, \ddot{\theta}, \ldots$, and $\left(\frac{\partial f}{\partial \theta}\right)_0$, $\left(\frac{\partial f}{\partial \phi}\right)_0, \ldots$ denote the values of $\frac{\partial f}{\partial \theta}, \frac{\partial f}{\partial \phi}, \ldots$ when $\theta = \theta_0$, $\phi = \phi_0, \ldots$.

Now this process can be carried further, and arranged as a process of approximation for expressing the values of $x, y, \ldots$ as series in ascending powers of the time. We have in fact as a first approximation $x = \frac{1}{2}\ddot{x}_0 t^2$, $y = \frac{1}{2}\ddot{y}_0 t^2$.

From such series we can deduce the initial curvatures of the paths of all the particles.

It will be easier to understand how this process is carried out after studying its application to a particular problem, and it will at the same time be seen how simplifications may at times suggest themselves. A complicated problem has been chosen intentionally.

247. Illustrative problem. *Two uniform rods AB, BC of masses m_1, m_2 and lengths a, b are freely hinged at B, and AB can turn about A in a vertical plane. The system starts from rest in a position in which AB is horizontal and BC vertical. It is required to determine the initial curvature of the path of any point of BC.*

Let AB make an angle θ with the horizontal, and BC an angle ϕ with the

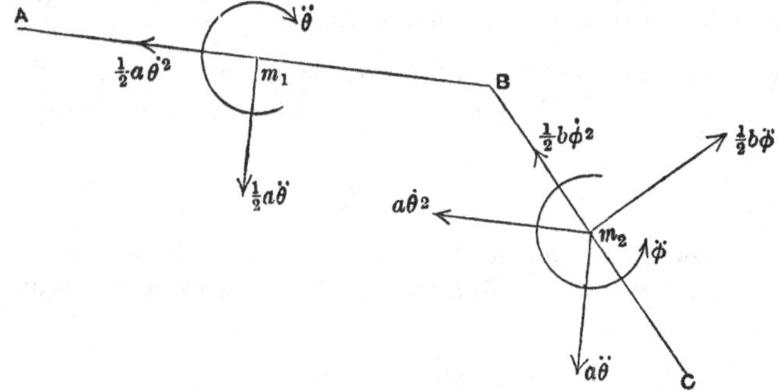

Fig. 80.

vertical at time t. Since B describes a circle of radius a about A, and since the centre of mass of BC describes a circle of radius $\frac{1}{2}b$ relative to B, the diagram of accelerations is that shown in Fig. 80.

By taking moments about B for BC, and about A for the system, we obtain the two equations

$$m_2(\tfrac{1}{4}b^2 + \tfrac{1}{12}b^2)\ddot{\phi} - m_2 a\ddot{\theta}\tfrac{1}{2}b\sin(\theta+\phi) - m_2 a\dot{\theta}^2\tfrac{1}{2}b\cos(\theta+\phi) = -\tfrac{1}{2}m_2 gb\sin\phi,$$

$$\begin{aligned} m_1(\tfrac{1}{4}a^2 + \tfrac{1}{12}a^2)\ddot{\theta} + m_2 a\ddot{\theta}\{a + \tfrac{1}{2}b\sin(\theta+\phi)\} + m_2 a\dot{\theta}^2\tfrac{1}{2}b\cos(\theta+\phi) \\ - m_2\tfrac{1}{12}b^2\ddot{\phi} - m_2\{\tfrac{1}{2}b + a\sin(\theta+\phi)\}\tfrac{1}{2}b\ddot{\phi} - m_2 a\cos(\theta+\phi)\tfrac{1}{2}b\dot{\phi}^2 \\ = \tfrac{1}{2}m_1 ga\cos\theta + m_2 g(a\cos\theta + \tfrac{1}{2}b\sin\phi). \end{aligned}$$

Adding the equations, and dividing out common factors, we have

$$(\tfrac{1}{3}m_1 + m_2)a\ddot{\theta} - \tfrac{1}{2}m_2 b\ddot{\phi}\sin(\theta+\phi) - \tfrac{1}{2}m_2 b\dot{\phi}^2\cos(\theta+\phi) = g\cos\theta(\tfrac{1}{2}m_1 + m_2)\ldots(1).$$

Also the first of the above equations is

$$\tfrac{1}{3}b\ddot{\phi}-\tfrac{1}{2}a\ddot{\theta}\sin(\theta+\phi)-\tfrac{1}{2}a\dot{\theta}^2\cos(\theta+\phi)=-\tfrac{1}{2}g\sin\phi \quad\ldots\ldots\ldots\ldots(2).$$

In the initial position $\theta=0$, $\phi=0$, $\dot{\theta}=0$, $\dot{\phi}=0$, and we have

$$\ddot{\phi}_0=0,\quad \ddot{\theta}_0=\frac{3m_1+6m_2}{2m_1+6m_2}\frac{g}{a}.$$

In any position we have, by Maclaurin's theorem,

$$\theta=\tfrac{1}{2}\ddot{\theta}_0t^2+\tfrac{1}{6}\dddot{\theta}_0t^3+\ldots,\qquad \phi=\tfrac{1}{2}\ddot{\phi}_0t^2+\tfrac{1}{6}\dddot{\phi}_0t^3+\ldots;$$

also

$$\dot{\theta}=\ddot{\theta}_0t+\tfrac{1}{2}\dddot{\theta}_0t^2+\ldots,\qquad \dot{\phi}=\ddot{\phi}_0t+\tfrac{1}{2}\dddot{\phi}_0t^2+\ldots.$$

Now, taking equation (2), we see that if $\dddot{\phi}_0$ were finite, ϕ would be of order t^3, and θ of order t^2, so that the terms would be respectively of orders 1, 2, 2, 3. This shows that $\dddot{\phi}_0$ must be zero. Again, if ϕ_0^{iv} is finite the equation can be reduced, by picking out the terms of order 2 in t, to

$$\tfrac{1}{6}b\phi_0^{\text{iv}}-\tfrac{1}{2}a\ddot{\theta}_0(\tfrac{1}{2}\ddot{\theta}_0)-\tfrac{1}{2}a\ddot{\theta}_0^2=0,$$

giving

$$\phi_0^{\text{iv}}=\frac{9a}{2b}\ddot{\theta}_0^2=\frac{9a}{2b}\left(\frac{3m_1+6m_2}{2m_1+6m_2}\frac{g}{a}\right)^2.$$

Again, taking equation (1), and observing that $\cos\theta=1-\frac{\theta^2}{2!}+\frac{\theta^4}{4!}-\ldots$, we see that the lowest power of t in this series is the fourth, and then it appears from equation (1) that the lowest power of t in $\ddot{\theta}$ is the fourth, so that the series for θ begins

$$\theta=\tfrac{1}{2}\ddot{\theta}_0t^2+\frac{1}{6!}\theta_0^{\text{vi}}t^6+\ldots.$$

Going back now to equation (2), it is clear that $\ddot{\phi}$ contains no term in t^3 but there is a term in t^4. In fact, picking out the terms in t^4 in equation (2) we have

$$\tfrac{1}{3}b\phi_0^{\text{vi}}\frac{t^4}{4!}-\tfrac{1}{2}a\ddot{\theta}_0\phi_0^{\text{iv}}\frac{t^4}{4!}=-\tfrac{1}{2}g\phi_0^{\text{iv}}\frac{t^4}{4!},$$

giving

$$\phi_0^{\text{vi}}=\frac{3}{2}\left(\frac{a}{b}\ddot{\theta}_0-\frac{g}{b}\right)\phi_0^{\text{iv}}=\frac{9}{4}\frac{m_1}{m_1+2m_2}\ddot{\theta}_0^3\frac{a^2}{b^2}.$$

Now, in the figure, taking as origin the initial position of B, and taking the axes of x and y horizontal and vertical, we can write for the coordinates of a point of BC distant r from B,

$$x=-a(1-\cos\theta)+r\sin\phi,\quad y=a\sin\theta+r\cos\phi;$$

expanding these we have approximately

$$x=-a\left(\frac{\theta^2}{2}-\frac{\theta^4}{24}\right)+r\left(\phi-\frac{\phi^3}{6}\right),\quad y=a\left(\theta-\frac{\theta^3}{6}\right)+r\left(1-\frac{\phi^2}{2}+\frac{\phi^4}{24}\right),$$

giving

$$\left.\begin{aligned} x&=-\frac{a}{8}t^4\ddot{\theta}_0^2+\frac{r}{24}t^4\phi_0^{\text{iv}}=\frac{a}{16}\ddot{\theta}_0^2t^4\left(\frac{3r}{b}-2\right),\\ y-r&=\tfrac{1}{2}at^2\ddot{\theta}_0,\end{aligned}\right\}$$

which are correct as far as t^2. Hence the initial path of the point is approximately a parabola

$$(y-r)^2 = 4ax\frac{b}{3r-2b},$$

and the radius of curvature of the path is $2ab/(3r-2b)$ unless $r=\frac{2}{3}b$.

If, however, $r=\frac{2}{3}b$, in order to get an approximate equation to the path, we must expand to a higher order. We find

$$x = \tfrac{2}{3}b\phi_0^{\text{vi}}\frac{t^6}{6!} = \frac{m_1}{m_1+2m_2}\frac{a^2}{b}\frac{\ddot{\theta}_0^3}{480}t^6,$$

correct as far as t^6, and thus the initial path is given by the approximate equation

$$(y-\tfrac{2}{3}b)^3 = 60abx\,(1+2m_2/m_1).$$

248. Examples.

1. Two equal uniform rods are freely jointed at common ends, the other end of the first is fixed so that the rods can turn about it, and the other end of the second is held at the same level as the fixed end of the first, so that the rods make equal angles α with the horizontal, and this end is let go. Prove that the initial angular accelerations of the rods are in the ratio

$$6-3\cos 2\alpha : 9\cos 2\alpha - 8.$$

2. Three equal uniform rods are freely jointed at B and C so as to form three sides of a quadrilateral $ABCD$, and the ends A and D can slide on a smooth horizontal rod. The system is initially held (by means of horizontal forces applied at A and D) in a symmetrical position with BC lowest and horizontal, and with AB and CD equally inclined at angles α to the horizontal. Prove that, when the ends A and D are released, the pressures at A and D are changed in the ratio $1+\sin^2\alpha : 5-3\sin^2\alpha$.

3. A uniform rod of length $2a$ is held at an inclination α to the horizontal in contact with a smooth peg at its middle point. Prove that, when the rod is let go, the initial radius of curvature of the path of a particle distant r from the middle point is $(a^2/r)\tan\alpha$.

4. Two equal uniform rods AB, BC each of length a are freely jointed at B, and can turn freely about A. Prove that, if the system is released from a horizontal position, the initial radius of curvature of the path of C is $\frac{2}{5}a$.

249. Small oscillations. *Illustrative problem.*

The following problem illustrates the application of the method of Art. 211.

A uniform rod is supported at its ends by two equal vertical cords suspended from fixed points. It is required to find the small oscillation in which the middle point moves vertically and the rod, remaining horizontal, turns round its middle point.

Let $2a$ be the length of the rod, l the length of either cord, z the distance through which the middle point has risen at time t, θ the angle through which the rod has turned in the same time. The depth of either end A or B below the corresponding point of support is $l-z$, and the distance AM or BN of an end from the equilibrium position of the corresponding cord is $2a \sin \frac{1}{2}\theta$. Hence we have

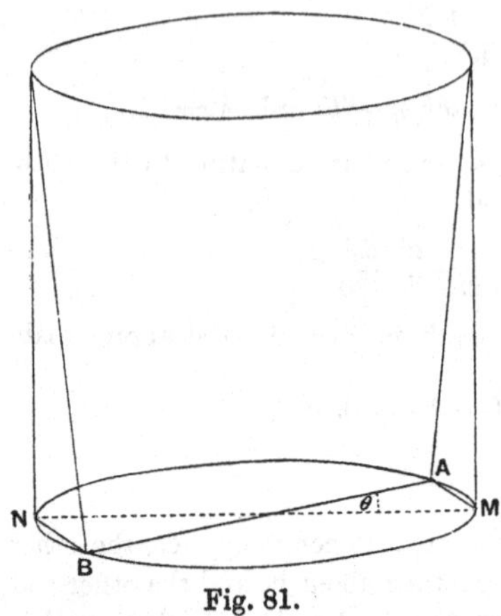

Fig. 81.

$$(l-z)^2 + 4a^2 \sin^2 \tfrac{1}{2}\theta = l^2;$$

this equation shows that when z and θ are small $z = \frac{1}{2}(a^2/l)\,\theta^2$ to the second order, and $\dot{z} = 0$ to the first order.

Now, if m is the mass of the rod, the kinetic energy in any position is

$$\tfrac{1}{2}m(\dot{z}^2 + \tfrac{1}{3}a^2\dot{\theta}^2),$$

and the potential energy is mgz, the lowest position being the standard position.

Hence, in the small oscillations, the kinetic energy is, with sufficient approximation,

$$\tfrac{1}{6}ma^2\dot{\theta}^2,$$

and the potential energy is, with sufficient approximation,

$$\tfrac{1}{2}mg\,(a^2/l)\,\theta^2.$$

The motion in θ is therefore the same as for small oscillations of a simple pendulum of length $\frac{1}{3}l$.

250. Examples.

1. A number of equal uniform rods each of length $2a$ are freely jointed at a common end and arranged at equal intervals like the ribs of an umbrella, and this cone of rods is placed in equilibrium over a smooth sphere so that the angle of the cone is 2α. Prove that, for small vertical oscillations of the joint, the length of the equivalent simple pendulum is

$$\tfrac{1}{3}a \cos\alpha\,(1+3\cos^2\alpha)/(1+2\cos^2\alpha).$$

2. Prove that the length of the equivalent simple pendulum for small oscillations of the handle of a garden roller rolling on a horizontal walk is

$$l - \frac{h}{1+M(k^2+a^2)/ma^2},$$

where a is the radius of the roller, M the mass of the roller alone, k its radius of gyration about its axis, m the mass of the handle, h the distance of the centre of mass of the handle from the axis of the roller, and l the length of the equivalent simple pendulum for the oscillations of the handle when the roller is held fixed.

3. Four equal uniform rods are freely jointed so as to have a common extremity, and four other like rods are similarly jointed; the other ends of the rods are then jointed in pairs so as to form eight edges of an octahedron. One of the joints where four rods meet is fixed and the other is attached to it by an elastic thread, so that in equilibrium the octahedron is regular and the thread vertical. Prove that the length of the equivalent simple pendulum for small vertical oscillations of the lowest point is $\frac{5}{8}(l-l_0)$, where l and l_0 are the equilibrium length and the natural length of the thread.

251. Stability of steady motions.

The principles of energy and momentum may frequently be applied to problems concerning the stability of steady motions. We shall illustrate the method by considering the steady motion of a spherical pendulum, that is a particle moving under gravity on the surface of a sphere so as to describe a horizontal circle.

Let θ be the angle which the radius vector from the centre of the sphere to the particle makes with the downwards vertical at time t, a the radius of the sphere, ϕ the angle contained between the plane through the particle and the vertical diameter and a fixed plane through the same diameter.

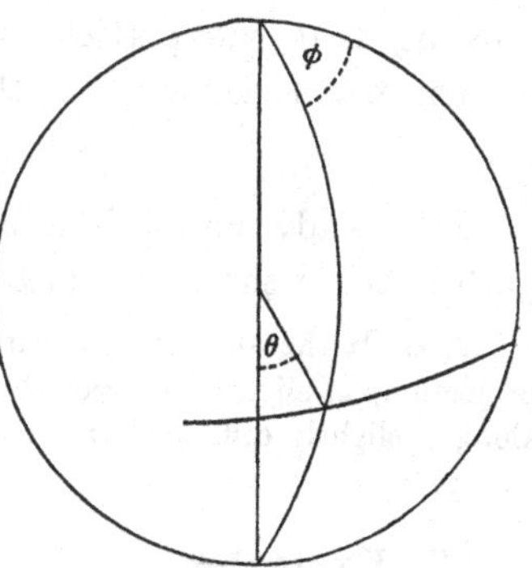

Fig. 82.

The energy equation is

$$\tfrac{1}{2}ma^2(\dot{\theta}^2+\sin^2\theta\dot{\phi}^2)+mga(1-\cos\theta) = \text{const.},$$

and the equation of constancy of moment of momentum about the vertical diameter is $ma^2\sin^2\theta\dot{\phi} = \text{const.}$

We wish to discover the condition that motion in a horizontal circle, $\theta=\alpha$, with angular velocity ω may be possible. We have

$$\dot{\phi}\sin^2\theta = \omega\sin^2\alpha,$$

so that the energy equation may be written

$$\tfrac{1}{2}a\left(\dot{\theta}^2+\omega^2\frac{\sin^4\alpha}{\sin^2\theta}\right)-g\cos\theta = \text{const.}$$

Differentiating with respect to the time we obtain the equation

$$\ddot{\theta}-\omega^2\frac{\sin^4\alpha\cos\theta}{\sin^3\theta}+\frac{g}{a}\sin\theta=0 \quad \text{..............(1).}$$

Now the steady motion is possible if ω is so adjusted that $\ddot{\theta}=0$ when $\theta=\alpha$. This gives us the condition

$$a\omega^2 = g \sec \alpha \quad \text{..........................(2).}$$

(Cf. Art. 79.)

If the particle is projected from a point for which θ is nearly equal to α, in a nearly horizontal direction, with an angular momentum $ma^2\omega \sin^2 \alpha$ about the vertical diameter, where ω is given by (2), then either it tends to remain always very near the circle $\theta=\alpha$, or to depart widely from it. Supposing it to remain near the circle, we may put $\theta=\alpha+\chi$, expand the terms of equation (1), and reject powers of χ above the first. We thus find

$$\ddot{\chi}+\chi\left[\frac{d}{d\theta}\left\{\frac{g}{a}\sin\theta-\omega^2\sin^4\alpha\frac{\cos\theta}{\sin^3\theta}\right\}\right]_{\theta=\alpha}=0,$$

or

$$\ddot{\chi}+\chi\frac{g}{a}\frac{1+3\cos^2\alpha}{\cos\alpha}=0,$$

showing that the particle oscillates about the state of steady motion in a period equal to that of a simple pendulum of length

$$a\cos\alpha/(1+3\cos^2\alpha).$$

The steady motion is stable if $\cos\alpha$ is positive, or the circular path is below the centre of the sphere.

Note. If the angular momentum (as well as the direction and point of projection) is slightly altered, the possible steady motion would take place along a slightly different circle; but the period of oscillation would be unchanged.

252. Examples.

1. Utilize the method of Art. 251 to show that the motion of a particle describing a circular orbit under a force $f(r)$ directed to the centre is stable if $[3+cf'(c)/f(c)]$ is positive, c being the radius of the circle. Deduce the results in Art. 106.

2. Prove that the steady motion with angular velocity ω of a conical pendulum of length l is stable, and that, if a small disturbance is made, oscillations take place in time

$$2\pi l\omega/\sqrt{(3g^2+l^2\omega^4)}.$$

3. A particle describes a horizontal circle of radius r on a smooth paraboloid of revolution whose axis is vertical and vertex downwards. Prove that, if it is slightly disturbed, its period of oscillation is

$$\pi\sqrt{\{(r^2+4a^2)/2ga\}},$$

where $4a$ is the latus rectum.

4. A circular wire of radius a and of negligible mass rotates freely about a vertical chord distant c from the centre; a small heavy ring can slide on the wire without friction. In the position of relative equilibrium the radius of the circle drawn through the ring makes an angle α with the vertical. Find the angular velocity with which the wire rotates, and prove that the length of the equivalent simple pendulum for small oscillations of the ring is

$$a \cos \alpha (c + a \sin \alpha) / \{c + a \sin \alpha (1 + 3 \cos^2 \alpha)\}.$$

Prove also that, if the wire is made to rotate uniformly, the period of small oscillations is the same as for a simple pendulum of length

$$a \cos \alpha \, (c + a \sin \alpha) / (c + a \sin^3 \alpha).$$

[In the second case energy is expended in keeping up the angular velocity of the wire, and an equation of motion of the ring must be formed by resolving along the tangent to the circle. The angular velocity in relative equilibrium is the same as before.]

5. An elastic circular ring of mass m and modulus of elasticity λ rotates uniformly in its own plane about its centre under no external forces. Prove that, if a is the radius in steady motion, and l is the radius when the ring is unstrained, the period of the small oscillations about the state of steady motion is

$$\surd \{2\pi lam / \lambda \, (4a - 3l)\}.$$

253. Illustrative problem. In further illustration of the principles of Energy and Momentum consider the following problem:

A uniform rod and a particle are connected by an inextensible thread attached to one end of the rod, the system is laid out straight, and the particle is projected at right angles to the thread. It is required to find the motion when there are no forces.

Let $2a$ be the length of the rod, l the length of the thread, χ the angle which the thread makes with the line of the rod produced at time t. Consider first the motion of the particle P relative to the centre of mass M of the rod AB.

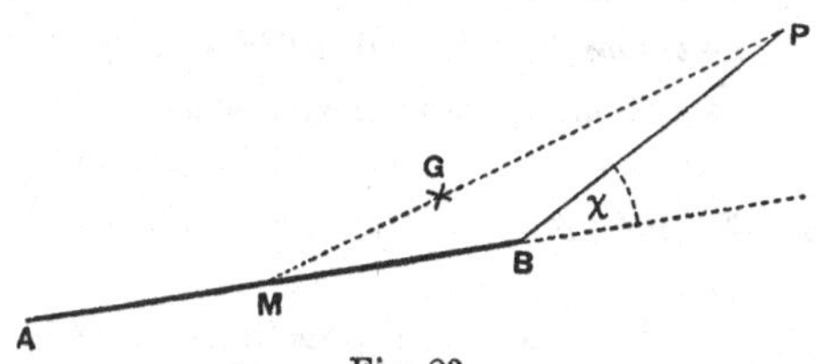

Fig. 83.

Let θ be the angle which AB makes at time t with its initial direction. Then the velocity of B relative to M is $a\dot{\theta}$ at right angles to AB, and, since BP makes an angle $\theta + \chi$ with a line fixed in the plane of motion, the velocity of P relative to B is $l(\dot{\theta} + \dot{\chi})$ perpendicular to BP. The velocity of P relative to M is the resultant of these two velocities. Its resolved parts along and perpendicular to AB are accordingly

$$-l(\dot{\theta} + \dot{\chi}) \sin \chi \quad \text{and} \quad a\dot{\theta} + l(\dot{\theta} + \dot{\chi}) \cos \chi.$$

Now the centre of mass G is always at the point dividing MP in the ratio of the masses of the particle and the rod; and, if these masses are p and m respectively, the velocity of M relative to G has components

$$\frac{p}{m+p}\,l\,(\dot{\theta}+\dot{\chi})\sin\chi \text{ and } -\frac{p}{m+p}\{a\dot{\theta}+l\,(\dot{\theta}+\dot{\chi})\cos\chi\}$$

along and perpendicular to AB, and the velocity of P relative to G has components

$$-\frac{m}{m+p}\,l\,(\dot{\theta}+\dot{\chi})\sin\chi \text{ and } \frac{m}{m+p}\{a\dot{\theta}+l\,(\dot{\theta}+\dot{\chi})\cos\chi\}$$

in the same directions.

Hence the moment of momentum in the motion relative to G is

$$\tfrac{1}{3}ma^2\dot{\theta}+\frac{mp}{m+p}[(a+l\cos\chi)\{a\dot{\theta}+l\,(\dot{\theta}+\dot{\chi})\cos\chi\}+l\sin\chi\,\{l\,(\dot{\theta}+\dot{\chi})\sin\chi\}],$$

or

$$\tfrac{1}{3}ma^2\dot{\theta}+\frac{mp}{m+p}[(a+l\cos\chi)\,a\dot{\theta}+(l+a\cos\chi)\,l\,(\dot{\theta}+\dot{\chi})];$$

also twice the kinetic energy in the motion relative to G is

$$\tfrac{1}{3}ma^2\dot{\theta}^2+\frac{mp}{m+p}[a^2\dot{\theta}^2+l^2(\dot{\theta}+\dot{\chi})^2+2al\dot{\theta}\,(\dot{\theta}+\dot{\chi})\cos\chi].$$

Now the centre of mass moves with uniform velocity in a straight line; and thus the kinetic energy of the whole mass placed at the centre of mass and moving with it is constant, and the moment about any fixed axis of the momentum of the whole mass placed at the centre of mass and moving with it is also constant. Also the kinetic energy of the system and its moment of momentum about any fixed axis are constants. Hence the moment of momentum in the motion relative to G and the kinetic energy in the same relative motion are constants.

Let V be the velocity with which the particle was initially projected at right angles to the thread; then the initial values of the moment of momentum and kinetic energy in the motion relative to G are

$$(a+l)\,Vmp/(m+p) \text{ and } \tfrac{1}{2}V^2mp/(m+p).$$

Hence throughout the motion we have the equations

$$\left.\begin{aligned}\tfrac{1}{3}(1+m/p)\,a^2\dot{\theta}+a\dot{\theta}\,(a+l\cos\chi)+l\,(\dot{\theta}+\dot{\chi})(l+a\cos\chi)&=(a+l)\,V,\\ \tfrac{1}{3}(1+m/p)\,a^2\dot{\theta}^2+a^2\dot{\theta}^2+l^2(\dot{\theta}+\dot{\chi})^2+2al\dot{\theta}\,(\dot{\theta}+\dot{\chi})\cos\chi&=V^2.\end{aligned}\right\}$$

254. Kinematical Note. It is sometimes convenient in calculating the velocities of points in a connected system to use the coordinates of a point referred to axes which do not retain the same directions. In the problem of Art. 253 we might have obtained the velocity of P relative to M by taking as axes lines through M along and perpendicular to AB. When we wish to calculate the velocity of a point in this way we have to attend to the fact that the component velocities parallel to the moving axes are not the differential coefficients (with respect to the time) of the coordinates referred to the same axes.

Consider the motion of a particle P whose coordinates at time t are x', y' referred to rectangular axes rotating in their own plane about the origin; let ϕ be the angle which the axis of x' makes with a fixed axis of x in the plane at time t, and x, y the coordinates of the particle referred to fixed rectangular axes of x and y. Also let u, v be component velocities of the particle parallel to the axes of x' and y'.

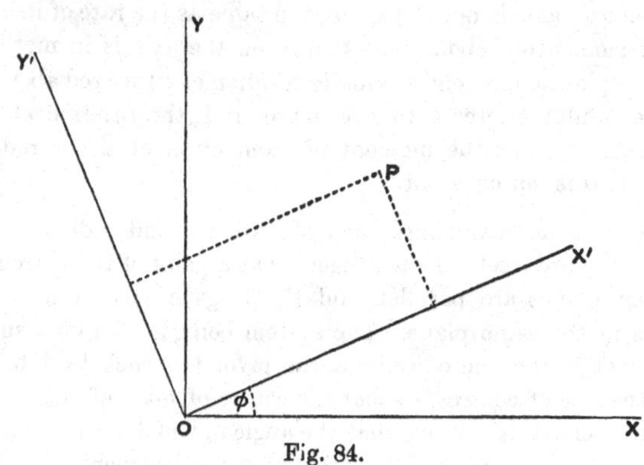

Fig. 84.

We have $$x = x' \cos\phi - y' \sin\phi, \quad y = y' \cos\phi + x' \sin\phi,$$
whence $$\left.\begin{aligned} \dot{x} &= (\dot{x}' - y'\dot{\phi})\cos\phi - (\dot{y}' + x'\dot{\phi})\sin\phi, \\ \dot{y} &= (\dot{y}' + x'\dot{\phi})\cos\phi + (\dot{x}' - y'\dot{\phi})\sin\phi. \end{aligned}\right\}$$
Also $$\dot{x} = u\cos\phi - v\sin\phi, \quad \dot{y} = v\cos\phi + u\sin\phi.$$
Hence we find $$u = \dot{x}' - y'\dot{\phi}, \quad v = \dot{y}' + x'\dot{\phi}.$$

Now, if we write ω for $\dot{\phi}$, ω is the angular velocity of the moving axes, and the resolved parts parallel to the moving axes of the velocity of the particle whose coordinates are x', y' are
$$\dot{x}' - \omega y' \text{ and } \dot{y}' + \omega x'.$$

We may prove in precisely the same way that, if α, β are the resolved parts of the acceleration of P parallel to the axes of x', y', then
$$\alpha = \dot{u} - \omega v \text{ and } \beta = \dot{v} + \omega u.$$

In the problem of Art. 253, we take axes through M along and perpendicular to AB. Then the angular velocity of the moving axes is $\dot{\theta}$, and the coordinates of P are $a + l\cos\chi$ and $l\sin\chi$. From these the component velocities of P relative to M which were obtained otherwise in that Article might be deduced.

255. Examples.

1. Two uniform rods AB, BC, freely jointed at B, move in one plane under no forces; it is required to find the motion.

We may use the figure and notation of Art. 253, taking P to be the middle point of BC, and writing m and p for the masses of AB and BC and $2a$ and $2l$ for their lengths. We have to add to the expression given in that

Article for the moment of momentum in the motion relative to G the term $\frac{1}{3}pl^2(\dot{\theta}+\dot{\chi})$, and to the expression there given for the kinetic energy in the motion relative to G the term $\frac{1}{3}pl^2(\dot{\theta}+\dot{\chi})^2$. The energy equation and the equation of constancy of moment of momentum determine the motion.

Note. This Example affords a good illustration of the result, to which attention was called in Art. 157, to the effect that the moment of kinetic reaction about an axis is not in general the same as the rate of increase of the moment of momentum about that axis when the axis is in motion. In the present Example the moment of kinetic reaction of either rod about B is zero, because the resultant force acting on either rod (the reaction at the hinge) passes through B; but the moment of momentum of either rod, or of the system, about B is not constant.

2. Two equal circular rings, each of radius a and radius of gyration k about its centre, are freely pivoted together at a point of their circumferences, so that their planes are parallel, and the rings are so thin they may be regarded as in the same plane. The system being at rest on a smooth table with the pivot in the line of centres, the pivot is struck by a blow perpendicular to the line of centres, so that the centre of mass of the system starts to move with velocity V. Prove that the angle θ, which either radius through the pivot makes with its initial direction at any subsequent time, is given by the equation

$$k^2(k^2+a^2\sin^2\theta)\dot{\theta}^2=V^2a^2.$$

3. A uniform straight tube of length $2a$ contains a particle of equal mass, and, the particle being close to the middle point, the tube is started to rotate about that point with angular velocity ω. Prove that, if there are no external forces, the velocity of the particle relative to the tube when it leaves it is $a\omega\sqrt{\frac{3}{5}}$.

4. Two horizontal threads are attached to a circular cylinder of negligible mass whose axis is vertical, are coiled in opposite directions round it, and carry equal particles which are initially at rest on two smooth horizontal planes. One of the particles is struck at right angles to its thread so that it starts off with velocity V and its thread begins to unwind from the cylinder. Prove that, if the initial length of the straight portion of the thread attached to the particle struck is c, its length r at time t is given by the equation

$$r^2=c^2+2aVt+\tfrac{1}{2}V^2t^2,$$

the cylinder being free to turn about its axis.

5. A thread is attached to a rigid cylinder of radius a and moment of inertia I about its axis, and carries a particle of mass m which is free to move on a smooth plane perpendicular to the axis, while the cylinder is free to rotate about the axis. The particle is projected on the plane at right angles to the thread with velocity V so that the thread tends to wind up round the cylinder. Prove that the length r of the straight portion at any subsequent time is given by the equation

$$(I+ma^2)\,r^2\dot{r}^2=\{I+m(r^2+a^2-c^2)\}\,a^2V^2,$$

where c is the initial value of r. Hence prove that

$$r^2 - c^2 = 2aVt + V^2t^2m/(M+m),$$

where $M = I/a^2$.

6. A cone of vertical angle 2α is free to turn about its axis, and a smooth groove is cut in its surface so as to make with the generators an angle β. A particle of mass m moves in the groove, and starts at a distance c from the vertex. Prove that, if at any subsequent time the particle is at a distance r from the vertex and the cone has turned through an angle θ, r and θ are connected by the equation

$$(I + mc^2 \sin^2 \alpha)\, e^{2\theta \sin \alpha \cot \beta} = (I + mr^2 \sin^2 \alpha),$$

where I is the moment of inertia of the cone about its axis.

7. An elliptic tube of latus rectum $2l$, eccentricity e, and moment of inertia I about its major axis, is rotating freely about its major axis, which is fixed, with angular velocity Ω, and contains a particle of mass m which is attracted to one focus by a force $\mu m/(\text{distance})^2$ and is initially at rest at the end of the major axis nearest the centre of force. Prove that, if the particle is slightly displaced, and if $\mu e(1+e)^2 < l^3\Omega^2$, it will come to rest relatively to the tube at an end of the nearer latus rectum, provided that

$$\Omega^2 l = 2\mu m e\,(1/ml^2 + 1/I).$$

8. Four equal uniform rods are freely hinged together so as to form a rhombus of side $2a$ and the system rotates about one diagonal, which is fixed in a vertical position, the highest point of the rhombus being fixed and the lowest being free to slide on the diagonal. Find the angular velocity in the steady motion in which each rod makes an angle α with the vertical, and prove that the period of the small oscillations about this state of steady motion is the same as for a simple pendulum of length

$$\tfrac{8}{3} a \cos\alpha\,(1 + 3\sin^2\alpha)/(1 + 3\cos^2\alpha).$$

Motion of a string or chain

256. Inextensible chain. When a chain moves in a straight line, the condition of inextensibility is that all the particles of it have at any instant the same velocity. When the chain forms a curve, and moves so as to be in contact with a given curve, the condition takes the form:—The velocity of a particle, resolved along the tangent to the curve at the position of the particle, is the same for all the particles.

257. Tension at a point of discontinuity. It often happens that two parts of a chain move in different ways, and that portions of the chain are continually transferred from the part that is moving in one way to the part that is moving in the other way. The tension at the place where the motion changes is then to be determined by the principle that the increase of momentum of a system in any interval is equal to the impulse of the force which acts upon it during that interval. (Art. 162.) This principle is to be applied to a hypothetical particle of the chain, supposed to pass during a very short interval from one state of motion to the other, and the mass of the hypothetical particle is to be taken to be the mass of the part of the chain which changes its motion during the interval. (Cf. Art. 189.) This principle is illustrated in the following problems.

It is important to observe that discontinuous motions such as are considered here in general involve dissipation of energy.

258. Illustrative Problems.

I. *A chain is coiled at the edge of a table with one end just hanging over. It is required to find the motion.*

At any time t let x be the length which has fallen over the edge, T the tension at the edge in the falling portion. There is no tension in the part coiled up. Let m be the mass per unit length of the chain.

During a very short interval Δt a length of the chain which can be taken to be $\dot{x}\Delta t$ is set in motion with velocity $\dot{x}$, and the impulse of the force by which it is set in motion can be taken to be $T\Delta t$. Hence we have the approximate equation

$$T\Delta t = m\dot{x}\Delta t \,.\, \dot{x},$$

which passes over into the exact equation

$$T = m\dot{x}^2.$$

The equation of motion of the falling portion is therefore

$$mx\ddot{x} = mxg - m\dot{x}^2.$$

Writing v for $\dot{x}$, this is $$xv\frac{dv}{dx} + v^2 = gx,$$

or $$\frac{d}{dx}(x^2v^2) = 2gx^2.$$

Integrating, and observing that v and x vanish together, we have

$$v^2 = \tfrac{2}{3}gx.$$

This equation gives the velocity of the falling portion when its length is x.

The time until the length is x is

$$\int_0^x \frac{dx}{\sqrt{(\frac{2}{3}gx)}} = \sqrt{\frac{6x}{g}}.$$

The potential energy lost while the free end falls through x is $\frac{1}{2}mgx^2$, and the kinetic energy gained is $\frac{1}{2}mxv^2$ or $\frac{1}{3}mgx^2$; and the amount of energy dissipated in the same time is $\frac{1}{6}mgx^2$.

II. *A chain, one end of which is held fixed, is initially held with the other end close to the fixed end, and the other end is then let go.*

Let $2l$ be the length of the chain, m the mass per unit length, $l+x$ the length of the part that has come to rest at time t, T the tension at its lower end.

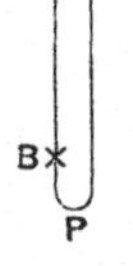

Fig. 85.

The free end has fallen through $2x$ under gravity, so that

$$2x=\tfrac{1}{2}gt^2, \quad \dot{x}=\tfrac{1}{2}gt,$$

and the falling portion is free from tension.

During a very short interval Δt a length approximately equal to $\frac{1}{2}gt \, . \, \Delta t$ passes from motion with velocity gt to rest, so that an impulse, which is approximately equal to $T\Delta t$, destroys an amount of momentum which is approximately equal to $\frac{1}{2}mg^2t^2\Delta t$. Hence we have the exact equation

$$T=\tfrac{1}{2}mg^2t^2.$$

Thus the motion and the tension at any time are determined.

259. Constrained motion of a chain under gravity. We shall suppose the chain to be in a rough tube, or in a groove cut on a rough surface, so that the line of it is a given curve. We shall take this curve to be in a vertical plane.

Let s be the distance, measured along the curve, of a point P of the curve from a fixed point, ρ the radius of curvature of the curve at P, ϕ the angle which the normal to the curve at P makes with the vertical. Let P' be a point near to P, for which s, ϕ become $s+\Delta s$, $\phi+\Delta\phi$. Between P and P' we may imagine a hypothetical particle of mass $m\Delta s$. Let v be the velocity of this particle, which we may take, with sufficient approximation, to be directed along the tangent to the curve at P. We may regard the particle as moving under the tensions T and $T+\Delta T$, which we may take to be directed along the tangents at P and P', the pressure of the curve, which we may take to be directed along the normal at P, and the friction, which we may take to be directed along the tangent at P. We denote the pressure and friction by $R\Delta s$ and $\mu R\Delta s$, so that R is the pressure per unit of length, and μ is the coefficient of friction.

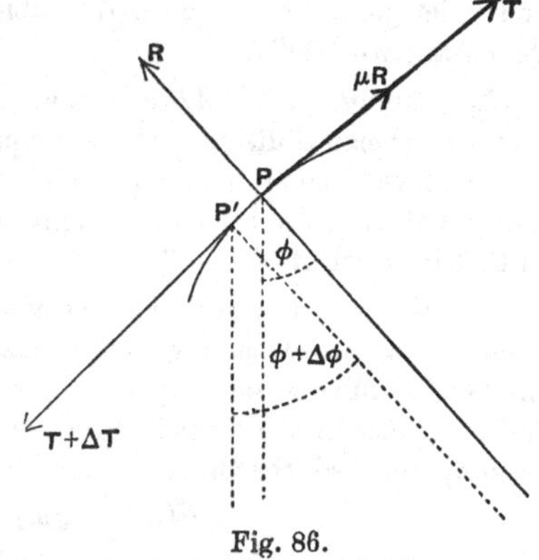

Fig. 86.

We form equations of motion by resolving along the tangent and normal at P. The equations are

$$m\Delta s\,.\,\dot{v} = mg\Delta s\,.\,\sin\phi + (T+\Delta T)\cos\Delta\phi - T - \mu R\Delta s,$$

$$m\Delta s\,.\,\frac{v^2}{\rho} = mg\Delta s\,.\,\cos\phi + (T+\Delta T)\sin\Delta\phi - R\Delta s.$$

On dividing by Δs, and passing to the limit, we have the exact equations of motion

$$m\dot{v} = mg\sin\phi + \frac{dT}{ds} - \mu R \quad\text{.................(1)},$$

$$m\frac{v^2}{\rho} = mg\cos\phi + \frac{T}{\rho} - R \quad\text{.....................(2)}.$$

If the curve is *smooth* we omit μR from the first equation. If, further, the ends of the chain are free, the velocity v can be determined by means of the energy equation, and the tension can be found by substituting for v in the equation (1). When the tension is known the pressure at any point can be found from the equation (2).

260. Examples.

1. A uniform chain of length a is laid out straight on a smooth table, and lies in a line at right angles to the edge of the table. One end is put just over the edge. Prove that, if the edge of the table is rounded off, so that the part of the chain which has run off at any time is vertical, the velocity of the chain as the last element leaves the table is $\sqrt{(ag)}$.

2. A uniform chain of length l and weight W is suspended by one end and the other end is at a height h above a smooth table. Prove that, if the upper end is let go, the pressure on the table as the coil is formed increases from $2hW/l$ to $(2h+3l)\,W/l$.

3. A uniform chain AB is held with its lower end fixed at B and its upper end A at a vertical distance above B equal to the length of the chain. The end A is released, and at the instant when it passes B the end B is also released. Prove that the chain becomes straight after an interval equal to three-quarters of that in which A fell to B.

4. Two uniform chains whose masses per unit of length are m_1 and m_2 are joined by a thread passing over a fixed smooth pulley. Initially the chains are held up in coils and they are released simultaneously without causing any finite impulse in the thread. Prove that, until one of the chains has become entirely uncoiled, the thread slips over the pulley with uniform acceleration

$$g\,(\sqrt{m_1} \sim \sqrt{m_2})/(\sqrt{m_1}+\sqrt{m_2}),$$

and that the portions of the chains which have become straight increase during the interval with uniform accelerations

$$2g\sqrt{m_2}/(\sqrt{m_1}+\sqrt{m_2}) \text{ and } 2g\sqrt{m_1}/(\sqrt{m_1}+\sqrt{m_2}).$$

5. A uniform chain of length l and weight W is placed on a line of greatest slope of a smooth plane of inclination α to the horizontal so that it just reaches to the bottom of the plane where there is a small smooth pulley over which it can run off. Prove that, when a length x has run off, the tension at the bottom of the plane is

$$W(1-\sin\alpha)\,x\,(l-x)/l^2.$$

6. A uniform chain is held with its highest point on the highest generator of a smooth horizontal circular cylinder, and lies on the cylinder in a vertical plane, subtending an angle β at the centre of the circular section on which it lies. Prove that, when the chain is let go, the lower end is the first part of it to leave the cylinder, and that this happens when the radius drawn through the upper end makes with the vertical an angle ϕ given by the equation

$$\tfrac{1}{2}\beta\cos(\phi+\beta)=\sin\beta+\sin\phi-\sin(\phi+\beta).$$

261. Chain moving freely in one plane. Kinematical equations. At any instant the chain forms a curve. Let A be the position on this curve of a chosen particle, P that of any other particle, and let s be the arc of the curve measured from A to P. If the chain is inextensible we may regard s as a parameter specifying the particle which is at the point P at time t. Let ϕ be the angle which the tangent at P to the curve, drawn in the sense of increase of s, makes with a fixed axis of x in the plane; ϕ is estimated as the angle through which a line coinciding with the axis of x must turn in the positive sense so as to coincide with the tangent. Also let ρ be the radius of curvature of the curve at P.

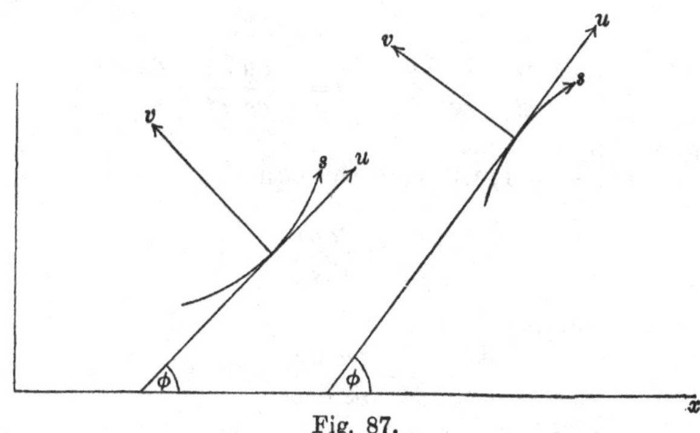

Fig. 87.

We resolve the velocity of the particle of the chain which is at P at time t into components u, v, of which u is directed along the tangent to the curve at P in the sense of increase of s, and v is directed along the normal. The sense of the normal is taken to be such that, if the curve is described in the sense of increase of s, the normal is drawn towards the left hand. If this sense is that of the normal drawn towards the centre of curvature, $\frac{\partial\phi}{\partial s}$ is positive; otherwise, it is negative. (See Fig. 87.) The absolute value of $\frac{\partial\phi}{\partial s}$, without regard to sign, is $\frac{1}{\rho}$.

Let x, y be the coordinates of P, *i.e.* of the position of the particle specified by s at time t. We have the equations

$$\cos\phi = \frac{\partial x}{\partial s}, \quad \sin\phi = \frac{\partial y}{\partial s},$$

in which the differential coefficients are partial, s and t being independent variables. From these equations we have

$$\frac{\partial^2 y}{\partial s^2} = \frac{\partial\phi}{\partial s}\frac{\partial x}{\partial s}, \quad \frac{\partial^2 x}{\partial s^2} = -\frac{\partial\phi}{\partial s}\frac{\partial y}{\partial s}.$$

Further the direction cosines of the normal drawn in the sense already chosen are $-\frac{\partial y}{\partial s}$ and $\frac{\partial x}{\partial s}$.

The velocity of the particle specified by s at time t has components u and v in the stated directions, and also has components $\frac{\partial x}{\partial t}, \frac{\partial y}{\partial t}$ parallel to the axes of coordinates. We have therefore the equations

$$u = \frac{\partial x}{\partial s}\frac{\partial x}{\partial t} + \frac{\partial y}{\partial s}\frac{\partial y}{\partial t}, \quad v = -\frac{\partial y}{\partial s}\frac{\partial x}{\partial t} + \frac{\partial x}{\partial s}\frac{\partial y}{\partial t}.$$

Since $\left(\frac{\partial x}{\partial s}\right)^2 + \left(\frac{\partial y}{\partial s}\right)^2 = 1$, we have the equation

$$\frac{\partial}{\partial t}\left[\left(\frac{\partial x}{\partial s}\right)^2 + \left(\frac{\partial y}{\partial s}\right)^2\right] = 0,$$

which is the same as

$$\frac{\partial x}{\partial s}\frac{\partial^2 x}{\partial s\partial t} + \frac{\partial y}{\partial s}\frac{\partial^2 y}{\partial s\partial t} = 0,$$

or

$$\frac{\partial}{\partial s}\left(\frac{\partial x}{\partial s}\frac{\partial x}{\partial t} + \frac{\partial y}{\partial s}\frac{\partial y}{\partial t}\right) - \frac{\partial^2 x}{\partial s^2}\frac{\partial x}{\partial t} - \frac{\partial^2 y}{\partial s^2}\frac{\partial y}{\partial t} = 0,$$

or

$$\frac{\partial u}{\partial s} - v\frac{\partial\phi}{\partial s} = 0.$$

This equation, combined with the statement that s and t are independent variables, expresses the condition of inextensibility of the chain.

The angular velocity $\frac{\partial \phi}{\partial t}$, with which the tangent to the curve at the position of the particle specified by s is turning, may be expressed in terms of u and v. We have the equation

$$\frac{\partial \phi}{\partial t} = -\sin\phi \frac{\partial}{\partial t}(\cos\phi) + \cos\phi \frac{\partial}{\partial t}(\sin\phi),$$

or

$$\frac{\partial \phi}{\partial t} = -\frac{\partial y}{\partial s}\frac{\partial^2 x}{\partial s \partial t} + \frac{\partial x}{\partial s}\frac{\partial^2 y}{\partial s \partial t}$$

$$= \frac{\partial}{\partial s}\left(-\frac{\partial y}{\partial s}\frac{\partial x}{\partial t} + \frac{\partial x}{\partial s}\frac{\partial y}{\partial t}\right) + \frac{\partial^2 y}{\partial s^2}\frac{\partial x}{\partial t} - \frac{\partial^2 x}{\partial s^2}\frac{\partial y}{\partial t}$$

$$= \frac{\partial}{\partial s}\left(-\frac{\partial y}{\partial s}\frac{\partial x}{\partial t} + \frac{\partial x}{\partial s}\frac{\partial y}{\partial t}\right) + \frac{\partial \phi}{\partial s}\left(\frac{\partial x}{\partial s}\frac{\partial x}{\partial t} + \frac{\partial y}{\partial s}\frac{\partial y}{\partial t}\right),$$

or

$$\frac{\partial \phi}{\partial t} = \frac{\partial v}{\partial s} + u\frac{\partial \phi}{\partial s}.$$

The two equations

$$\frac{\partial u}{\partial s} - v\frac{\partial \phi}{\partial s} = 0, \quad \frac{\partial v}{\partial s} + u\frac{\partial \phi}{\partial s} = \frac{\partial \phi}{\partial t}$$

are the kinematical conditions which must be satisfied at all points of the chain throughout the motion.

Note. If the chain is extensible, and s_0 is the natural length of the portion of it that is contained between a chosen particle A and any other particle P, the particle P is specified by the parameter s_0, and we may take s_0 and t as independent variables. We may then prove in the same way as in the above Article that the following kinematical equations must hold at all points of the chain:

$$\frac{\partial u}{\partial s_0} - v\frac{\partial \phi}{\partial s_0} = \frac{\partial \epsilon}{\partial t}, \quad \frac{\partial v}{\partial s_0} + u\frac{\partial \phi}{\partial s_0} = (1+\epsilon)\frac{\partial \phi}{\partial t},$$

where ϵ is the extension of the chain at the particle P.

262. Chain moving freely in one plane. Equations of motion. We form the equations of motion by resolving the kinetic reaction of a small element of the chain in the directions of the tangent and normal to the curve which it instantaneously forms. The component accelerations in these directions are obtained by the method of Art. 254 in the forms

$$\frac{\partial u}{\partial t} - v\frac{\partial \phi}{\partial t}, \quad \frac{\partial v}{\partial t} + u\frac{\partial \phi}{\partial t}.$$

The resultant of the tensions at the ends of the element is obtained in the same way as in Art. 259. If S and N denote the component forces per unit mass applied to the chain in the directions of the tangent and normal to the curve, the equations of motion are

$$m\left(\frac{\partial u}{\partial t}-v\frac{\partial \phi}{\partial t}\right)=\frac{\partial T}{\partial s}+mS,$$

$$m\left(\frac{\partial v}{\partial t}+u\frac{\partial \phi}{\partial t}\right)=T\frac{\partial \phi}{\partial s}+mN,$$

where m denotes the mass per unit of length.

263. Invariable form. Interesting cases of the motion of a chain arise in which the shape of the curve formed by the chain is invariable, but the chain moves along the curve. In discussing such cases it conduces to clearness to imagine the chain to be enclosed in a fine rigid tube, of the shape in question, and to move along the tube while the tube moves in its plane. The velocity of any point of the tube is then determined as the velocity of a point of a rigid body moving in two dimensions, and the velocity of any element of the chain will be found by compounding a certain velocity w relative to the tube with the velocity of any point of the tube. The direction of w is that of the tangent to the line of the tube at the point, and its magnitude is variable from point to point in accordance with the kinematical conditions.

Taking now the special case of a uniform chain moving under gravity, we show that the chain can move steadily in the form of a common catenary, the curve retaining its position as well as its form. The velocity w is in this case the velocity of an element of the chain, and, with the notation of Art. 261, we have

$$u=w, \quad v=0.$$

The kinematical conditions become

$$\frac{\partial w}{\partial s}=0, \quad \frac{\partial \phi}{\partial t}=w\frac{\partial \phi}{\partial s},$$

so that the chain moves uniformly along itself.

The equations of motion of Art. 262 are satisfied by

$$T=mgc\sec\phi+mw^2,$$

the curve being the catenary $s=c\tan\phi$.

264. Examples.

1. Prove that any curve which is a form of equilibrium for a uniform chain under conservative forces is a form which the chain can retain when moving uniformly along itself under the same forces, and that the tension is greater in the steady motion than in equilibrium by mw^2, where m is the mass per unit length of the chain, and w is the velocity with which the chain moves along itself.

2. A uniform chain moves over two smooth parallel rails distant $2a$ apart at the same level and is transferred from a coil at a distance h vertically below one rail to a coil at a distance $h+b$ vertically below the other. Prove that the portion between the rails can be a common catenary, provided that the velocity of the chain along itself is $\sqrt{(gb)}$.

3. A uniform chain moves in a plane under no forces in such a way that the curve of the chain retains an invariable form which rotates about a fixed point in the plane with uniform angular velocity ω, while the chain advances relatively to the curve with uniform velocity V. Prove that the general (p, r) equation of the curve must be of the form

$$(p+2V/\omega)\,r^2=ap+b,$$

where a and b are constants.

4. A uniform chain falls in a vertical plane under gravity. Prove that the square of the angular velocity of the tangent at any element is

$$\frac{1}{m}\left(\frac{T}{\rho^2}-\frac{\partial^2 T}{\partial s^2}\right).$$

5. A uniform chain hangs in equilibrium over a smooth pulley; one end is fixed at an extremity of the horizontal diameter, and portions hang vertically on both sides. Prove that, if the end is set free, the distance y of the lowest point from the horizontal diameter during the first part of the motion satisfies the equation

$$(l-y+\tfrac{1}{2}gt^2)\,\ddot{y}-(\dot{y}-gt)^2=g\,(y+\tfrac{1}{2}c),$$

where l is the length of the chain and $2c$ is the circumference of the circle.

6. A uniform chain of length $2L$ and mass $2L\mu$ has its ends attached to two points A, C, and passes over a smooth peg B between A and C and in the same horizontal line with them, the points A, B, C being so close together that the parts of the chain between them may be considered vertical. Elastic threads of natural lengths l and l' and moduluses λ and λ' are fastened to points P and P' of the chain on opposite sides of B and their other ends are fixed to points O and O' vertically below P and P'. The system oscillates so that the threads are always stretched, and the points P and P' are never for any finite time at rest. Prove that the time of a complete oscillation is

$$2\pi\,\sqrt{\{Ll'\mu/(\lambda l'+\lambda' l-\mu g l l')\}}.$$

7. A fine elliptic tube is constrained to rotate with uniform angular velocity ω about its major axis which is vertical, and contains a uniform chain whose length is equal to a quadrant of the ellipse. Prove that, if $\omega^2=4g/l$, where l is the latus rectum of the ellipse, the chain will be in stable relative equilibrium with one end at the lowest point.

8. A rough helical tube of pitch α and radius a is placed with its axis vertical, and a uniform chain is placed within it, the coefficient of friction between the tube and the chain being $\tan\alpha\cos\epsilon$. Prove that, when the chain has fallen a vertical distance ma, its velocity is $\sqrt{(ag\sec\alpha\sinh 2\mu)}$, where μ is determined by the equation

$$\cot\tfrac{1}{2}\epsilon\tanh\mu=\tanh(\mu\sin\epsilon+\tfrac{1}{2}m\cos\alpha\sin 2\epsilon).$$

265. Initial Motion. When the chain starts from rest in a position which is not one of equilibrium the initial velocities are zero, and the equations of motion are simplified by the omission of $\partial\phi/\partial t$. At the same time the kinematic conditions are altered in form. Since $\dfrac{\partial^2\phi}{\partial s\partial t}$ vanishes initially, the result of differentiating the equation $\dfrac{\partial u}{\partial s} = v\dfrac{\partial \phi}{\partial s}$ is

$$\frac{\partial^2 u}{\partial s\partial t} = \frac{\partial v}{\partial t}\frac{\partial \phi}{\partial s}.$$

We may write the equations of motion in the form

$$\left.\begin{aligned}\frac{\partial u}{\partial t} &= S + \frac{1}{m}\frac{\partial T}{\partial s},\\ \frac{\partial v}{\partial t} &= N + \frac{T}{m}\frac{\partial \phi}{\partial s}.\end{aligned}\right\}$$

Differentiating the first with respect to s, multiplying the second by $\dfrac{\partial\phi}{\partial s}$, and subtracting, we obtain an equation

$$\frac{\partial}{\partial s}\left(\frac{1}{m}\frac{\partial T}{\partial s}\right) - \frac{1}{m}\frac{T}{\rho^2} = -\frac{\partial S}{\partial s} + N\frac{\partial \phi}{\partial s}.$$

This equation serves to determine the initial tension at any point of the chain. To determine the arbitrary constants which enter into the solution of the equation we have to use the conditions which hold at the ends, or at other special points, of the chain. If, for example, one end of the chain is guided to move on a given curve, the acceleration of the extreme particle must be directed along the tangent to the curve.

Cases arise in which this method cannot be applied. In the case of a heavy chain with an end which moves on a smooth straight wire, not perpendicular to the tangent at the end, the equation of motion of an element at the end, found by resolving along the wire, cannot be satisfied if the acceleration of the element is finite (not infinite) and the tension is finite (not zero). The conclusion in such cases must be that the chain becomes slack at the end, and it may become slack throughout. In such cases it is usually convenient to suppose the end of the chain to be attached to a ring which can slide on the wire, and to take the mass of the ring, at first, to be finite; when the problem has been solved with this condition we can pass to the case above described by supposing the mass of the ring to be diminished without limit.

266. Impulsive Motion. The equations of impulsive motion of a chain which is suddenly set in motion are obtained at once by the method of Art. 262. We have only to regard S and N as the resolved parts of an impulse, reckoned per unit of mass, applied to an element, and T as impulsive tension. The equations are

$$\left.\begin{aligned} mu &= \frac{\partial T}{\partial s} + mS, \\ mv &= T\frac{\partial \phi}{\partial s} + mN. \end{aligned}\right\}$$

The kinematical conditions are the same as those which were obtained in Art. 261 for a chain in continuous motion.

In case no impulses are applied to the chain except at its ends, S and N vanish, and we can eliminate u and v, obtaining an equation for T in the form

$$\frac{\partial}{\partial s}\left(\frac{1}{m}\frac{\partial T}{\partial s}\right) - \frac{T}{m\rho^2} = 0.$$

The solution of this equation subject to the given terminal conditions gives the impulsive tension at any point of the chain.

267. Examples.

1. In the initial motion of a chain under gravity prove that the tension satisfies the equation

$$\frac{\partial}{\partial s}\left(\frac{1}{m}\frac{\partial T}{\partial s}\right) - \frac{T}{m\rho^2} = 0.$$

2. A uniform chain hangs under gravity with its ends attached to two rings which are free to slide on a smooth horizontal bar. Prove that, if the rings are initially held so that the tangents to the chain just below them make equal angles γ with the horizontal, and are let go, the tension at the lowest point is changed in the ratio $2M' : 2M' + M\cot^2\gamma$, where M is the mass of the chain, and M' that of either ring. [Cf. Ex. 5 in Art. 207.]

3. If the ends of the chain of Ex. 2 are free to move on smooth bars along the normals at the ends, and the chain is severed at its vertex, prove that the tension at a point where the tangent makes an angle ϕ with the horizontal immediately becomes

$$\tfrac{1}{2} Mg\phi \sec\phi \cos\gamma / (\cos\gamma + \gamma\sin\gamma).$$

4. Impulsive tensions T_α, T_β are applied at the ends of a piece of chain of mass M hanging in the form of a common catenary with terminal tangents inclined to the horizontal at angles α and β. Prove that the kinetic energy generated is

$$\frac{1}{2}\frac{\tan\alpha - \tan\beta}{M}\left\{\frac{(T_\alpha\cos\alpha - T_\beta\cos\beta)^2}{\alpha - \beta} + (T_\alpha^2 \sin\alpha\cos\alpha - T_\beta^2 \sin\beta\cos\beta)\right\}.$$

MISCELLANEOUS EXAMPLES

1. A rod of length $2a$ is held in a position inclined at an angle α to the vertical, and is then let fall on a smooth horizontal plane. Prove that, if there is no restitution, the end of the rod which strikes the plane will leave it immediately after impact provided that the height through which the rod falls is greater than

$$\tfrac{1}{18}a \sec \alpha \operatorname{cosec}^2 \alpha\,(1+3\sin^2\alpha)^2.$$

2. A sphere of radius a rolling on a rough table with velocity V comes to a slit of breadth b perpendicular to its path. Prove that, if there is no restitution, the condition that it should cross the slit without jumping is

$$V^2 > \tfrac{10}{7}ga\,(1-\cos\alpha)\sin^2\alpha\,(14-10\sin^2\alpha)/(7-10\sin^2\alpha)^2,$$

where $b=2a\sin\alpha$ and $17ga\cos\alpha > V^2+10ga$.

3. A sphere of mass m falls vertically and impinges with velocity V against a board of mass M which is moving with velocity U on a horizontal table. The coefficient of restitution between the sphere and the board is e, and the friction between the board and the table can be neglected. Prove that, if the coefficient of friction between the sphere and the board exceeds $2MU/(7M+2m)(1+e)V$, the kinetic energy lost in the impact is

$$\tfrac{1}{2}m(1-e^2)V^2+mMU^2/(7M+2m).$$

4. A ball is let fall upon a hoop, of which the mass is $1/n$ of that of the ball; the hoop is suspended from a point in its circumference, about which it can turn freely in a vertical plane. Prove that, if e is the coefficient of restitution, and α the inclination to the vertical of the radius passing through the point at which the ball strikes the hoop, the ball rebounds in a direction making with the horizontal an angle $\tan^{-1}\{(1+\tfrac{1}{2}n)\tan\alpha - e\cot\alpha\}$.

5. A plank of length $2a$ is turning about a horizontal axis through its centre of gravity, and a particle strikes the rising half, rebounds, and strikes the other half, the coefficient of restitution being unity. Prove that, if the motion indefinitely repeats itself, the inclination of the plank to the horizontal must never exceed α where $I(\pi+2\alpha)\tan\alpha = ma^2$, I being the moment of inertia of the plank about its axis, and m the mass of the particle.

6. Two equal rigid uniform laminæ, each in the shape of an equilateral triangle, rest with two edges in contact. They are struck at the same instant with equal blows P in opposite directions bisecting the common edge and one other edge of each, so that they are pressed together and begin to slide one over the other. Find the velocity v of the point of application of either blow resolved in its direction and prove that, if μ is the coefficient of friction, the kinetic energy generated in the system is $(1-\mu\sqrt{3})Pv$, assuming no restitution.

---

7. Two lengths $2a$ and $2b$ are cut from the same uniform rod of mass M and freely jointed at one end of each. The rods being at rest in a straight line, an impulse MV is applied at the free end of a. Prove that the kinetic energy when b is free is to that when the further end of b is fixed in the ratio

$$(4a+3b)(3a+4b)/12(a+b)^2.$$

8. An equilateral triangle, formed of three equal uniform rods hinged at their ends, is held in a vertical plane with one side horizontal and the opposite corner downwards. Prove that, if after falling through any height the middle point of the highest rod is suddenly stopped, the impulsive stresses at the upper and lower hinges will be in the ratio $\surd 13 : 1$.

9. A rectangle, sides $2a$ and $2b$, formed of four uniform rods of the same material and section, smoothly hinged at the ends, is moving without rotation on a smooth horizontal plane, when a side of length $2a$ impinges on a small rough peg (zero restitution). Prove that, for that side to acquire the greatest possible angular velocity, the point of impact must be at a distance $a\{(3b+a)/(3b+3a)\}^{\frac{1}{2}}$ from its centre. Prove also that the rectangle cannot begin to rotate as a rigid body unless the direction of motion before impact makes with the impinging side an angle greater than

$$\tan^{-1}\frac{a\,(3b+a)^{\frac{1}{2}}\,(3b+3a)^{\frac{1}{2}}}{b\,(2b+3a)}.$$

10. Twelve equal rods each of length $2a$ are so jointed together that they can be the edges of a cube, and the framework moves symmetrically through a configuration in which each rod makes an angle θ with the vertical; prove that, if u is the velocity of the centre of mass, the kinetic energy is $\frac{1}{2}M(\frac{7}{3}a^2\dot{\theta}^2+u^2)$, where M is the mass of the framework, and that, if the frame strikes the ground when $\dot{\theta}=0$, then u is reduced in the ratio

$$1/(1+\tfrac{7}{27}\operatorname{cosec}^2\theta).$$

11. Any number of equal uniform rods are jointed together so as to have a common extremity and placed symmetrically so as to be generators of a cone of vertical angle 2α; the system falling with velocity V strikes symmetrically a smooth fixed sphere of radius c (no restitution). Prove that the angular velocity with which each rod begins to turn is

$$V(c\cos\alpha \sim a\sin^3\alpha)/(\tfrac{4}{3}a^2\sin^2\alpha+c^2\cot^2\alpha-ac\sin 2\alpha).$$

---

12. Two equal uniform rods each of length $2a$ are freely hinged at one extremity, and their other extremities are connected by an inextensible thread of length $2l$. The system rests on two smooth pegs distant $2c$ apart in a horizontal line. Prove that, if the thread is severed, the initial angular acceleration of either rod is

$$(8a^2c-l^3)\,g/(\tfrac{8}{3}a^2l^2+32a^4c^2/l^2-8a^2cl).$$

13. A uniform circular disk is symmetrically suspended by two elastic cords of natural length c, inclined at an angle α to the vertical and attached to the highest point of the disk. Prove that, if one of the cords is cut, the initial radius of curvature of the path of the centre of the disk is

$$3b\,(b-c)/(c\sin 4\alpha-b\sin 2\alpha),$$

where b is the equilibrium length of each cord.

14. A uniform rod of length $2a$ and weight W rests on a rough horizontal plane with its pressure on the plane uniformly distributed. A horizontal force P, large enough to produce motion, is suddenly applied at one end perpendicularly to the length of the rod. Prove that the rod begins to turn about a point distant x from its middle point, where x is the positive root of the equation

$$x^3 - (\tfrac{1}{3} - 2P/\mu W)\, a^2 x - \tfrac{2}{3} Pa^3/\mu W = 0,$$

and μ is the coefficient of friction.

15. A uniform circular disk (mass M) rotates in a horizontal plane with angular velocity ω. Close round it moves a ring of mass m and radius c rotating about its centre with angular velocity $\nu\,(<\omega)$. The ring carries a massless smooth spoke along a radius, and a bead of mass p can move on the spoke under the action of a force to the centre of the ring equal to $\mu/(\text{distance})^2$, and the bead is in relative equilibrium at a distance a from the centre. Prove that, if a slight continuous action now begins between the disk and the ring, of the nature of friction and proportional to the relative angular velocity, the distance of the bead from the centre, and the angular velocity of the ring, will at first increase, and their values after a short time t will be

$$a + \tfrac{1}{3} t^3 a\nu\lambda\, (\omega - \nu)/(mc^2 + pa^2),$$

and

$$\nu + t\lambda\, (\omega - \nu)/(mc^2 + pa^2) - \tfrac{1}{2}\lambda t^2\, [\lambda\, (\omega - \nu)/(mc^2 + pa^2)]\, [2/Mc^2 + 1/(mc^2 + pa^2)],$$

where $\lambda\dot{\theta}$ is the frictional couple when the relative angular velocity is $\dot{\theta}$.

16. A series of $2n$ uniform equal rods, each of mass m, are hinged together and held so that they are alternately horizontal and vertical, each vertical rod being lower than the preceding; the highest rod is horizontal and can turn freely round its end which is fixed. Prove that, when the rods are let go, the horizontal component X_{2r} and the vertical component Y_{2r} of the initial action between the $2r$th and the $(2r+1)$th rods are given by

$$X_{2r} = B\,(-5 + 2\sqrt{6})^r + C\,(-5 - 2\sqrt{6})^r,$$
$$Y_{2r} = B'\,(-5 + 2\sqrt{6})^r + C'\,(-5 - 2\sqrt{6})^r,$$

the constants B, C, B', C' being determined by the equations

$$X_{2n} = 0,\quad Y_{2n} = 0,\quad X_2 + 2X_0 = 0,\quad 2Y_2 + 16Y_0 - 5mg = 0.$$

17. A chain is formed of n equal symmetrical rods, each of length $2a$ and radius of gyration k about its centre of mass. One end is fixed and the whole is supported in a horizontal line. Prove that, if the supports are simultaneously removed, the free end begins to move with acceleration

$$g\,[1 + (-)^{n+1}\,\text{sech}\,\log\,(\tanh^n \tfrac{1}{2}\theta)],\ \text{where}\ \theta = \log\,(a/k).$$

---

18. A particle of mass M rests on a smooth table, and is connected with a particle of mass m by an inextensible thread passing through a hole in the table. Prove that, if m is released from rest in a position in which its polar

coordinates are a, a referred to the hole as origin and the vertical as initial line, then in the initial motion

$$(M+m)\,\ddot{r}_0 = mg\cos a,\quad a\ddot{\theta}_0 = -g\sin a,$$

$$a\,(M+m)\,r_0^{\text{iv}} = 3mg^2\sin^2 a,\quad a^2\theta_0^{\text{iv}} = g^2\sin a\cos a\,(M+6m)/(M+m).$$

Also prove that the initial radius of curvature of the path of m is

$$3\,(\ddot{x}_0^2+\ddot{y}_0^2)^{\frac{3}{2}}/(x_0^{\text{iv}}\ddot{y}_0 - y_0^{\text{iv}}\ddot{x}_0),$$

where $\quad \ddot{x}_0 = \ddot{r}_0,\ \ddot{y}_0 = a\ddot{\theta}_0,\ x_0^{\text{iv}} = r_0^{\text{iv}} - 3a\ddot{\theta}_0^2,\ y_0^{\text{iv}} = a\theta_0^{\text{iv}} + 6\ddot{r}_0\ddot{\theta}_0.$

19. A garden roller is at rest on a horizontal plane which is rough enough to prevent slipping; and the handle is so held that the plane through the axis of the cylinder and the centre of mass of the handle makes an angle a with the horizontal. Show that, if the handle is let go, the initial radius of curvature of the path described by its centre of inertia is

$$cn^{-2}\,(\sin^2 a + n^2\cos^2 a)^{\frac{3}{2}},$$

where $\quad (n-1)\,M\,(K^2+a^2) = ma^2,$

c is the distance of the centre of mass of the handle from the axis of the cylinder, m its mass, and MK^2 the moment of inertia of the cylinder about its axis, the cylinder being homogeneous and of radius a.

20. A rough plank of mass M is free to turn in a vertical plane about a horizontal axis distant c from its centre of mass, and a uniform sphere of mass m is placed on the plank at a distance b from the axis on the side remote from the centre of mass, the plank being held horizontal. Prove that, when the plank is let go, the initial radius of curvature of the path of the centre of the sphere is $21b\theta/(5-11\theta)$, where $\theta = (mb - Mc)/(mb + Ma)$, and Mab is the moment of inertia of the plank about the axis.

---

21. An elastic circular ring of which the radius when unstrained is a rests on a smooth surface of revolution, whose axis is vertical, in the form of a circle of radius r. Prove that the period of the small oscillations in which each element moves in a vertical plane is the same as for a simple pendulum of length l, where $1/l = \sin a\cos a/(r-a) - \sec a/\rho$, ρ being the radius of curvature of the meridian curve at a point on the ring, and a the inclination of the normal to the vertical.

22. Two equal spheres, each of radius a and moment of inertia I about an axis through its centre, have their centres connected by an elastic thread passing through holes in their surfaces, and are set to vibrate symmetrically, so that the spheres turn through equal angles about their centres and the thread remains in one plane. Prove (i) that, if in equilibrium the tension of the thread is T, then the time of an oscillation of small amplitude is

$$2\pi\sqrt{(I/Ta)},$$

and (ii) that, if the natural length of the thread is $2a$ and λ is its modulus of elasticity, then the period of a small oscillation of amplitude a is

$$\frac{4}{a}\sqrt{\left(\frac{2I}{\lambda a}\right)}\int_0^{\frac{\pi}{2}}\frac{d\theta}{\sqrt{(1-\frac{1}{2}\sin^2\theta)}}.$$

[There are no forces besides the tension of the thread and the pressure between the spheres.]

23. A particle is placed on one of the plane faces of a uniform gravitating circular cylinder at a very small distance from the centre of the face; prove that it will make small oscillations in a period

$$2\,(a^2+h^2)^{\frac{1}{4}}\,(\pi/\gamma\rho h)^{\frac{1}{2}},$$

where a, h, ρ are the radius of the cylinder, its height, and the density of its material.

24. A uniform rod rests in equilibrium on a rough gravitating uniform sphere under no forces but the attraction of the sphere. Prove that, if slightly displaced, it will oscillate in time

$$2\pi l\,(a^2+l^2)^{\frac{3}{4}}/a\sqrt{(3\gamma m)},$$

where m is the mass of the sphere, a its radius, and $2l$ the length of the rod.

---

25. A particle can move in a smooth plane tube, which can rotate about a vertical axis in its plane. In a position of relative equilibrium ω is the angular velocity, a the distance of the particle from the axis, ρ the radius of curvature of the tube at the point occupied by the particle, and α the angle which the normal at this point makes with the vertical. Prove that the period of a small oscillation is

$$\frac{2\pi}{\omega}\sqrt{\left(\frac{\rho\sin\alpha}{a-\rho\sin\alpha\cos^2\alpha}\right)},\quad\text{or}\quad\frac{2\pi}{\omega}\sqrt{\left(\frac{\rho\sin\alpha}{a+3\rho\sin\alpha\cos^2\alpha}\right)},$$

according as the angular velocity is maintained constant, or the tube rotates freely.

26. A thread of length l has its ends attached to two points distant c apart on a vertical axis, and a bead can slide on the thread; the system rotates freely about the vertical axis with angular velocity ω. Prove that, if

$$\omega^2 > 2gl/(l^2-c^2),$$

the time of a small oscillation about a position of relative equilibrium is

$$2\pi\sqrt{\{A\,(l^4-A^2c^2)/2l^2g\,(l^2+3A^2)\}},$$

where $$A=2gl^2/\omega^2\,(l^2-c^2).$$

27. A particle describes a circle uniformly under the influence of two centres of force which attract inversely as the square of the distance. Prove that the motion is stable if $3\cos\theta\cos\phi<1$, where θ, ϕ are the angles which a radius of the circle subtends at the centres of force.

28. A uniform rod of length $2b$ can slide with its ends on a smooth vertical circular wire of radius a and the wire is made to rotate about a vertical diameter with uniform angular velocity ω. Prove that the lowest horizontal position is stable if

$$9\omega^2(a^2-\tfrac{4}{3}b^2)<g\sqrt{(a^2-b^2)}.$$

29. One end of a rigid uniform rod of length $2a$ formed of gravitating matter is constrained to move uniformly in a circle of radius c with angular velocity ω, and the rod is attracted to a fixed particle of mass m at the centre of the circle. Prove that the rod can move steadily projecting inwards towards the centre, and that this steady motion is stable if

$$\gamma m>\omega^2 c(c-2a)^2.$$

---

30. A bead is free to slide on a rod of negligible mass whose ends slide without friction on a fixed circle. Prove that, if there are no external forces, the bead moves relatively to the rod as if repelled from the middle point with a force varying inversely as the cube of the distance.

31. Two equal uniform rods AB, BC each of mass m and length $2a$ are freely jointed at B and have their middle points joined by an elastic string, and the system moves in one plane under no forces. Prove that, if θ is the angle between the string and either rod at any time, ϕ the angle which the string makes with a fixed line, and V the potential energy of the stretched string, then throughout the motion

$$(\tfrac{1}{3}+\cos^2\theta)\,\dot\phi=\text{const.},$$

$$ma^2\{(\tfrac{1}{3}+\cos^2\theta)\,\dot\phi^2+(\tfrac{1}{3}+\sin^2\theta)\,\dot\theta^2\}+V=\text{const.}$$

32. Two equal uniform rods AC, CB, hinged at C, and having their extremities A, B connected by a thread so that ACB is a right angle, are revolving in their own plane with uniform angular velocity about the angle A which is fixed. Prove that, if the thread is severed, the reaction at the hinge is instantaneously changed in the ratio $\sqrt{5}:4$.

33. A uniform rod of mass m and length $2a$ moves at right angles to itself on a smooth table, and impinges symmetrically on a uniform circular disk of mass m' and radius a spinning freely about its centre. Prove that, if there is no restitution, and the edge of the disk is rough enough to prevent slipping, the bodies will separate after an interval in which the unmolested disk would have turned through an angle whose circular measure is

$$(m'+3m)/(m'+m).$$

34. A uniform cube, of mass M, and radius of gyration k about an axis through its centre, rests on a smooth horizontal plane, and a smooth circular groove of radius a is cut on the upper face and passes through the centre O of that face. A particle of mass m is projected along the groove from O with

velocity V. Prove that, if $a\phi$ is the arc traversed by the particle, and θ the angle turned through by the block in any time,

$$\cot\frac{\sqrt{(1+\beta^2)}}{\beta}(\tfrac{1}{2}\phi-\theta)=\frac{\beta}{\sqrt{(1+\beta^2)}}\cot\tfrac{1}{2}\phi,$$

where $$\beta^2=k^2(M+m)/4ma^2.$$

35. Two rough horizontal cylinders each of radius c are fixed with their axes inclined to each other at an angle 2α; and a uniform sphere of radius a rolls between them, starting with its centre very nearly above the point of intersection of the highest generators. Prove that the vertical velocity of its centre in a position in which the radii to the two points of contact make angles ϕ with the horizontal is

$$\sin\alpha\cos\phi\sqrt{\{10g(a+c)(1-\sin\phi)/(7-5\cos^2\alpha\cos^2\phi)\}}.$$

36. A particle of mass m is placed in a smooth straight tube which can rotate in a vertical plane about its middle point, and the system starts from rest with the tube horizontal. Prove that the angle θ which the tube makes with the vertical when its angular velocity is a maximum and equal to ω is given by the equation $4(mr^2+I)\omega^4-8mgr\omega^2\cos\theta+mg^2\sin^2\theta=0$, where I is the moment of inertia of the tube about its middle point, and r is the distance of the particle from that point.

37. A square formed of four similar uniform rods freely jointed at their extremities, is laid on a smooth horizontal table, one of its corners being fixed; show that, if angular velocities ω, ω' in the plane of the table are communicated to the rods that meet at this corner, the greatest value of the angle between them in the subsequent motion is

$$\tfrac{1}{2}\cos^{-1}\{-\tfrac{5}{6}(\omega-\omega')^2/(\omega^2+\omega'^2)\}.$$

38. Two equal homogeneous cubes are moving on a smooth table with equal and opposite velocities V in parallel lines, and impinge so that finite portions of opposing faces come into contact; show that, so long as they remain in contact, the line joining their centres meets the opposing faces at a distance x from the centres of the faces which satisfies the equation

$$\dot{x}^2(x^2+\tfrac{2}{3}a^2)(x_0^2+\tfrac{2}{3}a^2)=V^2x_0^2(a^2+x^2-x_0^2),$$

where $2a$ is a side of either cube, and x_0 is the initial value of x.

Prove further that, if the line joining the centres at the instant of impact cuts the opposing faces at an angle $\tfrac{1}{3}\pi$, then while the faces are in contact they slip with uniform relative velocity, and separate after an interval $(1+\sqrt{3})\,a/V$ after turning through an angle

$$2\sqrt{\tfrac{3}{5}}\{\tan^{-1}\sqrt{\tfrac{3}{5}}+\tan^{-1}\sqrt{\tfrac{1}{5}}\}.$$

---

39. A string without weight is coiled round a rough horizontal uniform solid cylinder of mass M and radius a which is free to turn about its axis. To the free extremity of the string is attached a uniform chain of mass m and length l; if the chain is gathered close up and then let go, prove that the angle θ, through which the cylinder has turned after a time t, before the chain is fully stretched, satisfies the equation $Mla\theta=m(\tfrac{1}{2}gt^2-a\theta)^2$.

40. A great length of uniform chain is coiled at the edge of a horizontal platform, and one end is allowed to hang over until it just reaches another platform distant h below the first. The chain then runs down under gravity. Prove that it ultimately acquires a finite terminal velocity V, that its velocity at time t is $V \tanh(Vt/h)$, and that the length of chain which has then run down is $h \log \cosh(Vt/h)$.

41. Two buckets each of mass M are connected by a chain of negligible mass which passes over a fixed smooth pulley. On the bottom of one of them lies a length l of uniform chain, whose mass is μl, one end of which is attached to a fixed point just above the bottom of the bucket. Prove that, if the system starts to move from rest, the velocity of the bucket when there remains upon it a length y of chain is V, where

$$V^2 = 2g(l-y) - \frac{4Mg}{\mu} \log \frac{2M+\mu l}{2M+\mu y}.$$

42. Two scale-pans each of mass M are supported by a cord of negligible mass passing over a smooth pulley, and a uniform chain of mass m and length l is held by its upper end above one of the scale-pans so that it just reaches the pan. Find the acceleration of the pan when a length x of chain has fallen upon it, and prove that the whole chain will have fallen upon it after an interval $\sqrt{\{\frac{1}{2}l(4M+m)/Mg\}}$.

---

43. A chain of length l slides from rest down a line of greatest slope on a smooth plane of inclination a to the horizontal, the end of the chain hanging initially just over the edge. Prove that the time of leaving the plane is $\sqrt{\{l/g(1-\sin a)\}} \log(\cot \frac{1}{2}a)$.

44. One end of a uniform chain of length l and mass m is fixed to a horizontal platform of mass $(2k-1)m$; the chain passes over a smooth fixed pulley, and is coiled on the platform. As the platform descends vertically, the chain uncoils, rises vertically and passes over the pulley. Prove that, at any time t before the chain is completely uncoiled, the depth x of the platform satisfies an equation of the form $\dot{x}^2 = a + \beta x + \gamma e^{-kx/l}$, where a, β, γ are constants.

45. A chain whose density varies uniformly from ρ at one end to 3ρ at the other end is placed symmetrically on a small smooth pulley and is then let go. Prove that it leaves the pulley with velocity $\frac{1}{3}\sqrt{(11lg)}$, where $2l$ is its length.

46. An elastic string (modulus λ, mass ma, unstretched length a) is confined within a straight tube to one end of which it is fastened, and the tube rotates about that end with uniform angular velocity ω in a horizontal plane. Show that the length of the string in equilibrium is

$$a \frac{\tan \theta}{\theta}, \text{ where } \theta = a\omega \sqrt{\frac{m}{\lambda}}.$$

---

47. A uniform chain falls in a vertical plane with uniform acceleration f retaining an invariable form, while the chain advances along itself with a velocity which at any instant is the same for all points of the chain. Prove that the angle ϕ which the tangent at any point of the chain makes with the horizontal, considered as a function of the time t and of the arc s measured up to this point from some definite point of the chain, satisfies the two partial differential equations

$$(f-g)\left\{\cos\phi\frac{\partial^2\phi}{\partial s^2}+2\sin\phi\left(\frac{\partial\phi}{\partial s}\right)^2\right\}+\frac{\partial^2\phi}{\partial t^2}\frac{\partial\phi}{\partial s}-\frac{\partial\phi}{\partial t}\frac{\partial^2\phi}{\partial s\partial t}=0,$$

$$\frac{\partial\phi}{\partial s}\frac{\partial^2\phi}{\partial s\partial t}-\frac{\partial\phi}{\partial t}\frac{\partial^2\phi}{\partial s^2}=0.$$

48. The ends of a chain of variable density are held at the same level, and the chain hangs in the form of an arc of a circle subtending an angle $2\theta\,(<\pi)$ at the centre. Prove that, if equal tangential impulses are applied at the ends, the initial normal velocities at the lowest point and at either end are in the ratio $1 : \cos\theta$.

49. An endless uniform chain, lying in the form of a circle, receives a tangential pluck at one point A, which gives it an impulsive tension T_0 at that point; prove that the impulsive tension at any point P is

$$T_0\sinh(2\pi-\theta)\,\text{cosech}\,2\pi,$$

θ being the angle which AP subtends at the centre. Prove also that P starts to move in a direction making an angle ϕ with the tangent, where

$$\tan\phi=(e^{4\pi}-e^{2\theta})/(e^{4\pi}+e^{2\theta}).$$

50. A uniform chain is suspended from two points in the same horizontal line so that the tangents at the ends make angles a with the horizontal; and the ends can slide on fixed straight wires which are at right angles to the tangents at the ends. Prove that, if the wire supporting one end is removed, that end starts to move in a direction making with the horizontal an angle θ, where

$$\tan\theta=(1+\sin^2 a+2a\tan a)/\sin a\cos a.$$

Prove also that the tension at the other end is diminished in the ratio

$$1 : 1+\tfrac{1}{2}a^{-1}\cot a.$$

CHAPTER X†

THE ROTATION OF THE EARTH

268. It is a fact of observation that there is a relative motion of the Earth and the stars by which every star moves relatively to the Earth continually from East to West, or, what is geometrically the same thing, by which any part of the Earth's surface moves relatively to the stars continually from West to East. This motion can be precisely described by saying that the Earth rotates about its polar axis. The time in which the Earth turns through four right angles is called a "sidereal day." The rotation is such that, if the polar axis is supposed to be drawn from South to North, the sense of this axis and the sense of the rotation are related like the senses of translation and rotation of a right-handed screw.

269. Sidereal Time and Mean Solar Time. This process of relative rotation has for ages been accepted as a "time-measuring process," that it to say it has been regarded as taking place *uniformly*. Time measured by this process is called "sidereal time."

Now we have said (Article 3) that the process used for measuring time is the average rotation of the Earth relative to the Sun. To explain this statement, consider in the first place the motion of the Sun relative to a frame whose origin is the centre of the Earth and whose lines of reference go out thence to stars so distant as to have no observable annual parallax. The path and motion of the Sun relative to this frame are the same as the motion (in a planetary orbit) of the Earth, relative to a frame whose origin is in the Sun and whose lines of reference go out thence to the same stars (cf. Ex. 3 of Art. 44). The Sun's path relative to this frame of Earth and stars is very nearly the same as if his motion were an elliptic motion about a focus at the centre of the Earth. The sense in which the Sun describes his orbit is the same as the sense in which any particular meridian plane of the Earth turns about the polar axis, that is to say, the Sun is always moving from stars which have a more westerly position towards stars which have a more easterly position in the plane of his path. The elements of the elliptic orbit are not quite constant; in particular the apse line has a small progressive motion in the sense in which the orbit is described, and the line of intersection of the plane of the orbit with the plane of the Earth's equator

† Articles in this Chapter which are marked with an asterisk (*) may be omitted in a first reading.

(known as the line of nodes) has a small progressive motion in the opposite sense. The Sun passes the line of nodes at the Equinoxes, and the periodic time in the orbit is a *year* (technically a "tropical year"). Now it is to be observed that, relatively to a frame fixed in the Earth, the Sun makes about 365¼ revolutions round the Earth in a year, and the stars make about 366¼ revolutions, but the time of revolution of the Sun is not a constant multiple of the time of revolution of the stars. The variability arises in the first place from the fact that the motion of the Sun in his path, relative to the frame of Earth and stars, is much more nearly elliptic motion about a focus than uniform circular motion, and in the second place from the fact that the plane of the Sun's path is inclined to the equator. To define the measurement of time by the average rotation of the Earth relative to the Sun, we imagine a point to move (relatively to the frame of Earth and stars) in the Sun's path, with a uniform angular motion about the centre of the Earth (*i.e.* so that the time of describing any angle is a constant multiple of the time in which the Earth turns through the same angle), and at such a rate as always to coincide with the Sun at the nearer apse of his path; then we imagine a second point to move in the plane of the Earth's equator with a uniform angular motion about the centre of the Earth, and at such a rate as always to coincide with the first point at the node corresponding to the Vernal Equinox. This second point is called the *Mean Sun.* We may determine a frame of reference by taking the centre of the Earth as origin, the line joining the origin to the Mean Sun as a line of reference, and the plane through this line and the polar axis as a plane of reference. Relatively to this frame the Earth rotates about its polar axis in an interval called a mean solar day; this rotation can be used instead of the rotation relative to the stars as time-measuring process, and time so measured is *mean solar time.* The unit of time is the time in which the Earth rotates relatively to this frame through an angle equal to 1/86400 of four right angles, and this unit is the *mean solar second.*

270. The law of gravitation. When we say that the Earth is rotating, we imply that a body at rest relative to it is moving round the polar axis. Any particle of the body is describing a circle about a centre on the axis, and therefore has an acceleration directed towards the centre of this circle. If we refer the motion to axes which rotate with the Earth the particle has no such acceleration. The specification of the acceleration of the particle, and therefore of the forces acting on the body, depends upon the axes to which the motion is referred.

The law of gravitation is a statement concerning the forces that act upon the particles of bodies. It implies that the motion is referred to some axes or other. For a complete statement of the law the origin and axes to which the motion is referred ought to be specified. In other words, the law implies that a *frame of reference* has been chosen; and a complete statement of the law would involve the specification of this frame of reference.

When the law is applied to the motions of bodies within the Solar System an adequate frame of reference can be specified by the statements: (i) The origin is the centre of mass of the system. (ii) The axes are determined by stars so distant as to have no observable annual parallax.

Relatively to this frame the Earth as a whole has certain motions. Of these the most conspicuous are the orbital motion about the Sun and the rotation about the polar axis.

271. Gravity. The acceleration denoted by g, and described as the "acceleration due to gravity," is specified by reference to axes fixed in the Earth. It may be precisely defined as the *initial* acceleration, relative to such axes, of a particle starting from rest, relative to such axes, in a position near the Earth's surface.

This acceleration is not identical with the acceleration produced in the particle by the field of the Earth's gravitation. The latter is denoted by g'. (Cf. Ch. VI.)

Let Ω denote the angular velocity of the Earth's rotation, so that $2\pi/\Omega$ is the number of mean solar seconds in a sidereal day. Let p denote the distance of a particle from the polar axis. Let f denote the acceleration of the Earth's centre of mass referred to the frame specified by the centre of mass of the solar system and the "fixed" stars. The acceleration of a body, treated as a particle, which is at rest relatively to the Earth, is compounded of the accelerations f and $p\Omega^2$; the acceleration $p\Omega^2$ is directed towards the point where the polar axis cuts a plane which is at right angles to it and passes through the position of the particle.

Let m be the mass of the body, as determined by the law of gravitation (Ch. VI). The forces acting upon it are the force mf due to the field in which the Earth moves, the force mg' due to the Earth's gravitational field, and a force W which keeps the particle in relative equilibrium.

We disregard in this statement the difference in the values of the intensity f of the external field at the centre and surface of the Earth. (See Art. 274.)

The direction of W is that of a plumb-line at the place; in other words it is the "vertical" at the place. The sense of W is upwards.

The kinetic reaction of the particle is compounded of mf in the direction of the acceleration f and $mp\Omega^2$ in the direction of the acceleration $p\Omega^2$.

Hence the resultant of W and mg' is equal to $mp\Omega^2$ in the direction of the acceleration $p\Omega^2$.

If the particle is released, its initial acceleration is compounded of f, $p\Omega^2$ and g. The forces acting upon it are then those specified by mf and mg'. Hence $W = mg$; and the line of action of W is directly opposed to that of the acceleration g.

In obtaining the relation $W = mg$ in Chapter III we neglected the rotation of the Earth. It now appears that, when g is defined as above, the relation is unaffected by taking account of this rotation.

272. Variation of gravity with latitude. Let l be the angle which the vertical at a place makes with the plane of the equator. Then l is the (Astronomical) latitude of the place.

Let λ be the angle which the direction of the Earth's gravitational field at the place makes with the plane of the Equator.

Consider a body at rest relative to the Earth. Its kinetic reaction consists of vectors mf, $mp\Omega^2$; and the forces acting upon the body are mf, mg', W. The directions and senses of all these vectors have been specified.

Form an equation of motion by resolving in the direction of the polar axis. The equation is

$$0 = mg' \sin\lambda - W \sin l.$$

Form an equation of motion by resolving parallel to the direction of the acceleration $p\Omega^2$. The equation is

$$mp\Omega^2 = mg' \cos\lambda - W \cos l.$$

Since $W = mg$, we have

$$\frac{g'}{\sin l} = \frac{g}{\sin\lambda} = \frac{p\Omega^2}{\sin(l-\lambda)} \quad\ldots\ldots\ldots\ldots\ldots\ldots(1).$$

Of the quantities in these equations g, Ω, l are known by observation and p is known in terms of l when the figure of the Earth is known. The equations determine λ and g'.

If the Earth is regarded as spherical, and as made up of concentric spherical strata of equal density, the line of action of the force mg' passes through the centre, and we have

$$g' = \gamma E/R^2, \quad p = R\cos\lambda,$$

where R is the radius of the Earth, and E is its mass.

Hence

$$\frac{\sin(l-\lambda)}{\sin\lambda\cos\lambda} = \frac{R\Omega^2}{g}, \quad g = \frac{\gamma E}{R^2}\frac{\sin\lambda}{\sin l} \quad \text{............(2).}$$

Now $R\Omega^2/g$ is a small fraction equal to $\frac{1}{289}$ nearly, and therefore $l-\lambda$ is a small angle, approximately equal to $\frac{1}{289}\sin l\cos l$ radians. This angle is called the "deviation of the plumb-line." Also g is approximately equal to

$$\frac{\gamma E}{R^2}(1 - \tfrac{1}{289}\cos^2 l).$$

With the above assumptions as to the figure and constitution of the Earth, λ becomes the "geocentric" latitude of the place. The assumptions enable us to account for the variation of g with latitude. There is a small correction to the formula for g on account of the spheroidal figure of the Earth.

273. Mass and weighing. When two bodies are found to be of the same weight, by weighing them in a common balance, it is verified that the forces required to support them in equilibrium relative to the Earth are equal at the same place. Hence the product mg is the same for both. Now the ratio $g : g'$ is $\sin\lambda : \sin l$, where l is the Astronomical latitude of the place, and λ is the angle which the direction of the Earth's gravitational field at the place makes with the plane of the Equator. It follows that the product mg' is the same for the two bodies. But the ratio of two masses, as determined by the law of gravitation, is the ratio of the forces with which they are attracted by a gravitating body when they occupy, successively, the same position with respect to that body. Hence the masses of the two bodies, as determined by the law of gravitation, are equal.

The determination of the mass of a body by weighing it in a common balance may therefore be regarded as a particular case of the determination of mass by means of mutual action, on the basis of the law of gravitation, as was stated in Chapter VI.

274. Lunar deflexion of gravity. In the above discussion we have treated the external field as uniform, or as having the same intensity at the centre of mass of the Earth and at any point on its surface.

The external field arises from the gravitational attractions of the Sun, Moon and Planets. Its intensity varies slightly from centre to surface. This variation is most marked in the case of the Moon on account of its comparatively small distance from the Earth.

Let f denote, as before, the intensity of the external field at the Earth's centre of mass, and let f' denote the intensity at a point on the surface. A force compounded of mf', in the sense of f', and mf, in the sense of f reversed, is available for producing motion of the body m relative to the Earth.

The effect of this force is to make the direction of the plumb-line at a place deviate slightly from the direction which it would take if f' were the same as f. Since the difference between f and f' arises mainly from the attraction of the Moon, this effect is generally referred to as the "lunar deflexion of gravity."

The direct measurement of this effect is extremely difficult*. The theoretical value can, however, be determined. Cf. Ex. 5 in Art. 275.

The force which produces the lunar deflexion of gravity is the same as that which produces the tides, at least in so far as these depend upon the Moon. The force which arises, as above, from the difference between f and f' is the tide-generating force.

275. Examples.

[In these examples the Earth is regarded as a homogeneous sphere.]

1. If the Earth were to rotate so fast that bodies at the equator had no weight, prove that, in any latitude, the plumb-line would be parallel to the polar axis.

2. If the acceleration due to gravity at the Poles is g_0 and at the Equator g_e, prove that in (geocentric) latitude λ the value of g is

$$\sqrt{(g_0^2 \sin^2 \lambda + g_e^2 \cos^2 \lambda)},$$

and that the deviation of the plumb-line from the (geometrical) vertical is

$$\tan^{-1} \{(g_0 - g_e) \sin \lambda \cos \lambda / (g_0 \sin^2 \lambda + g_e \cos^2 \lambda)\}.$$

3. Prove that a pendulum which beats seconds at the Poles will lose approximately $30m \cos^2 l$ beats per minute in latitude l, where $1+m : 1$ is the ratio of the values of g at the Poles and at the Equator.

4. A train of mass m is travelling with uniform speed v along a parallel of latitude in latitude l. Prove that the difference between the pressures on the rails when the train travels due East and when it travels due West is $4mv\Omega \cos l$ approximately.

5. Assuming that the mass of the Moon is $\frac{1}{80}$ of that of the Earth, and that the Moon's distance is 60 times the Earth's radius, prove that, owing to the Moon's attraction, a seconds' pendulum at the Earth's surface will be losing at a rate $\frac{1}{400}(3 \sin^2 a - 1)$ seconds per day, where a is the altitude of the Moon at the place of observation.

* See G. H. Darwin, *The Tides and kindred phenomena in the Solar system*, London, 1898.

***276. Motion of a free body near the Earth's surface.** We form first the equations of motion of the body referred to axes fixed in the Earth. As in Art. 272 we take the Earth to be spherical. We take the origin to be at the centre of the Earth, the axis of z to be the polar axis (from South Pole to North Pole), the axis of x to be the intersection of the plane of the equator and the meridian plane near which the motion takes place, the positive sense of the axis of x being *from* the centre *to* the meridian in question; also we take the axis of y to be at right angles to this meridian plane and directed towards the East. This system is a right-handed system. By the results of Art. 254, the component velocities of the body parallel to these axes are not $\dot{x}$, $\dot{y}$, $\dot{z}$, but they are

$$\dot{x} - \Omega y, \quad \dot{y} + \Omega x, \quad \dot{z},$$

and the component accelerations are

$$\frac{d}{dt}(\dot{x} - \Omega y) - \Omega(\dot{y} + \Omega x), \quad \frac{d}{dt}(\dot{y} + \Omega x) + \Omega(\dot{x} - \Omega y), \quad \ddot{z}.$$

Hence the equations of motion of the body are

$$\left.\begin{aligned} m(\ddot{x} - 2\Omega\dot{y} - \Omega^2 x) &= -(\gamma m E/R^2)\cos\lambda, \\ m(\ddot{y} + 2\Omega\dot{x} - \Omega^2 y) &= \quad 0, \\ m\ddot{z} \qquad\qquad &= -(\gamma m E/R^2)\sin\lambda, \end{aligned}\right\}$$

where λ is the angle which the radius of the Earth drawn through the body makes with the plane of the equator. Now, as the body remains near a place, we may take λ to be constant, and we may in the terms containing Ω^2, put $x = R\cos\lambda$ and $y = 0$. Then, using equations (2) of Art. 272, we find

$$\begin{aligned} \ddot{x} - 2\Omega\dot{y} &= -g\cos l, \\ \ddot{y} + 2\Omega\dot{x} &= \quad 0, \\ \ddot{z} \qquad &= -g\sin l. \end{aligned}$$

Since these equations contain only differential coefficients of x, y, z with respect to the time, we may, without making any alteration, suppose the origin to be on the Earth's surface in the latitude and longitude near which the motion takes place.

We shall now, taking the origin as just explained, transform to the horizontal drawn southwards as axis of x', the horizontal drawn eastwards as axis of y', and the vertical drawn upwards as axis of z'. We have

$$x' = x\sin l - z\cos l, \quad y' = y, \quad z' = z\sin l + x\cos l.$$

We thus obtain the equations

$$\left.\begin{aligned} &\ddot{x}' - 2\Omega\dot{y}' \sin l = 0, \\ &\ddot{y}' + 2\Omega(\dot{x}' \sin l + \dot{z}' \cos l) = 0, \\ &\ddot{z}' - 2\Omega\dot{y}' \cos l = -g; \end{aligned}\right\} \quad\ldots\ldots\ldots\ldots(1),$$

these equations determine the motion of the body relative to the axes at the place of observation.

***277. Initial motion.** Suppose the body to fall from rest relative to the Earth. Then the initial velocities relative to the axes at the place of observation are given by the equations

$$\dot{x}' = 0, \; \dot{y}' = 0, \; \dot{z}' = 0,$$

and we shall suppose that the initial value of the coordinate y' is zero. The motion is determined by equations (1) of Art. 276. Integrating the first of these, we have

$$\dot{x}' = 2\Omega y' \sin l \quad\ldots\ldots\ldots\ldots(1),$$

and integrating the third equation, we have

$$-\dot{z}' = gt - 2\Omega y' \cos l \quad\ldots\ldots\ldots\ldots(2),$$

where t is the time from the beginning of the motion. Substituting in the second equation, and neglecting $\Omega^2 y'$, we have, on integration,

$$\dot{y}' = \Omega g t^2 \cos l,$$

so that

$$y' = \tfrac{1}{3}\Omega g t^3 \cos l \quad\ldots\ldots\ldots\ldots(3).$$

Substituting in equations (1) and (2), and neglecting terms of the same order as before, we have, on integration,

$$\left.\begin{aligned} x' &= x_0', \\ z' &= z_0' - \tfrac{1}{2}gt^2, \end{aligned}\right\}$$

where x_0' and z_0' are the initial values of x' and z'.

In the beginning of the motion the acceleration relative to axes fixed on the Earth is directed vertically downwards, and it is what we have called g. To the order of approximation here adopted the vertical component of acceleration remains constant throughout the motion.

It appears that the body falls a little to the East of the starting point, the eastward deviation in a fall through a height h being very approximately

$$\tfrac{2}{3}\Omega \sqrt{(2h^3/g)} \cos l.$$

This result accords well with observed facts.

***278. Motion of a Pendulum.** Let a simple circular pendulum of length L be free to move about its point of support, which is fixed relatively to the Earth, and let T be the tension of the suspending fibre.

Let x', y', z' be the coordinates of the bob referred to the system of axes described in Art. 276, the origin being at the equilibrium position; then the line of action of T makes with the axes angles whose cosines are

$$-x'/L, \ -y'/L, \ (L-z')/L,$$

and we have the relation

$$x'^2 + y'^2 + (L-z')^2 = L^2 \qquad \text{.....................(1).}$$

Now the equations of motion are, by Art. 276,

$$\left.\begin{aligned} m\ddot{x}' - 2m\Omega\dot{y}' \sin l &= -T(x'/L), \\ m\ddot{y}' + 2m\Omega(\dot{x}' \sin l + \dot{z}' \cos l) &= -T(y'/L), \\ m\ddot{z}' - 2m\Omega\dot{y}' \cos l &= -mg + T(L-z')/L. \end{aligned}\right\} \qquad \text{......(2).}$$

We shall integrate these equations on the assumption that the pendulum makes small oscillations. On this assumption we have approximately

$$z' = \tfrac{1}{2}(x'^2 + y'^2)/L \qquad \text{.......................(3).}$$

Multiply the equations (2) in order by $\dot{x}', \dot{y}', \dot{z}'$, and add. The terms containing T vanish identically by (1), the terms containing Ω also vanish identically, and the equation can be integrated. Omitting $\dot{z}'^2$ in the integral equation, and substituting for z' from (3), we have

$$\tfrac{1}{2}m(\dot{x}'^2 + \dot{y}'^2) = \text{const.} - \tfrac{1}{2}mg(x'^2 + y'^2)/L \qquad \text{.........(4).}$$

Again, multiplying the first of equations (2) by $-y'$, and the second by x', adding, and omitting the term in $x'\dot{z}'$, we have on integration

$$x'\dot{y}' - y'\dot{x}' = -\Omega \sin l\,(x'^2 + y'^2) + \text{const.} \qquad \text{.........(5).}$$

Introducing polar coordinates in the horizontal plane given by

$$x' = r\cos\theta, \ y' = r\sin\theta,$$

from equations (4) and (5) we obtain equations of the form

$$\left.\begin{aligned} \dot{r}^2 + r^2\dot{\theta}^2 &= A - r^2(g/L), \\ r^2\dot{\theta} &= B - r^2\Omega\sin l, \end{aligned}\right\}$$

and, if we put

$$\theta + \Omega t \sin l = \phi \qquad \text{..........................(6),}$$

we shall have

$$\left.\begin{aligned}\dot{r}^2+r^2\dot{\phi}^2&=(A+2\Omega B\sin l)-r^2\{(g/L)+\Omega^2\sin^2 l\},\\ r^2\dot{\phi}&=B.\end{aligned}\right\}\dots(7).$$

These equations determine the motion. It is to be noticed that r and ϕ are polar coordinates referred to an initial line which rotates about the vertical from East to West with an angular velocity $\Omega \sin l$.

***279. Foucault's Pendulum.** When the pendulum can turn freely about its point of support and is set oscillating so as to pass through its equilibrium position, the system is known as a Foucault's Pendulum.

Since r can vanish, it follows by the second of equations (7) of the last Article that B must vanish, and thus $\dot{\phi}$ vanishes throughout the motion. Hence the pendulum oscillates so that its plane of vibration turns round the vertical relatively to the Earth with angular velocity $\Omega \sin l$ from East to West.

The first of equations (7) of the last Article then becomes, if we neglect $\Omega^2 \sin^2 l$ in comparison with g/L,

$$\dot{r}^2 = A - r^2 (g/L),$$

showing that the horizontal motion in the plane of vibration is simple harmonic motion of period $2\pi\sqrt{(L/g)}$.

If a is the amplitude of the simple harmonic motion, so that the pendulum has no velocity in the plane of vibration when $r = a$, it will not move as here described unless its angular velocity relative to the Earth is $\Omega \sin l$ from East to West. To start the pendulum, therefore, it is not sufficient to hold it aside from its equilibrium position; it must be projected at right angles to the vertical plane containing it with velocity $a\Omega \sin l$. When thus set going it moves like a simple pendulum of the same length in a plane which turns about the vertical from East to West with angular velocity $\Omega \sin l$.

This result accords well with observed facts.

***280. Examples.**

1. A projectile is projected from a point on the Earth's surface with velocity V at an elevation α in a vertical plane making an angle β with the meridian (East of South). Prove that after an interval t it will have

moved southwards through x, eastwards through y, and upwards through z, where

$$\left.\begin{aligned} x &= Vt\cos a\,\{\cos\beta+\Omega t\sin l\sin\beta\},\\ y &= Vt\,\{\cos a\sin\beta-\Omega t\,(\sin l\cos\beta\cos a+\cos l\sin a)\}+\tfrac{1}{3}\Omega g t^3\cos l,\\ z &= Vt\,\{\sin a+\Omega t\cos l\sin\beta\cos a\}-\tfrac{1}{2}gt^2, \end{aligned}\right\}$$

approximately, $\Omega^2 y$ being neglected.

2. Prove that, if the bob of a pendulum of length L is let go from a position of rest relative to the Earth when its displacement from its equilibrium position is a, and the vertical plane through it makes an angle β with the meridian (East of South), its path is given approximately by the equation

$$(\beta-\theta)=\Omega\sqrt{(L/g)}\sin l\,\{\sqrt{(a^2-r^2)}/r-\cos^{-1}(r/a)\},$$

powers of $L\Omega^2/g$ above the first being neglected.

3. A particle is observed to move, relatively to a certain frame, with a simple harmonic motion of period $2\pi/n$ in a line, which turns uniformly about the mean position of the particle in a plane fixed relatively to the frame with angular velocity ω; prove that the acceleration of the particle when at distance r from its mean position is compounded of a radial acceleration $(n^2+\omega^2)\,r$, and a transverse acceleration $2\omega\dot{r}$ in the sense in which the line turns.

CHAPTER XI

SUMMARY AND DISCUSSION OF THE PRINCIPLES OF DYNAMICS

GALILEO discovered by experiment that the velocity of a falling body is proportional to the time during which it has been falling, and he was thus led to the notion of acceleration. He recognized in the motion of a body on a very smooth horizontal plane that a body, which could be regarded as subject to no forces, moved uniformly in a straight line; and he was thus led to connect the existence of force with the production of acceleration.

Newton found that the notion of acceleration, thus introduced by Galileo, availed for the description of the motions of the bodies of the Solar System equally with the motion of falling bodies near the Earth's surface, and he made the idea of force, as that which produces acceleration, the cardinal notion in his philosophy. Newton also introduced the notion of mass, as distinct from weight, and stated that the mass of a body is the quantity of matter which it contains. He formulated his theory in a series of definitions, in the three celebrated Laws of Motion, which he called *Axiomata sive Leges Motus*, and in the Scholia attached thereto. We give here a translation of the three Laws of Motion:

"*First Law.* Every body remains in its state of rest or of uniform "motion in a straight line, except in so far as it is compelled by "impressed forces to change its state."

"*Second Law.* Change of motion is proportional to the impressed "moving force, and takes place in the direction in which that force "is impressed."

"*Third Law.* Reaction is always equal and opposite to action; "or the actions of two bodies one on the other are always equal and "oppositely directed."

The definitions preceding the laws introduce the notions of mass, and of impressed moving force as an action on a body by which its state of motion is changed, and as proportional to what we now call momentum generated in a given interval. The scholia attached to the laws contain a demonstration of the theorem of the parallelogram of forces, and an account of the determination of masses by direct experiment with the ballistic balance. The latter is given as a verification of the Third Law.

In the course of this book the theoretical aspect of the science has been developed from two principles which are essentially the same as Newton's laws of motion, but are expressed in a form that is more convenient for application. They are

I. The kinetic reaction of a particle has the same magnitude, direction and sense as the resultant force acting on the particle (Art. 64).

II. The magnitude of the force exerted by one particle on another is equal to the magnitude of the force exerted by the second particle upon the first, the lines of action of both the forces coincide with the line joining the particles, and the forces have opposite senses (Art. 142).

These principles correspond precisely to the second and third of Newton's laws. The first law may be regarded as a particular case of the second; for, if there is no impressed force, there is no change of motion, and the motion goes on unchanged. In Newton's time this particular principle was so subversive of current ideas that it was necessary to state it explicitly.

The first step in the formulation of the principles of Mechanics* is the recognition of the vectorial character of such quantities as velocity and acceleration. The statement that velocity is a vector is the proposition that is often called the "parallelogram of velocities." It is not a physical law, nor is it a mathematical proposition capable of mathematical proof from definitions, postulates and axioms, but it is a definition arrived at by gradually increasing the precision of a notion already formed. This notion is the notion of velocity as *rate of displacement per unit of time.*

The discussion, given in many books as a "proof," by means of the motion of a ball in a moving tube, is valuable as an illustration; but the process that

* Discussions of the principles of Mechanics will be found in the works cited on p. 300 below, and also in H. Hertz's *Principles of Mechanics*, Translation, London, 1899.

it illustrates is not the composition of two velocities relative to the same frame, but the composition of a velocity relative to one frame with the velocity of that frame relative to another frame. The analytical formulation of this latter process is very simple (see Art. 27).

We make a step which has physical significance when we recognize the existence of a *field of force*. The establishment of this notion was one of the services rendered to science by Galileo. He showed that we could say of a free body near the Earth's surface that it has such and such an acceleration, no matter how its motion is started. In Newton's hands the principle was carried further. It was found to be possible to say of a body anywhere in the Solar System, and free from contact with other bodies, that it had a definite acceleration.

It is hardly necessary to say that neither Galileo nor anyone else has ever experimented upon a free body. Galileo found how to isolate the effect which we now call the "acceleration due to gravity," and he demonstrated the existence and nature of this effect conclusively.

It is inferred that there is some action of the Earth upon bodies in its neighbourhood, or of one body of the Solar System on another, by which the acceleration is produced. This hypothetical action is called *force*.

When we draw this inference we go beyond the facts. The occurrence of definite accelerations in definite places is a physical fact. The inference that some "action" or "force" produces them may, or may not, be legitimate. In so far as the analytical formulation of the facts is concerned it is unnecessary. In our Chapter II it has not been introduced. In our Chapter IV it is introduced merely for the purpose of stating results in the same terms as in subsequent Chapters.

We make another step which has physical significance when we recognize that the motion of bodies in a field of force is modified when they are in contact with other bodies. A book placed on a table rests on the table, instead of falling through to the floor. A ball thrown into the air does not move in a parabolic path, but the trajectory is steeper in falling than in rising. It does appear to be a legitimate inference that there is an action of some sort, due to the table, or due to the air, whereby the acceleration that a free body would have is modified. When we infer such action we assert the existence of *force*.

The existence of *pressure* between bodies in contact seems obvious to common sense. Nevertheless it is to be noted that the pressure is just as much *inferred* from an observation about the motion of the bodies as the action between gravitating bodies is inferred from the motions of these bodies. Yet action at a distance appears to common sense to be absurd. We shall make a mistake if we suppose that the existence of any action between bodies is verified by our muscular sensations, although it was from these sensations that the notion of such action grew up. In like manner it is not verified, nor is its measure determined by the use of the spring balance (Art. 58). The result that, under suitable conditions, the extension of the spring, by a body hung on to it, is proportional to the weight of the body (as determined by the common balance) is a fact about the elasticity of the spring.

Another point to be noted is that the notion of force is not really necessary to the analytical formulation of those parts of the science in which we pay attention to the motion of one body at a time. For example, in our Chapters III and V (as well as in II and IV), nearly all the questions discussed could be expressed without using the notion of force. We might, for instance, discuss the motion of a particle which moves in a given field of force, and has, in addition to the acceleration of a free body, an acceleration directed along the tangent of its path, whatever that tangent may be, in the sense opposite to the velocity, and proportional to a power of that velocity, whatever the magnitude and sense of the velocity may be. We should have the method and results of Art. 138. In the same parts of the science the notion of mass is irrelevant. We have introduced it in Chapter III solely in order that the statement of the results may take the same form as in the subsequent parts of the theory.

It would appear from this discussion that the *action of one body on another* is a concept—something conceived by us—in terms of which we describe the motions of bodies. We infer the existence of the action from observed *accelerations,* which we regard as produced by the actions. It would appear also that we are at liberty to define "force" in the way that we find most convenient. We define it as a particular measure of the action of one body on another, and we state how it is to be measured.

The definition can be given in most precise terms when the body acted upon can be treated as a particle. We *define* the magnitude of any force, acting on a particle, as the product of the mass of the particle and the acceleration that is produced in it by the corresponding action.

The definition is incomplete until we state what the nature of the dependence of force upon direction is to be taken to be. We *define* the force acting on a particle as a *vector localized at a point.*

From this point of view the "parallelogram of forces" becomes part of a conventional definition. The "proofs" and "verifications" given in most books may be regarded as verifications that the definition is, as a matter of fact, convenient. One way in which the definition may be arrived at has been sketched in Art. 61.

The definition of force remains incomplete until we explain what is meant by the "mass" of a body, or of a particle.

To do this we must introduce the Law of Reaction. As has been explained in Chapter VI, this Law is equivalent to the statement that the accelerations, which are produced in two bodies by their mutual actions, have a ratio which is always the same so long as the bodies remain the same. The reciprocal of this ratio is the ratio of the masses of the two bodies.

There are two quite distinct sets of circumstances in which we can observe accelerations or changes of velocity; and in accordance with our concept of force, these changes of velocity are regarded as produced by mutual actions. We may consider, in the first place, the mutual actions of the bodies and the Earth; and we are thus led to the mass-ratio of two bodies, as the ratio of their weights when weighed in a common balance. In the second place, we may let the bodies collide, and determine their mass-ratio by the method of the ballistic balance. The fact that the result is the same, sometimes expressed by the phrase "identity of gravitational and inertial mass," seems to the present writer to be the central fact of Mechanics.

As has been already pointed out, the notions of force and mass are not essential to the analytical formulation of those parts of the science in which we study the motion of one body at a time (the body being treated as a particle). They are essential as soon as we begin to discuss the motions of several bodies forming a connected system.

In the course of this discussion we introduce two subsidiary principles, both of which were introduced by Newton: the law of gravitation, and the conception of a body as a system of particles. We have already worked out in considerable detail the consequences of these principles, when applied to bodies which may be treated as rigid. It may be stated here that no new principle is required for the more complete discussion of the motions of rigid bodies, or for the discussion of the motions of deformable solid bodies or of fluids.

The conception of bodies as made up of particles, and the conception of the mutual actions of bodies, as made up of forces between particles, are, as a matter of historical fact, the two conceptions upon which the existing science of Mechanics is based. They possess further the advantages, (1) that it is possible to found upon them a strictly logical deductive theory, (2) that this theory provides an adequate abstract formulation of the rules obeyed by the motions of the bodies of the Solar System, and of matter in bulk under ordinary conditions. They have thus historically developed into a scheme which successfully coordinates an immense number of disconnected observations concerning matters of fact. Accordingly this theory constitutes a science—a logically valid and practically valuable method of representing observed facts by abstract formulas.

We must be on our guard against identifying the "particles" of the mechanical theory with the atoms and molecules of chemistry and the kinetic theory of gases, or with the electrons and corpuscles of modern physical speculation. The mechanical conception of the constitution of bodies is independent of the chemical and electrical conceptions; and the problem of bringing the various conceptions into harmony with each other has not been solved. There is no reason for thinking that it is incapable of solution*.

It appears to be desirable to explain how it may be possible for internal forces between the hypothetical particles of a body, or a set of bodies, to be adjusted so that the motion of the particles may represent the motions of the bodies.

It has been already explained in Chapter VI how the masses of the hypothetical particles can be assigned.

In the case of a free body, the external forces are gravitational attractions between the particles of the body and the particles of other bodies, and so they can be regarded as known.

A body which is not free is in contact with some other body. We regard all the bodies which are thus in contact as forming a single "system of bodies."

Let the body, or the system of bodies, be replaced by a system of particles. The masses of the particles, and the external forces acting on them, are known.

* See the remarks on the 'Beneke Preis-stiftung' in *Göttingen Nachrichten*, 1901 ("Geschäftliche Mitteilungen"), and cf. H. M. Macdonald, *Electric Waves, Appendix B*, Cambridge, 1902, and J. G. Leathem, *Volume and surface integrals used in Physics*, Cambridge, 1905.

To make the motion of the particles represent the motion of the body, or system of bodies, each particle must have a suitable acceleration. Thus the kinetic reactions of the particles can be regarded as known.

Let there be n particles in the system. The $3n$ components of kinetic reaction can be regarded as given. The magnitudes of the internal forces between them are $\frac{1}{2}n(n-1)$ quantities. The $\frac{1}{2}n(n-1)$ unknown quantities are connected with the known quantities by $3n$ equations, which are the equations of motion of the particles.

The $3n$ equations are of the form

$$m_1\ddot{x}_1 = X_1 + X_1',$$

in which $m_1\ddot{x}_1$ and X_1 are known, and X_1' is of the form

$$F_{12}\cos\theta_{12} + F_{13}\cos\theta_{13} + \ldots + F_{1n}\cos\theta_{1n},$$

where the angles $\theta_{12}, \ldots$ are those which the lines joining the particles make with the axis of x, and F_{12} denotes the force exerted on the particle m_1 by the particle m_2, and so on.

These quantities are such that, if θ_{21} is the same as θ_{12}, then $F_{21} = -F_{12}$, and therefore the equations of the types

$$\Sigma X' = 0, \quad \Sigma(yZ' - zY') = 0$$

are satisfied.

But the equations of the types

$$\Sigma m\ddot{x} = \Sigma X, \quad \Sigma m(y\ddot{z} - z\ddot{y}) = \Sigma(yZ - zY)$$

also are satisfied identically, since the accelerations and the external forces are supposed to be adjusted correctly.

We conclude that, if the particles are sufficiently numerous, the $\frac{1}{2}n(n-1)$ quantities F_{12} can be adjusted in an infinite number of ways so that the $3n$ equations may be satisfied.

It appears that the forces between the hypothetical particles are largely indeterminate. This result offers no difficulty so long as we do not attempt actually to assign these forces. We conclude that the motion of the body, or system of bodies, can be represented by the motion of a system of particles.

The method that is actually adopted involves a restriction upon the hypothetical forces, which, nevertheless, leaves them largely indeterminate. The method involves the introduction of the notion of *stress*.

Consider a body resting on a horizontal plane in the field of the Earth's gravity. Let the body be imagined to be divided into two parts by a horizontal plane. When we represent the body by a system of particles we may suppose that none of the particles are in the plane. Consider the forces acting upon those particles which are above the plane. Those forces which are due to the Earth's gravity act vertically downwards. Those which are due to the mutual gravitation between the particles below the plane and those above it have horizontal components and vertical components, but the vertical components are directed downwards. If these were all the internal forces the centre of mass of the particles which are above the plane would have an acceleration, of which the vertical component would be different from zero and would be directed downwards. Since the centre of mass of the particles does not move, the particles below the plane must be regarded as exerting upon those above the plane forces which, on the whole, counteract the gravitational attractions, and thus the internal forces between the two sets of particles must be regarded as consisting of other forces besides these attractions.

Since the law of gravitation is assumed to hold for all distances that are measurable by ordinary means (Art. 146), we must regard the additional forces as being exerted only between particles which are very near together.

In general let a plane surface pass through a point O of a body, and draw on the plane a closed curve C of area S containing the point O. Some of the lines of action of forces between neighbouring particles on the two sides of the plane cross the plane within the curve C. We consider the forces thus exerted upon the particles which lie on a chosen side of the plane. Let ξ, η, ζ denote the sums of the components of these forces parallel to the axes. Then ξ, η, ζ are the components of a vector quantity, which is called the "resultant stress" or "resultant traction" across the area S of the plane. The quantities ξ/S, η/S, ζ/S are the components of a vector quantity which is called the "average stress" or "average traction" across the area S of the plane. We suppose that as the area S is diminished, by contracting the curve C towards the point O, the components of the average stress tend to definite finite limits; then these limits are the components of the "stress" or the "traction" across the plane at the point O.

Let S now denote any closed geometrical surface drawn in the body, X_ν, Y_ν, Z_ν the components of the stress or traction across the tangent plane at any point of S. Then the part of the body within S is to be regarded as a system of particles which move under forces, and the sums of the components parallel to the axes, and the sums of the moments about the axes, of the forces which arise from actions between neighbouring particles on the two sides of S are expressed by such formulæ as

$$\iint X_\nu dS, \qquad \iint (yZ_\nu - zY_\nu)\, dS,$$

where the integration extends over the surface.

This specification of the internal forces by means of stress is found to be adequate for the description of the motions of extended bodies.

The stress across a plane at a point of a body is a measurable quantity which can sometimes be determined theoretically and in some cases measured practically. The simplest examples are pressure in a fluid and tension in a string or chain. This tension is the resultant of the tractions across a plane which is normal to the line of the chain.

The introduction of the notion of stress carries with it a distinction between two classes of forces:—*body forces* and *surface tractions.*

Gravitational forces are proportional to the masses of the particles on which they act. The sum of the components, parallel to any fixed direction, of all the gravitational forces which act upon the part of a body within any small volume is proportional to the volume. For theoretical purposes we regard such forces as examples of a possible class of forces which we call "body forces." They may be specified by the force per unit of volume, or per unit of mass.

The resultant traction across a portion of a geometrical plane, drawn through a body, is an example of another class of forces, which we call "surface tractions." These forces act across surfaces, and are proportional to the areas of the surfaces across which they act, when these areas are small enough. They may be specified by the force per unit of area, or, what is the same thing, by the traction across a plane at a point.

In the course of this book the energy equation has been regarded as one of the first integrals of the equations of motion of a conservative system. This mode of treatment appears to the writer to be the most natural when the science is based upon Newton's laws of motion, or any equivalent statements; but modern Physics would assign to the energy equation a much more important *rôle*. This comes about through the doctrine of the conservation of energy. The energy equation in Mechanics is seen to be but an example of a general principle applicable to all kinds of physical processes.

Attempts have been made to discard the notion of force, and to develope the theory of Mechanics from the notions of mass and energy. It has been proposed also to discard the conception of bodies as made up of particles at the same time as the notion of force. One difficulty in the way of this method of formulation is the difficulty of giving any account of the retained notion of mass. In the Newtonian Mechanics we have, on the basis of the Law of Reaction, a clear and definite meaning for the term "mass." Another difficulty in the way of the "energetic" method of formulation is the difficulty of giving any adequate account of potential energy, or of work. These difficulties may perhaps be overcome in the future. In the present state of science we may make a compromise between the two methods, by taking the notions of *kinetic energy* and *work* from the Newtonian system, and destroying the scaffolding of *forces* and *particles* by which they are reached. The *masses* that occur in this intermediate method of formulation are then regarded as coefficients in the expression for the kinetic energy.

The possibility of this intermediate method depends upon an analytical transformation of the equations of motion, as developed in accordance with the Newtonian method. This analytical transformation proceeds by way of generalization of the *principle of virtual work*. Just as all the equations of equilibrium of a system can be deduced from an equation of the form

$$\Sigma\left[(X+X')\dot{x}'+(Y+Y')\dot{y}'+(Z+Z')\dot{z}'\right]=0,$$

as has been explained in Art. 208, so all the equations of motion of the system can be deduced from an equation of motion of the form

$$\Sigma\left[m(\ddot{x}\dot{x}'+\ddot{y}\dot{y}'+\ddot{z}\dot{z}')\right]=\Sigma\left[(X+X')\dot{x}'+(Y+Y')\dot{y}'+(Z+Z')\dot{z}'\right], \quad \ldots\ldots(\mathrm{A})$$

which may be obtained by the method of Art. 208. The important result is that the terms of the equations of motion which, in the Newtonian method, represent what have been called in this book "kinetic reactions*" are expressible in terms of the kinetic energy.

To explain this statement we consider the case in which the position of the system at any time can be expressed in terms of a finite number of independent geometrical quantities. Let these quantities be denoted by θ, ϕ, ... Then the kinetic energy T can be expressed as a homogeneous quadratic function of the corresponding velocities $\dot{\theta}$, $\dot{\phi}$, ... ; and the left-hand member of equation (A) can be expressed in the form

$$\left\{\frac{d}{dt}\left(\frac{\partial T}{\partial \dot{\theta}}\right)-\frac{\partial T}{\partial \theta}\right\}\dot{\theta}'+\left\{\frac{d}{dt}\left(\frac{\partial T}{\partial \dot{\phi}}\right)-\frac{\partial T}{\partial \phi}\right\}\dot{\phi}'+\ldots,$$

in which $\dot{\theta}'$, $\dot{\phi}'$, ... represent any set of velocities with which the system might pass through the position denoted by θ, ϕ, ... This result is due to Lagrange.

It appears from this discussion that, if we can find, for any system, an expression for the kinetic energy and an expression for the rate at which work is done, we can obtain the equations of motion of the system without introducing any considerations of "forces" or "particles."

The formulation of the principles of Mechanics implies that choice is made of the frame of reference and of the time-measuring process. This statement remains true whether the formulation is carried out in terms of mass and force, or in terms of kinetic energy and work; for the two methods require the specification of accelerations and velocities. When we say that a particle at a certain place has a certain acceleration, the place and the acceleration must be specified by reference to some frame or other, and the specification of the acceleration involves also the use of some method or other of measuring time. A similar statement holds for velocities.

* In some books they are called "effective forces."

For many theoretical purposes it is unnecessary to specify either the frame of reference or the time-measuring process; it is sufficient to suppose that they have been chosen. But in any problem concerning observable motions of actual bodies, the description of the motion is incomplete until the reference system, both for space and time, is specified. We may ask two questions: (1) How is the system specified? (2) How ought the system to be specified? It is a little difficult to answer briefly either of these questions; and it is comparatively easy to answer the slightly different question: What reference-systems are inadmissible? The answer is that no system ought to be admitted which conflicts with the principles of Mechanics, or the law of gravitation, or the principle of the conservation of energy.

A system of reference which satisfies the conditions of this question and answer may be described as "kinetic*." A frame of reference which satisfies the conditions will be called a "kinetic frame," and time measured in accordance with the conditions will be called "kinetic time."

To illustrate this question, and the answer, let us consider the motion of the Earth. The principles of Mechanics require that the Earth should be regarded as a body having a certain mass and a certain centre of mass. Observations of falling bodies and Astronomical observations lead us, in accordance with the concept of

* W. H. Macaulay, in the Article 'Motion, Laws of' in *Ency. Brit.* 10th Edition, vol. 30 (1902), describes what is here called a "kinetic frame" as a "Newtonian base." In regard to the general question of the relativity of motion, reference should be made to Newton's original argument in the *Principia*, Lib. 1, 'Scholium' attached to the 'Definitiones,' and to the following more recent works:—J. C. Maxwell, *Matter and Motion* (London, 1882), new edition by J. Larmor (1920), Thomson and Tait, *Natural Philosophy*, Part I (Cambridge, 1879), E. Mach, *The Science of Mechanics*, Translation (Chicago, 1893), C. Neumann, *Ueber die Principien der Galilei-Newton'sche Theorie* (Leipzig, 1870), K. Pearson, *The Grammar of Science* (London, 1900), H. Poincaré, *La science et l'hypothèse* (Paris, N.D.), the Article by W. H. Macaulay cited above and the Article by A. Voss in *Ency. d. math. Wiss.* Bd. IV, Teil 1, Art. 1 (Leipzig, 1901). In regard to the reference system of Astronomy see the Article by E. Anding in *Ency. d. math. Wiss.* Bd. VI, Teil 2, Art. 1 (Leipzig, 1905). It need hardly be said that the view adopted from Newton by Maxwell and by Thomson and Tait, viz. that we have knowledge of absolute direction but not of absolute position, differs from that stated in the text. Since the question is not of practical importance, it has seemed to the present writer to be desirable to set forth, as clearly as may be, a view which seems to him to be logically defensible, rather than to emphasize the divergence of this view from those held by others.

force, to regard the Earth as exerting forces on other bodies, and the law of reaction states that these bodies exert forces on the Earth, and, therefore, that the centre of mass of the Earth has certain component accelerations. Thus we cannot choose as a frame of reference axes fixed in the Earth, and at the same time maintain the law of reaction. The change from the geocentric astronomy of Ptolemy to the heliocentric astronomy of Copernicus may be regarded as an instance of the discarding of an unsuitable frame of reference.

As an illustration of the restrictions limiting the choice of the time-measuring process we may consider the forces that can affect the rotation of the Earth. The system of Earth and Moon, with the fluid ocean on the Earth, executes various internal relative motions, among which the tides are conspicuous. Such internal relative motions generally involve dissipation of energy* in a system, for they do not take place without friction. We are thus led to expect that the kinetic energy of the Earth's rotation is being dissipated at a finite rate, or that the period of the diurnal rotation (the length of the day) is gradually increasing. On the basis of the law of gravitation and the principle of the conservation of energy, but without fixing beforehand what the time-measuring process is to be, astronomers have shown that one of the inequalities in the motion of the Moon could be explained by the supposition that such a gradual slackening in the speed of the Earth's rotation is taking place. This result implies that the time-measuring process is not the rotation of the Earth, or, in other words, that sidereal time is not kinetic time.

The result is usually stated in the form that the Earth as a time-keeper is losing at the rate of so many seconds in a century†.

The processes by which we reach a kinetic frame of reference and a kinetic time-measuring process are approximative. It has always proved to be possible to correct a choice previously made so as to harmonize the observations of the motions of actual bodies with the principles of Mechanics—at least very approximately;

* That is to say a change in the form of the energy by which less of it is rendered available, as, for example, in the conversion of kinetic energy into heat.

† The rate is variously estimated. Two estimates are 22 seconds per century and 8·3 seconds per century. See Thomson and Tait, *Nat. Phil.* Part II, Appendix G (contributed by G. H. Darwin).

there are certain small discrepancies. By means of the law of gravitation we can determine, to a certain order of approximation, the masses of the bodies which compose the Solar System and the position relative to these bodies of the centre of mass of the system. It has proved to be sufficient to take this centre of mass as origin, and to take, as lines of reference, lines drawn to "fixed" stars which have no appreciable proper motion or annual parallax.

In regard to the measurement of time we have no natural system of reference such as the "fixed" stars provide for the determination of direction; but we can proceed in a different fashion by means of the familiar process of changing the independent variable. Let t denote sidereal time, that is to say time determined by the rotation of the Earth relative to the stars; t is, of course, measured from some particular epoch, the instant of the occurrence of some assigned event, and we may take the interval t to denote t sidereal days. During this interval the Earth turns through $2\pi t$ radians. Let the Earth as a time-keeper be losing at the rate of ϵ seconds per day. We know that ϵ is a very small fraction. Let a new variable τ be introduced by the equation

$$\tau = t - \frac{\epsilon}{86400}\frac{t^2}{2}.$$

If we measure time by τ instead of t, the quantity τ measures kinetic time so far as it has been necessary as yet to determine its measure.

This discussion suggests also a method by which we might dispense with the "fixed" stars in the choice of a frame of reference. We may construct a frame, of which the origin is the centre of mass of the Sun, by means of three lines drawn from the origin. We may take these lines arbitrarily; for instance, we may draw two of them to the centres of mass of the Earth and Jupiter, and the third, in a chosen sense, at right angles to the plane of these two. This frame does not, of course, continue to be a kinetic frame; but we can take it to coincide with a kinetic frame at some instant. It will then move relatively to the kinetic frame, and the kinetic frame will move relatively to it. If the relative motion of the two frames were known, we could determine the position of the kinetic frame in the system after a short interval of time; and thus we might by a continued approximation, determine the position of the kinetic frame at any time. This method has no practical value; but it appears to have some theoretical interest. This interest will be more apparent if we reflect that, according to the law of universal gravitation, there are gravitational forces acting between the bodies of the Solar System and the stars. However small the forces which thus act upon the bodies of our system may be, it remains true that the centre of mass of the system cannot, in the long run, be a proper origin for a kinetic frame. The frame which we now adopt, with origin at the centre of mass of the Solar System, and axes pointing to fixed stars, may be taken to coincide with a kinetic frame at some instant. Then we are able to state that the relative motion of the two frames is so small that it has not been detected by any observations.

Finally it must be said that the choice of a kinetic frame and of kinetic time, instead of any other frame and time, is a convention. We have set out to describe the motions of bodies; and we wish to utilize the results that have been accumulated during three centuries by scientific investigators who, for the most part, paid little attention to the question of systems of reference. To achieve our object we must state, as precisely as we can, what our system of reference is, and how actual bodies move with reference to it. We do this when we say that the system of reference is what we have called "kinetic," and when we explain how a kinetic frame can be found and how kinetic time can be determined, with, at any rate, sufficient approximation for our purpose.

APPENDIX

MEASUREMENT AND UNITS

(*a*) *Measurement.* The mathematical theory of measurement rests on the assumed possibility of dividing an object into an integral number of parts which are identical in respect of some property. Thus, to measure the length of a segment of a line, we must suppose the segment divided into a number of equal segments, where the test of equality of length is congruence; to measure the mass of a body we must suppose it capable of division into a number of bodies of equal mass, where equality of mass is tested by weighing; to measure an interval of time we measure the angle turned through by the Earth in the interval; this requires the division of an angle into a number of equal angles, and the test of equality of angles is congruence.

The measurement of an object in respect of any property requires (1) a unit or standard of comparison, and (2) a mode of referring to the standard. The standard must be an object which possesses the property in question. The mode of referring to the standard must be such that it determines a positive number (integral, rational but not integral, or irrational) which is the measure of the object in respect of the property. The number is determined by the following rules :—

(α) When the object can be divided into an integral number n of parts, each of which is identical with the standard in respect of the property in question, the measure of the object in respect of that property is n.

(β) When the object and the standard can be divided into p and q parts respectively (p and q being integers), such that all the parts are identical in respect of the property in question, the measure of the object in respect of that property is the rational fraction p/q.

Here it is to be noted (1) that the rule (α) is the case of the rule (β) for which $q=1$, and (2) that in practice the integer q may be taken so large that an integer p may be found for which the fraction p/q measures the object within the limits of experimental error.

In the mathematical theory of measurement the case where no rational fraction p/q can measure the object may not be so simply dismissed. It may happen that however great q is taken there is no corresponding number p, but that, while the fraction p/q would measure an object somewhat smaller than that to be measured, the fraction $(p+1)/q$ would measure an object somewhat greater than that to be measured. When this is the case the measure sought is an irrational number. We may in fact separate all rational numbers into two classes—a "superior" class and an "inferior" class—so that all the numbers in the superior class are too large to be the measure of the object, and all those in the inferior class are too small. Every rational number without exception falls into one or other of the two classes, and the separation between them is marked by an irrational number which is the measure of the object.

Suppose, for example, that we wish to measure the diagonal of a square whose side is the unit of length. We may separate all rational numbers into two classes—those whose squares are greater than two, and those whose squares are less than two. Every rational number without exception falls into one or other of the two classes. The separation between the two classes is marked by the irrational number $\sqrt{2}$, and this irrational number is the required measure.

(*b*) *Number and Quantity.* When the unit is stated the magnitude of an object is precisely determined by its measure in terms of the unit, and this measure is always a number. The "object" may be anything which we can think of as measurable in respect of any property, and the phrase "magnitude of an object" is thus coextensive in meaning with the word "quantity." The quantity does not change when the unit chosen to measure it changes, and thus the quantity is not identical with the number expressing it.

A number can express a quantity only when the unit of measurement is stated or understood. When the unit is stated or implied the number expresses the quantity.

Mathematical equations, and inequalities, are relations between numbers, expressing that a certain number which has been arrived at in one way is equal to, greater than, or less than, a certain number which has been arrived at in another way.

Mathematical equations, and inequalities, between numbers expressing quantities are valid expressions of relations between the quantities, as distinct from the numbers, only if they hold good for all systems of units.

(*c*) *Fundamental and derived Quantities.* The fundamental Physical quantities are lengths, times, and masses. In Dynamics, as considered in this book, all the other quantities which occur are derived from these. Thus, velocity is measured by a fraction of which the numerator is a number expressing a length and the denominator is a number expressing an interval of time; acceleration is measured by a fraction of which the numerator is a number expressing a velocity and the denominator is a number expressing an interval of time; force is measured by the product of a number expressing a mass and a number expressing an acceleration; and all the other magnitudes that occur are in similar ways dependent upon lengths, times, and masses.

(*d*) *Dimensions.* A number which expresses a quantity is said to be of one "dimension" in that quantity. If the unit of measurement is altered so that the new unit is a certain multiple x of the old, the number expressing the quantity in terms of the new unit is the quotient by x of the number expressing the quantity in terms of the old unit.

The number expressing a derived quantity is, in every case, the product of three numbers A, B, C, of which A is a homogeneous expression of some degree p in numbers expressing lengths, B is a homogeneous expression of some degree q in numbers expressing intervals of time, and C is a homogeneous expression of some degree r in numbers expressing masses. We say that the quantity is of p dimensions in length, q dimensions in time, and r dimensions in mass. We express this shortly by saying that the *dimension symbol* of the quantity is $[L]^p [T]^q [M]^r$. The numbers p, q, r may be positive or negative, integral or fractional, or zero.

If the units of length, time, and mass are changed so that the new units are respectively x, y, z times the old, the measure of any quantity in terms of the new units is obtained from its measure in terms of the old units by dividing by $x^p y^q z^r$, where $[L]^p [T]^q [M]^r$ is the dimension symbol of the quantity.

The condition that a mathematical equation or inequality between numbers expressing quantities may be a valid expression of a relation between the quantities is that every term in it must be of the same dimensions.

(*e*) *Physical Quantities.* We give here a list showing the principal derived quantities that occur in Dynamics and their dimension symbols.

Velocity	$[L]^1[T]^{-1}$.
Acceleration	$[L]^1[T]^{-2}$.
Momentum, Impulse	$[L]^1[T]^{-1}[M]^1$.
Moment of Momentum, Impulsive Couple	$[L]^2[T]^{-1}[M]^1$.
Kinetic Reaction, Force	$[L]^1[T]^{-2}[M]^1$.
Kinetic Energy, Work	$[L]^2[T]^{-2}[M]^1$.
Power	$[L]^2[T]^{-3}[M]^1$.
Density	$[L]^{-3}[M]^1$.
Constant of Gravitation	$[L]^3[T]^{-2}[M]^{-1}$.

(*f*) *Method of Dimensions.* We can frequently determine the *form* of a result by consideration of the dimensions of the quantities involved. This will be made clear by the consideration of some examples. Thus, if we assume that the period of oscillation of a pendulum can depend only on its mass, its length, and the acceleration due to gravity, we can prove that it is proportional to the square root of the length. Since the quantity to be expressed is an interval of time its expression cannot involve any power of a mass, and we have assumed that no mass but the mass of the body can enter into the expression; the period is therefore independent of the mass of the body. Now g has dimension symbol $[L]^1[T]^{-2}$, and therefore $1/\sqrt{g}$

has dimension symbol $[T]^1[L]^{-\frac{1}{2}}$, hence the only way in which the expression of the period can contain the length l of the pendulum is by being proportional to its square root. This argument would prove that the period is a numerical multiple of $\sqrt{(l/g)}$. Again, to take another example, consider the ellipticity of the Earth supposed to depend on the angular velocity of rotation ω, the mean density ρ, and the constant of gravitation γ. The product $\gamma\rho$ has dimension symbol $[T]^{-2}$, and thus $\omega^2/\gamma\rho$ is a number (angles being measured in radians); the ellipticity being a number, must be a function of $\omega^2/\gamma\rho$.

The method of dimensions supplies also a useful means of verification. In any piece of mathematical reasoning where the numbers represent quantities all the terms in each equation must be of the same dimensions.

INDEX

The numbers refer to pages

Printed in the United States
By Bookmasters